P9-DJA-196

Enabling Context-Aware Web Services

Methods, Architectures, and Technologies

Contents

Preface

Over the years, the Web has gone through many transformations, from traditional linking and sharing of computers and documents (i.e., "Web of Data"), to current connecting of people (i.e., "Web of People"). With the recent advances in radio-frequency identification (RFID) technology, sensor networks, and Web services, the Web will continue the transformation and will be slowly evolving into the so-called "Web of Things and Services". Indeed, this future Web will provide an environment where everyday physical objects such as buildings, sidewalks, and commodities are readable, recognizable, addressable, and even controllable using services via the Web. The capability of integrating the information from both the physical world and the virtual one not only affects the way we live, but also creates tremendous new Web-based business opportunities such as support of independent living of elderly persons, intelligent traffic management, efficient supply chains, and improved environmental monitoring.

Context-aware Web services are emerging as an important technology to underpin the development of new applications (user centric, highly personalized) on the future ubiquitous Web. This book compiles the newest developments and advances in context awareness and Web services from world's leading researchers in this field. It offers a comprehensive and systematic presentation of methodologies, architectures, and technologies that enable the development of context-aware Web services. The whole book is organized into three major parts: Methods, Architectures, and Technologies. The Methods part focuses on the principle of context awareness in Web service and various ways to model context-aware Web services at the specification level. The Architectures part focuses on the infrastructures, frameworks and standards for building context-aware Web services. The Technologies part focuses on the various techniques adapted from general research areas e.g., semantic Web, database, artificial intelligence, and formal methods in the development of context-aware Web services.

This book is the first of its kinds to bridge the gap between two previously separated research and development areas: context-awareness and Web services. It serves well as a valuable reference point for researchers, educators, and engineers who are working in Internet computing, service-oriented computing, distributed computing, and e-Business, as well as graduate students who wish to learn and spot the opportunities for their studies in this emerging research and development area. It is also of general interest to anyone using the service paradigms for software development, particularly on developing context-aware applications. It is our hope that the work presented in this book will stimulate new discussions and generate original ideas that will further develop this important area.

We would like to thank the authors for their contributions and the reviewers for their expertise to improve the manuscripts. Moreover, we are grateful to CRC Press for the opportunity to publish this book. Our special thanks go to Li-Ming Leong, Yong-Ling Lam, and Marsha Pronin of Taylor & Francis Group for their support and professionalism during the whole publication process of this book.

Quan Z. Sheng, Jian Yu, Schahram Dustdar
August 2009

Contributors

Jose M. del Alamo
Universidad Politecnica de Madrid
Madrid, Spain

Carlos Baladron
University of Valladolid
Valladolid, Spain

Kosala Yapa Bandara
Dublin City University
Dublin, Ireland

Philippe Beaune
Ecole Nationale Superieure des Mines
Saint-Etienne, France

Yazid Benazzouz
France Telecom R&D Meylan & ENSM-SE
Saint-Etienne, France

Xavier Blanc
University of Lille 1
Lille, France

Olivier Boissier
Ecole Nationale Superieure des Mines
Saint-Etienne, France

Tom Broens
Telematica Instituut
The Netherlands

Alejandro Cadenas
Telefonica I+D
Madrid, Spain

Belen Carro
University of Valladolid
Valladolid, Spain

Sophie Chabridon
Institut Telecom, Telecom & Management SudParis
Paris, France

Chii Chang
Monash University
Melbourne, Australia

Denis Conan
Institut Telecom, Telecom & Management SudParis
Paris, France

Stefan Dietze
The Open University
Milton Keynes, United Kingdom

Laurence Duchien
University of Lille 1
Lille, France

Schahram Dustdar
Vienna University of Technology
Vienna, Austria

John Domingue
The Open University
Milton Keynes, United Kingdom

Paolo Falcarin
Politecnico di Torino
Turin, Italy

Babak Farshchian
SINTEF
Trondheim, Norway

Stefan Forsstrom
Mid Sweden University
Sundsvall, Sweden

Stefania Galizia
Innova Spa
Rome, Italy

A.E.S. Goh
Nanyang Technological University
Singapore

Laurent-Walter Goix
Telecom Italia
Turin, Italy

Pierre Grenon
The Open University
Milton Keynes, United Kingdom

Alessio Gugliotta
Innova Spa
Rome, Italy

Aart van Halteren
Philips Research
The Netherlands

Dimitra I. Kaklamani
National Technical University of Athens
Athens, Greece

Theo Kanter
Mid Sweden University
Sundsvall, Sweden

Georgia M. Kapitsaki
National Technical University of Athens
Athens, Greece

Victor Kardeby
Mid Sweden University
Sundsvall, Sweden

Zakia KaziAoul
Institut Telecom, Telecom & Management SudParis
Paris, France

Shonali Krishnaswamy
Monash University
Melbourne, Australia

Luca Lamorte
Telecom Italia
Turin, Italy

Rafael Leano
University of Lille 1
Lille, France

S.S.G. Lee
Nanyang Technological University
Singapore

Kewen Liao
University of Adelaide
Adelaide, Australia

Sea Ling
Monash University
Melbourne, Australia

Georgios V. Lioudakis
National Technical University of Athens
Athens, Greece

Michael Mrissa
Universite de Lyon
Lyon, France

Massimo Maresca
University of Padova and M3S
Genova, Italy

Isabel Ordas
Telefonica I+D
Madrid, Spain

Patrik Osterberg
Mid Sweden University
Sundsvall, Sweden

Claus Pahl
Dublin City University
Dublin, Ireland

Carlos Parra
University of Lille 1
Lille, France

Pravin Pawar
University of Twente
The Netherlands

Carlos Pedrinaci
The Open University
Milton Keynes, United Kingdom

Nicolas Pessemier
University of Lille 1
Lille, France

Stefan Pettersson
Mid Sweden University
Sundsvall, Sweden

George N. Prezerakos
Technological Education Institute of Piraeus
Piraeus, Greece

Hung Keng Pung
National University of Singapore
Singapore

Fano Ramaparany
France Telecom R&D Meylan & ENSM-SE
Saint-Etienne, France

Alvaro Martinez Reol
Telefonica I+D
Madrid, Spain

Daniel Romero
University of Lille 1
Lille, France

Romain Rouvoy
University of Lille 1
Lille, France

Antonio Sanchez-Esguevillas
Telefonica I+D
Valladolid, Spain

Lionel Seinturier
University of Lille 1
Lille, France

Quan Z. Sheng
University of Adelaide
Adelaide, Australia

Michele Stecca
University of Padova and M3S
Genova, Italy

Chantal Taconet
Institut Telecom, Telecom & Management SudParis
Paris, France

P.S. Tan
Singapore Institute of Manufacturing Technology
Singapore

Ruben Trapero
Universidad Politécnica de Madrid
Madrid, Spain

Hong-Linh Truong
Vienna University of Technology
Vienna, Austria

Nikolaos D. Tselikas
University of Peloponese
Peloponnese, Greece

Bert-Jan van
Beijnum Telematica Instituut
The Netherlands

Iakovos St. Venieris
National Technical University of Athens
Athens, Greece

Victor Villagra
Technical University of Madrid
Madrid, Spain

Katarzyna Wac
University of Geneva
Geneva, Switzerland

Jamie Walters
Mid Sweden University
Sundsvall, Sweden

MingXue Wang
Dublin City University
Dublin, Ireland

Stefan Wesner
High Performance Computing Centre
Stuttgart, Germany

Hoi S. Wong
University of Adelaide
Adelaide, Australia

Jian Yu
University of Adelaide
Adelaide, Australia

Jian Zhu
National University of Singapore
Singapore

Part I

Methodology

Chapter 1

Context-Aware Web Service Development: Methodologies and Approaches

Georgia M. Kapitsaki, George N. Prezerakos, and Nikolaos D. Tselikas

Abstract The development of context-aware Web services is an interesting issue in service provision nowadays. This book chapter delves into the literature by presenting the main visible trends. The description of the proposed approaches is divided into three categories: programming language extensions, which intervene into the language level in order to add or integrate context-awareness constructs, model-driven techniques, which exploit model-driven development principles, and semantic technologies, which rely mainly on ontologies and reasoning operations. Various example methodologies are depicted for each category giving the reader the possibility to choose the most appropriate direction based also on the short evaluation provided at the end.

1.1 Introduction

Context-awareness plays a vital role in service provision nowadays as service providers are focusing on providing personalized services to end-users. At the same time Web services are constantly gaining ground for the construction of context-aware applications. They can be found as part of desktop Web applications and in mobile computing where various mobile devices offer Web service based applications. The development process of such context-aware Web services is an important aspect prior to the service provision, since context-awareness requirements need to be taken into account and be incorporated during the service development. Developers can profit from techniques that facilitate the introduction of context handling during the service development phase.

In this book chapter we deal with development approaches that lead to the construction of context-aware Web services. The term "methodology," however, is not used in its traditional meaning. We do not focus on traditional software development methodologies (like the waterfall model or agile development), which usually divide the development process into distinct steps (requirements, design, coding, testing, deployment), but rather on general approaches that can assist the development work in the framework of several software lifecycle models. Moreover, these approaches are usually targeting specific architectures or middleware platforms proposed for the provision of context-aware Web services.

The inclusion of context handling issues in the development process may be performed directly at the code level of the service or earlier during the service design phase. This can be achieved either by popular programming or development techniques (e.g., model-driven development, semantic Web, aspects) and in this sense a significant number of research papers on context-aware service development falls into the three categories described below.

Category name	**Short description**
Programming languages extensions	Addition of programming language constructs focusing on context handling in Web services.
Model-driven development	Combination of Web service and context models towards the automatic production of service code.
Approaches based on semantic technologies	Ontological description of context data that allow extensive reasoning capabilities.

The category of *programming languages extensions* refers to attempts where specific modifications or constructs are introduced in a programming language in order to add context-aware capabilities. The service logic is enriched with code fragments responsible for performing the context adaptation. This can be achieved for example by separating the main business code from the context-sensitive code parts or by introducing different program layers that are either activated or deactivated based on the execution context. Uses of Aspect Oriented Programming (AOP) also fall in this category, whereas extensions to different languages that can be used in the construction of Web services are proposed (e.g., Java, Python). This way developers can introduce code level context manipulation schemes to the development of Web services.

Model-Driven Development (MDD) is related to the transformations between meta-models that capture domain characteristics and may eventually lead to the full or partial generation of platform specific code of the application under development through the necessary transformations between models and code. Model-driven engineering can be exploited in the stages of design and implementation. This is usually achieved by introducing a context information model during the design stage, so that the service can be created based on the adaptation to context reflected in the context model. The language most widely used in the design phase is the Unified Modeling Language (UML), although cases of other Domain Specific Languages constructed for specific development tools can also be found in the literature. MDD methodologies can be exploited for the development of context-aware Web services, where Web service properties are taken into account during the modeling phase and may be used for the generation of the service code.

Semantic technologies provide means for the sophisticated representation of information including reasoning and inference capabilities. Ontologies are usually exploited for the modeling of context information in specific domains of interest. However, it can be noted that in many cases the ontology specification drives the development process. Approaches that make use of semantic technologies either concentrate the development effort around the system ontologies or exploit the semantic capabilities of Web services to allow developers to reuse existing services in specific domains.

In several cases the presented approaches refer specifically to context-aware Web services. However, the majority of cases constitute generic attempts that can be applied to the field of Web services with adequate adaptations. Nevertheless, it is important to be able to reuse existing techniques for context-aware Web service development that have proven their usability in service development paradigms other than Web services.

1.2 Exploiting Programming Languages Extensions

Approaches related to extensions to programming languages or additions of new language constructs can be adopted in the field of Web services for adaptation at the service implementation level. Nevertheless, in order to be able to exploit these principles, the language used for the Web service implementation must be the one proposed in the chosen solution.

A usual group of approaches is denoted with the term "Context-oriented Programming" or COP. Generally, COP aims in incorporating context-related issues in the structure of a software system. The first solution proposed under this general term for the programming language Python is found in (14). In this approach the code is separated to context-free skeleton and context-dependent stubs. The skeleton refers to the main implementation, which includes gaps referred to as "open terms" that are to be adapted to context information. More precisely, open terms consist of:

- *goals*: express the goal of an entity (e.g., greet the user)
- *context*: represents contextual information
- *event*: describes an event that triggers the open term execution (optional part)

On the other hand, stubs represent a specific execution scheme applicable to a specific goal under defined context conditions. The context behavior is injected to the main logic by the context-filling procedure, which binds open terms with the appropriate stub and provides, thus, the desired context-dependent behavior during service execution. This process is performed through an external entity (the matchbox), which corresponds to a stub repository with registered entries for the available stubs. The relevant code fragments are retrieved from the stub repository, when specific context-related conditions are met. Stubs and the program skeleton are expressed in Extensible Markup Language (XML) as depicted in Figure 14.1, whereas context-filling takes place by a call to the *fillgap* routine. The figure presents a very simple service that greets the user based on the user's native tongue and current location.

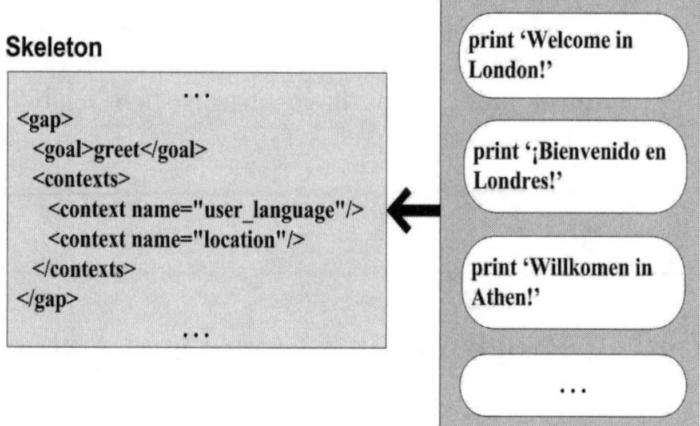

Figure 1.1: Context-oriented Programming for the Python language.

Another approach to COP introduces the notion of "layer" into the program execution in order to express behavior variations based on contextual properties (11). Layers correspond to sets of partial class and method definitions that become activated depending on context information. They are treated as first-class constructs and can be triggered, i.e., activated or deactivated, from any part of the application code. This way context dependencies are kept separate from the base program definition. Applicable variations depend either on an actor entity interacting with the program (or system), the system properties or external environmental conditions. Viewed together, these properties depict the contextual situation as a whole. Layers can be inserted in the corresponding implementations for different programming languages: Java (ContextJ), Squeak/Smalltalk (ContextS) and Common Lisp (ContextL), allowing the program to be modified dynamically at runtime in a context-aware fashion. Regarding the Java version, a subset of ContextJ applicable to standard Java environments is available from the current work, namely ContextJ*. In ContextJ* the class `Layer` is used for the construction of new layer objects depicting context attribute conditions. Each class that wants to support context-dependent behavioral variations needs to include an interface that captures the method definitions that include context dependencies. In order to choose which layer should be activated a call is forwarded to the `Layer Definitions` container, which is parameterized based on the interface mentioned above. These constructs are demonstrated for a simple case in Figure 14.2, where the method *greeting* of the class *GreetingService* depends on current context information. Different versions of the method call are given depending on the activated layer (none, Location or Greeting).

A similar idea to Context-oriented programming is illustrated by the Isotope Programming Model (IPM) (19) proposed as an extension to object-oriented programming (OOP). Objects are defined by a number of attributes and a default behavior described in default methods – forming the main element – and a number of isotope elements. Instead of using one unique file for the method definition as in OOP, IPM splits the method into several isotope elements depicting different contextual conditions. Apart from the method definition, each isotope is accompanied by a context block illustrating the relevant context conditions (usually in the form of logical constraints) that activate the selection of the respective behavior part as shown in Figure 14.3. At execution time, the most appropriate isotope is selected based on the evaluation of its context block and its recency, as indicated by the timestamp information included in the isotope element. If no suitable isotope element is identified, the default behavior contained in the main element is executed.

An additional group of approaches stems from the paradigm of AOP, whereas some AOP features are also visible in the layered approach of COP. Indeed context can be seen as a crosscutting concern spanning through various program parts and activities. Through the definition of appropriate pointcuts, advices injecting context-adapted behavior can be applied during the service execution. Generic AOP languages such as AspectWerkz[1] can be exploited for the injection of context-related behavior into different Web service parts.

[1] http://aspectwerkz.codehaus.org/

```
import static be.ac.vu.prog.contextj.ContextJ.*;

//File: Layers.java
public class Layers {
  public static final Layer Location = new
Layer();
  public static final Layer German = new
Layer();
  . . .
}

//File: IGreetingService.java
public interface IGreetingservice {
  public String greeting();
}

//File: GreetingService.java
public class Greetingservice implements iGreetingService{
  String userName; String cityName;

  private LayerDefinitions<IGreetingService> layers =
    new LayerDefinitions<IGreetingService>();

  public String greeting() {
    return layers.select().greeting();
  }

  {
    layers.define(RootLayer,
        new IGreetingService() {
            public String greeting() {
                  return "Hello " + userName + "!";
            }
        }
    );

    layers.define(Layers.Location,
       new IGreetingService() {
         public String greeting() {
           return layers.next(this)+".Welcome in " + cityName;
         }
       }
    );

    layers.define(Layers.German,
       new IGreetingService() {
         public String greetUser() {
             return "Willkommen in " + cityName;
         }
       }
    );
  . . .
  }
}
```

Figure 1.2: Example of language constructs for COP using ContextJ*.

However, a number of languages more specific to the notion of context that build on top of AOP principles also exist in the literature and are shortly presented here. A generic aspect language CSLogicAJ (Context-aware Service-oriented LogicAJ) for the definition of such pointcuts is proposed in (25). The solution targets Service-Oriented Architectures (SOA) and is, therefore, applicable to Web services that express the most popular SOA implementation. CSLogicAJ is built on LogicAJ (24), an aspect-oriented extension for Java. CSLogicAJ contains logic metavariables that exploit notations from the Prolog language and can be used uniformly in pointcuts and advices (e.g., ?latitude). It allows the selection of join points and context information on service level. Context-sensitive service aspects are applied synchronously by intercepting calls to service interfaces or asynchronously as reactions to context changes. A simple pointcut expression for service call logging and sorting

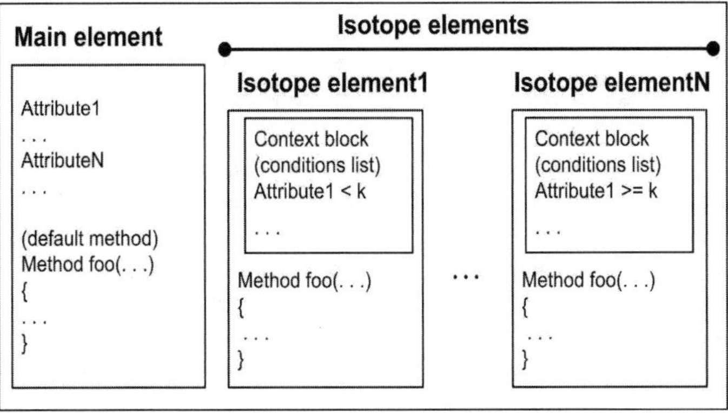

Figure 1.3: Object structure in the Isotope Programming Model.

actions on the service elements (e.g., returned service list) activated when the user position changes is shown in Figure 14.4.

```
onchange updateSorting(?lat ?lng):
   position(?GPSSender, ?lat, ?lng)
   current_service(?service) &&
   service_attr(?service, "service.id", ?id)
{
   log("called method: " + ?"m" + " on service "+ "?id");
   sorter.sort(new Double("?lat").doubleValue(),
            new Double("?lng").doubleValue());
   ...
}
```

Figure 1.4: CSLogicAJ pointcut expression for service call logging.

The adaptation of aspects to context information is discussed in (30) resulting to context-aware aspects. Context information is separated from the aspects, whose execution is, however, driven by context. The aspects are either applied or not based on their execution conditions. These context conditions, which are specified together with the aspect definition, can be composed, i.e., aggregate different context situations, or parameterized, i.e., bound with context properties that obtain a range of values (e.g., location name, discount rate). An important feature of the solution lies in context snapshotting, which is related with the assignment of timestamps to context information. This way, the program behavior can be affected not only by current context values, but also by past ones. Context-aware aspects have been implemented in an open extension of Java, the AOP framework Reflex. Reflex allows first-class pointcuts for dynamic crosscutting. Hooksets of Reflex can specify conditions of when to send a specific message to a meta object, achieving this way AOP-compliant execution of code fragments and also context-aware service behavior. In order to demonstrate the above a context definition example is illustrated in Figure 1.5 for the case of an electronic shop applying specific discounts to client purchases. `PromotionCtx` is bound to occurrences of `MsgReceive` operations and is parameterized by the discount rate (attribute *rate*). The activation state of the context is depicted through the *getState* method. Readers interested in more information on AOP frameworks that can be exploited for context-awareness purposes can refer to the comparative study of (5).

```
//File: PromotionCtx.java
class PromotionCtx extends Context
{
  double rate; /* with setter */

  // cflow(execution(* WebServiceRequest+.*(..)))
  CFlow cf = CFlowFactory.get(new Hookset(MsgReceive.class,
              new NameCS("WebServiceRequest", true),
              new AnyOS()));

  ContextState getState() {
    if(!cf.in()) return null;
    return new PromotionState(rate);
  }
}

//File: PromotionState.java
class PromotionState extends ContextState
{
   double rate; // init in constructor
   double getRate() {
     return rate;
   }
}
```

Figure 1.5: Parameterized context definition for context-aware aspects.

1.3 Model-Driven Development

The principles of MDD can be applied in the field of context-aware Web service development by introducing appropriate context-aware service models or independent context models that can be combined with the corresponding Web service models in order to express context adaptation conditions. In many cases, the development approaches are close to the principles of OMG's Model-Driven Architecture (MDA) (20) using Platform Independent and Platform Specific Models (PIMs and PSMs respectively), whereas others focus mainly on executable code results.

As already mentioned, the most widely adopted modeling language towards model-driven development is UML. Therefore, the approaches presented in this section follow mainly the UML notation. Different UML metamodels or profiles have been proposed towards model-driven application development. Metamodeling refers to the extension of the UML metaclasses for adaptation to different domains of usage, whereas profiling is a special metamodeling technique comprised of *stereotypes* that extend existing metaclasses and *tags* that are standard meta-attributes. A generic UML profile not specific to Web services is presented in (2), where the application modeling is performed through class and sequence diagrams. Thereinafter, the application model can be mapped to different platforms through appropriate transformations. The profile is shown in Figure 1.6, where the values in brackets indicate the UML extended metaclasses. Its elements can be divided into two main parts:

- *context elements* (Figure 1.6a): a number of stereotypes and tags are introduced to model context properties (`«Context»`), the periodic or event-based collection process of context information (stereotypes `«Periodic Collection»` and `«EventCollection»` respectively), precision and other quality attributes (`«ContextQuality»`) and the context situations that can affect the application behavior (`«ContextState»`) combined through `«And»` and `«Or»` constructs.

- *context-awareness mechanisms* (Figure 1.6b): refer to the ways the context information affects the main application and include three adaptation types. Structural adaptation is related with adding or removing attributes and methods from the application objects and is denoted through various versions of the class that may be modified (`«Variable Structure»`). Architectural adaptation refers to optional object instantiation through the `«Optional»` stereotype. Finally, behavioral adaptation extends the UML sequence diagram in order to support the inclusion of several sequences that can be optionally activated. The set of context-dependent interactions is depicted using `«Variable Sequence»`, which includes different interaction versions (`«SequenceVariant»`). Optional objects participating in the sequence diagram are shown using the stereotype «OptionalLine» in the corresponding lifelines.

In the model-driven solution of (8) context is divided into state-based, that corresponds to context attributes at a specific point in time, and event-based, that represents changes to the state of an entity. Context-awareness is modeled through constraints, which can be state constraints for specific points in time or event constraints, when they are identified through a series of events. The application is adapted to context information in two ways: through context-aware bindings that associate context values to application entities and context-aware inserts that modify the application structure (e.g., by adding a new application attribute) or behavior (e.g., by executing additional code). The main part of the proposed profile is depicted in Figure 1.7. During the application design UML activity diagrams and the entities «ContextMonitor» and «ContextAdapter» are also exploited. The first entity is responsible for the monitoring of the contextual conditions and the detection of cases, when the specified constraints are met, whereas the second is responsible for the actual adaptation. The application, which contains many AOP characteristics, is mapped to executable code in the AOP language AspectJ,[2] where Adapters and Monitors are implemented as aspects.

A significant modeling attempt that is specific to Web services can be found in ContextUML (27) and is illustrated in Figure 1.8. Different constructs are introduced in the metamodel, in order to model: Web service properties adaptable to context («CAObject» and related elements), context information containing plain or aggregated values («Context» and its subclasses) and context sources («ContextSource») or service groups that provide appropriate access to context resources («ContextServiceCommunity»). Context dependencies are depicted through the two subtypes of the «CAMechanism» class: «ContextBinding» refers to a direct mapping between a context-aware element and a context value (e.g., user age or temperature value in Celsius degrees), whereas «ContextTriggering» is related to the modification of the service execution through an appropriate «Action» executed, when specific conditions are met («ContextConstraint»).

The ContextUML metamodel is extended in (23) towards the direction of further decoupling the context model from the service model design. In the proposed scheme the elements of the extended ContextUML profile are combined with UML activity diagrams in order to model applications consisting of Web services. AOP techniques are also exploited in the platform specific level for the service adaptation to the contextual environment. A similar approach is proposed in (12). Different UML profiles are introduced at the design level resulting in the modeling and, finally, in the executable code and configuration files of a fully functional Web application consisting of different context-adapted Web services. In this solution context management is

[2]http://www.eclipse.org/aspectj/

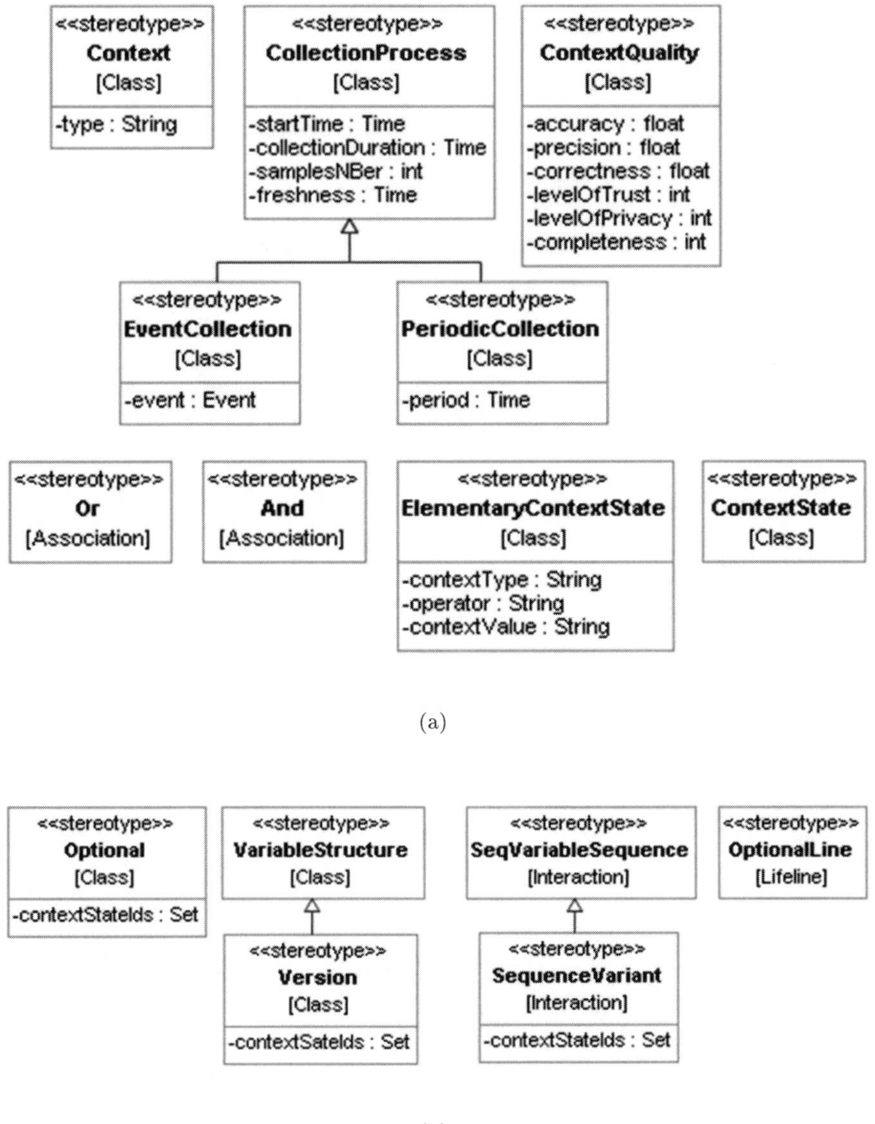

Figure 1.6: Ayed profile for context properties and context-awareness.

decoupled from the main application logic in all development stages. The Eclipse Modeling Framework (EMF) (4) and dedicated tools are exploited

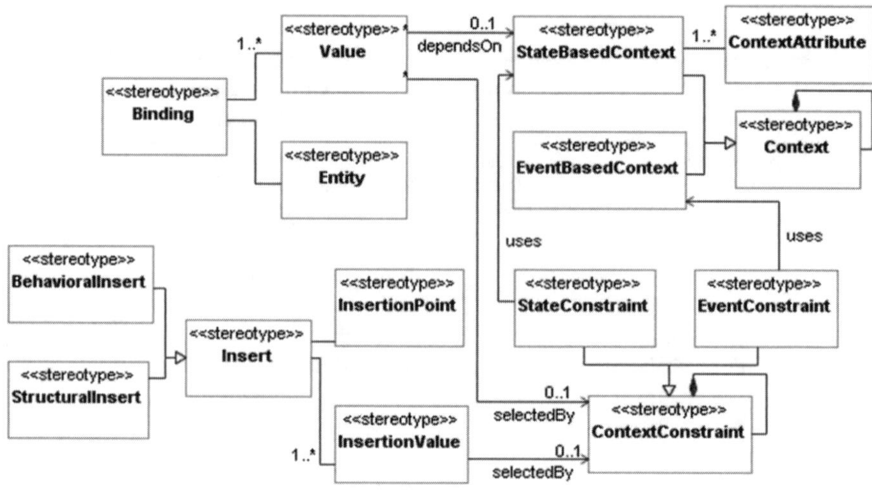

Figure 1.7: Main elements of the Grassi and Sindico profile.

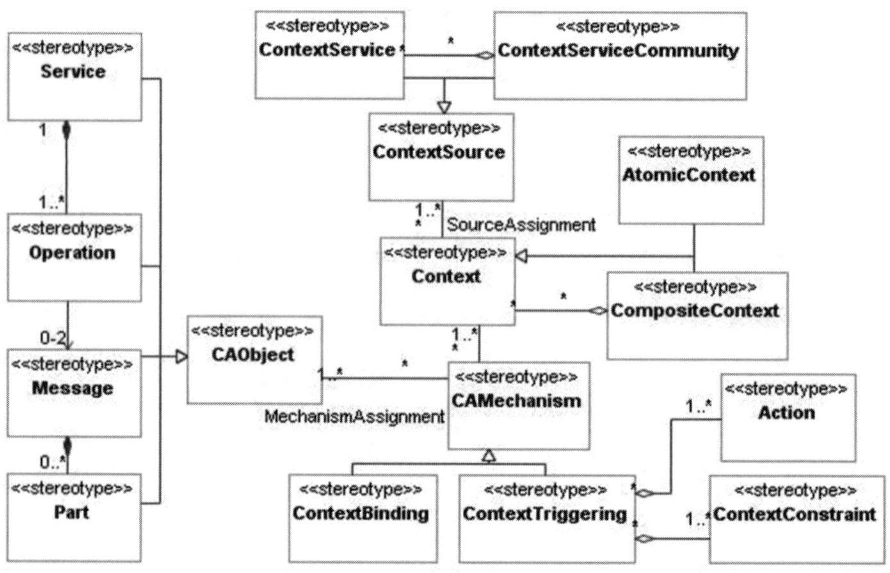

Figure 1.8: The metaclasses of the ContextUML metamodel.

for the model parsing functions, whereas templates of the Apache Velocity engine[3] drive the transformation process. The proposed technique allows the reuse of existing Web services through a reverse transformation between UML models and WSDL descriptions (the direct process is described in (9)). Web services are distinguished into business Web services that participate in the main service logic and context Web services that expose the functionality of context sources, whereas the application flow is modeled by a state transition diagram. The steps of the methodology are illustrated in Figure 1.9.

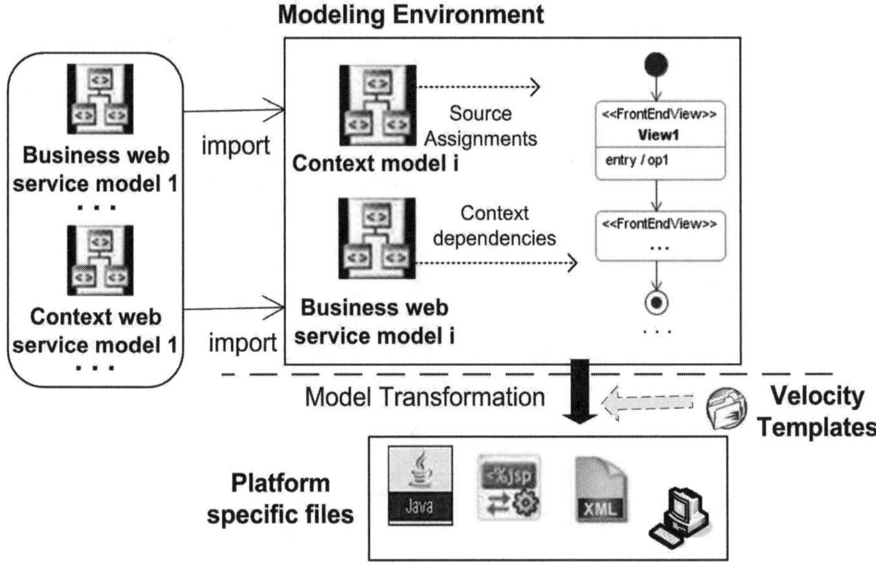

Figure 1.9: Steps of model-driven development exploiting state machines and platform specific implementation mapping.

In Context-Aware Service Oriented Architecture (CASOA) presented in (32) the context view is also separated from the business view throughout the application development. Context information is modeled using ontologies, which have been transferred in UML notation for use with MDA abstraction levels (Context View). In order to bind contextual properties and business entities, a Composition View is introduced. It consists of *Business Process Components* responsible for business logic functions and *Context Processing Components* aiming at performing adaptation actions to the business behavior; the two component groups are bound through the *Component Manager*

[3] http://velocity.apache.org/

element. Multiple business and context components are put together through the *Composite Component Manager* at Adaptation View, whereas Service View is denoted to the representation of Web service properties (operation, messages, etc.). These different views are depicted in Figure 1.10.

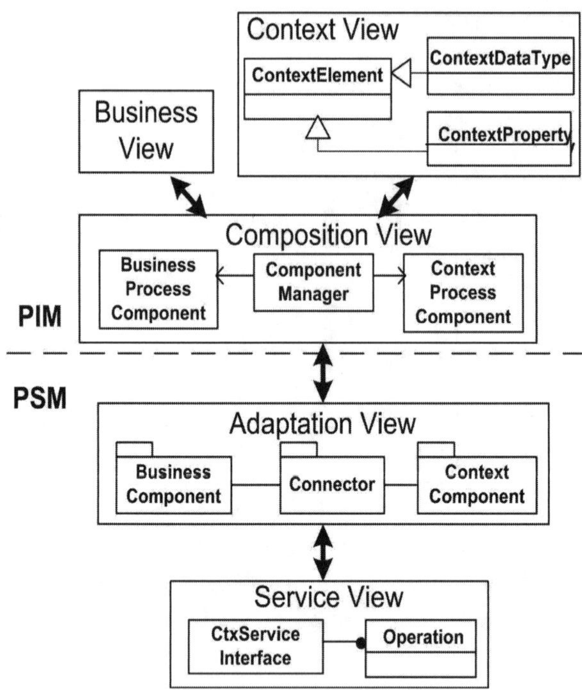

Figure 1.10: Views and components of Context-Aware Service-Oriented Architecture.

Another model-driven approach built around three layers with different abstraction degrees is presented in (1):

- *First layer – service specification level*: the context-aware service and its environment are modeled from an external perspective in terms of contextual changes, information retrieval, and appropriate actions neglecting information about implementation platforms and context retrieval mechanisms.

- *Second layer – platform-independent service design*: the model view is moved to the internal perspective expressing how the different application parts communicate.

- *Third layer – platform-specific service design*: corresponds to the PSM service model and refers to different middleware including Web services.

The procedure is built mainly around the abstract service platform A-MUSE,[4] which includes context sources and action providers, whereas a specialization of Interaction System Design Language (ISDL),[5] namely Events-Conditions-Actions Domain Language (ECA-DL), is exploited for modeling purposes along with UML class diagrams and Object Constraint Language (OCL) (21) expressions. The service creation process is preceded by a preparation phase, where the service developers identify the modeling properties, such as modeling language, abstract platforms, and transformations.

1.4 Approaches Based on Semantic Technologies

Semantic technologies are used to encode dependencies between pieces of information separately from data and content providing means for the sophisticated representation of information including reasoning and inference capabilities. Moreover, the Web Ontology Language (OWL) has been finalized since 2004 by W3C as one of the standard formats for ontology representation (34). Ontologies are usually exploited for the modeling of context information in specific domains of interest. However, in many cases the ontology specification drives the development process. Thus, in this section we distinguish two main subcategories that make use of semantic technologies. In the first subcategory the development effort is concentrated around the system ontologies, whereas in the second the exploitation of the semantic capabilities of Web services is undertaken.

In the first subcategory, where ontology definitions or service semantic properties are driving the development process in order to allow developers to reuse existing services in specific domains, Knublauch presents the architecture of an application that finds appropriate holiday destinations and activities for customers based on OWL ontologies (15). The functionality of the application is made available to software agents through a Web service interface and to end-users through conventional Web browsers. Input to these services is in both cases a collection of data objects/structures about a customer (e.g., age, available budget, etc.), while the output is a list of suitable destinations along with a list of suggested activities (e.g., sightseeing, relevant sports, bars, etc.) and corresponding contact addresses. Both input and output data structures are represented in OWL, while OWL parsing and/or mapping into object-oriented languages (e.g., Java) can be performed through frameworks

[4]http://www.freeband.nl

[5]http://isdl.ctit.utwente.nl/

like Jena (6). In (15), apart from input/output data structures, ontologies are also used to represent the background knowledge, which is necessary to the application in order to fulfill its task. These are two separate yet linked layers, namely the Semantic Web Layer, which makes ontologies and interfaces available to the public, and the Internal Layer, which consists of control and reasoning mechanisms (Figure 1.11). On the one hand, core ontologies define the basic knowledge structure through base classes, which can be extended or instantiated arbitrarily by external ontology providers on the semantic Web and must be hardwired into the executable system. On the other hand, the knowledge encoded in the external ontologies can only be used by generic reasoning engines like rule execution engines or description logic classifiers (3). Protege with OWL Plugin (16) follows the above rationale in semantic Web application development. Protege is an ontology-driven development tool providing intelligent guidance to detect mistakes as a debugger in a programming environment, serving also as a rapid prototyping environment.

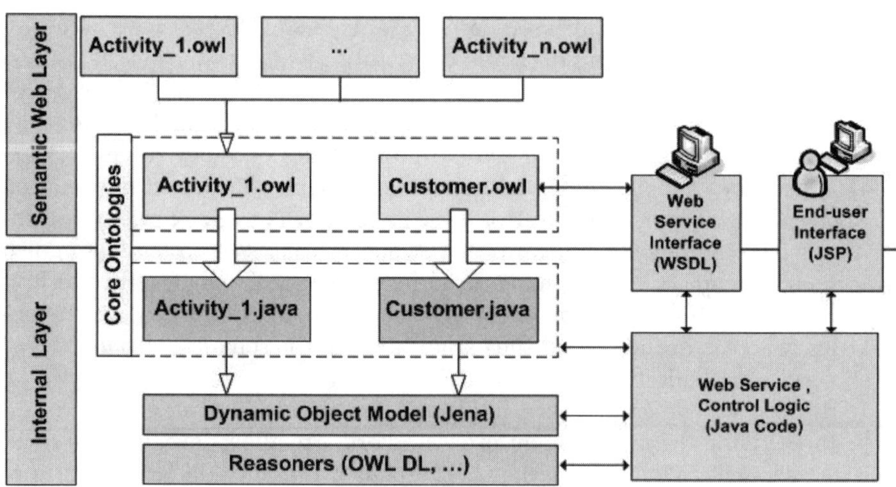

Figure 1.11: Layers in semantic Web applications.

In (29), Sun et al. apply software engineering methods and tools to visualize, simulate, and verify Semantic Markup for Web Service (OWL-S) process models. They present the software engineering language Live Sequence Charts (LSCs) and its tool Play-Engine (10) to visualize and simulate OWL-S process model ontologies, which capture the essential information about how a service is to be invoked and executed, as well as the expected outputs. LSCs are a powerful visual formalism serving as an enriched requirements' specification language. Two kinds of charts are distinguished in LSCs: existential charts are mainly used to describe possible interactions between participants in early system design stages, while at a later stage universal charts are used to specify behaviors that should always be exhibited. A chart typically consists of multiple instances, which are represented as vertical lines. Along with each line there is a finite number of locations that are the joint points of instances and messages. A location carries the "temperature annotation" for progress within an instance and messages passing between instances are represented as horizontal lines. "Cold conditions" are used as assistants specifying complex control structures like "do-while," "guarded-choice," etc. "Hot conditions" are asserted to assure critical properties at certain points of execution. Figure 1.12 depicts an LSC universal chart capturing the required interactions between a Serving-Using Agent and a Budget-Checking Agent cooperating in the Check Holiday Budget service.

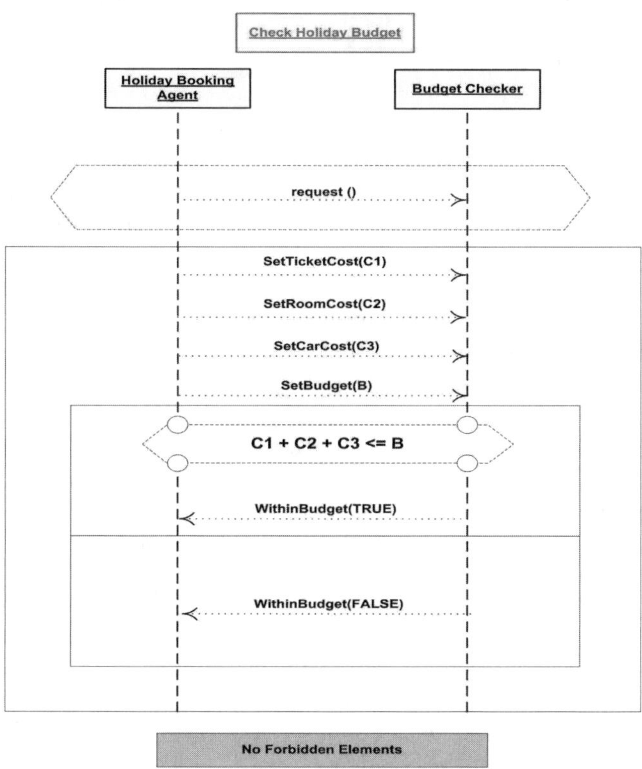

Figure 1.12: An LSC example: Check Holiday Budget.

When the Service-Using Agent requests the "Check Budget" service, necessary information like *budget_ticket_cost*, *budget_room_cost* and *budget_car_cost* is provided by the Service-Using Agent. The Budget-Checking Agent returns true if the budget is more or equal to the sum of the ticket/room/car-rental cost and false otherwise. One of the significant advantages of using LSCs is that LSC descriptions can be executed by Play-Engine without implementing the underlying object system. Play-Engine is a tool supporting a "play-in" - "play-out" approach to the specification, validation, analysis, and execution of LSCs. Behavior is "played-in" directly from the system's user interface, and as this is being done the Play-Engine is constructing LSCs. Later, behavior can be "played-out" independently from the user interface, and the tool executes the LSCs directly, driving this way the system's behavior. With "playing-out" Play-Engine computes a "maximal response" to a user-provided event. During the computing of such an event "hot-conditions" are evaluated. If any "hot condition" evaluates to false, a violation is caught. Otherwise, simulation continues with the user-provided events. This way, users may detect undesired behaviors allowed by the specification early in the development.

Another Semantic Web Service Development Tool – called ODE SWS - is presented in (7). ODE-SWS is a development environment for the design of Semantic Web Services (SWS) at the knowledge level. It describes the service following a problem-solving approach, in which the Semantic Web Services are modeled by "tasks" and "methods." More specifically, "tasks" describe the functional feature of a service, i.e., service's input/output roles, pre/post conditions and effects. Each task describes, on the one hand, the roles and conditions required to execute the service (i.e., input roles and pre-conditions) and on the other hand the results of the service execution (i.e., output roles, post-conditions and effects). "Methods" describe how a task can be solved. In other words, they specify the control of the reasoning steps needed to solve a given task. A method is defined by a set of input/output roles, the preconditions, which must be verified for method execution, and the postconditions, which describe the final state after the execution. A method describes also the internal components that are executed to solve a task. These components are tasks (called subtasks), each solved by other methods that can be composed of other sub-tasks, and so forth. Figure 1.13a shows the representation – as an eclipse – of the task *Task_Book Restaurant*, which defines the functional features of the service *BookRestaurant*. Figure 1.13b depicts an example of the internal description of a method (rounded corner square): the task *Task_BookRestaurant* is solved by the method *Method_BookRestaurant*, which is composed of four subtasks *Task_Find Restaurant*, *Task_CheckWorkingHours*, *Task_SelectTable*, and *Task_ BookTable*. The description of the internal structure of the method is completed with the definition of the execution coordination of its subtasks; that is, a method specifies the control flow among its subtasks. Figure 1.13c presents graphically the coordination of the subtasks of the method *Method_Book Restaurant*: the tasks *Task_FindRestaurant* and

Task_ CheckWorkingHours are executed first in order to select a restaurant in a given day and time. Then, the task *Task_ SelectTable* is executed in a loop until the user selects the desired table and, finally, the task *Task_ BookTable* books the table.

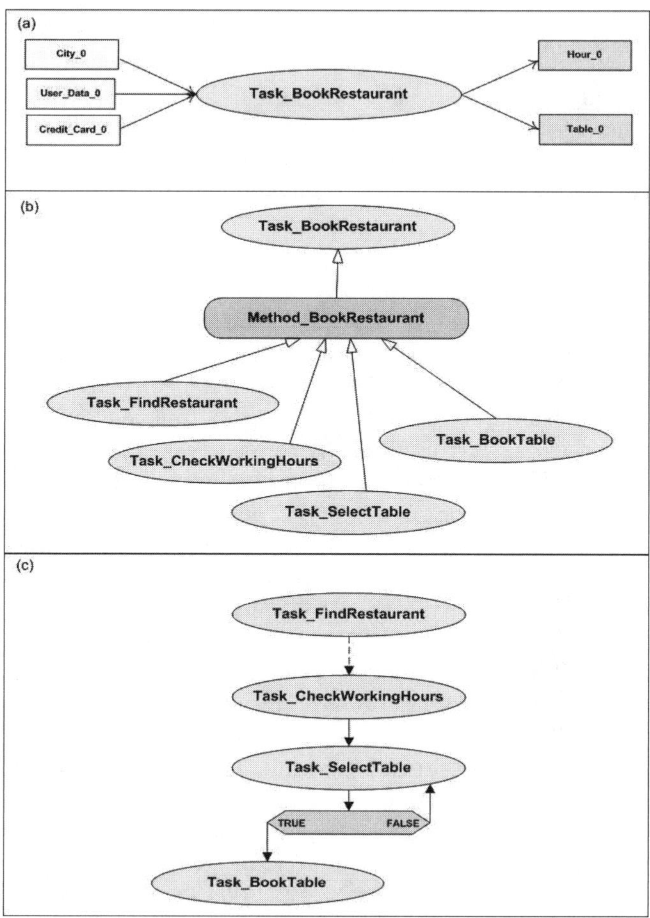

Figure 1.13: Tasks & Methods: The Book Restaurant example.

Another ontology-based approach for the specification and reconciliation of Web service context is presented in (28). Maamar et al. have also highlighted the importance of dealing with the composition of Web services at three connected levels (18). The lower level is dedicated to the messages exchanged among the interactive Web services of a composite service. The mid level is about the semantics of the content that these messages transfer and the need for common semantics becomes intensified especially when Web services coming from different providers take part in the composition. Finally, the higher level is related to the context, in which the Web service composition takes place. Since Web services may belong to different providers, their context definition can be different in terms of structure, name, number, and meaning of arguments, etc. Subramanian et al. present a two-step process in order to address the potential context heterogeneity of Web services. The first step consists of specifying contexts using a dedicated language: OWL-C (Ontology Web Language for Context ontologies), which is detailed in (18). The second step consists of fixing the heterogeneity of contexts using mediation mechanisms. These mechanisms are supported by a prototype called ConWeS (Context-based semantic Web Services).

In the second subcategory, ontologies and context are used to discover reusable WSs for development purposes. Furthermore, the studies within this subcategory describe the extraction of ontologies from WS descriptions in order to identify the context conditions where the WS can be reused. In (31), the term "context" is defined from two perspectives: one from service requesters' and another one from Web services'. From the former perspective, context is defined as the surrounding environment affecting requesters' service discovery and access, such as requesters' preferences, activities and accessible network and devices, while from the latter perspective, context is defined as the surrounding environment affecting Web service delivery and execution, such as networks and protocols for service binding, devices, and platforms for service execution, and so on. In this study an ontology-based context model towards a formal definition of contextual descriptions pertaining to both service requesters and Web services is presented. Yang et al. have implemented the ontology-based context model into a rule-base system using Java Expert System Shell (JESS)[6] for context elicitation by using an algorithm – called Rete – to match rules. This procedure is much faster than a simple set of cascading "if-then" statements adopted in conventional programming languages. Based on this context elicitation system, the authors present a Context-Aware Services-Oriented Architecture (CA-SOA) for providing context-aware services requests, publication, and discovery. OWL-S is utilized as the vehicle to carry contextual information, since OWL-S is a known tool from the semantic Web society tailored for Web services. Nevertheless, one of the main advantages is that both context model and CA-SOA are open and not limited to OWL-S. Other proprietary (e.g., Web Service Level Agreement from IBM) or standard (e.g., Web Services Agreement Specification from Global Grid Forum) solutions can be used to carry the knowledge.

An approach for the automatic retrieval of Web services is proposed in (13). In contrast with other approaches, this approach uses a context on-

[6] www.jessrules.com

tology as the basis of the capability descriptions of Web services and when the capability descriptions of both Web services and the Web requests are grounded onto the same ontology, a matchmaking method is given for judging the equivalency between them. Jin and Liu argue that the real meaning of Web services should be captured by the corresponding controllable entities that Web services can interact with and can bring effects to. Thus, the Web service context is actually related to the effects that the Web service and its meaning can impose to its content and the way that the Web service imposes the effects. Furthermore, these controllable entities are domain-relevant and independent to any Web services. Consequently, the conceptualization of the controllable resources becomes the shared domain knowledge for domain Web services. Based on their previous work, they propose a two-level context ontology for specifying both entities and effects. The ontology on the first level declares the context entities of Web services, while the second level introduced (the domain ontology) is sharable and helps Web services find each other automatically.

Another approach based on domain ontologies is proposed in (26). Sabou et al. cite the problem of semi-automatically learning Web service domain ontologies and target a better understanding of the ontology learning task in the context of Web services and the identification of technologies that could potentially be used. They have designed a framework for performing ontology learning in the context of Web services addressing several issues that constrain the development of an ontology learning solution in two ways. First, the framework exploits the peculiarities of Web service documentations to extract information used for ontology building. More specifically, the sublanguage characteristics of these texts lead to the identification of a set of heuristics. These heuristics are implemented as pattern-based extraction rules defined on top of linguistic information. Second, the learned ontologies are suited for Web service descriptions as they contain both static and procedural knowledge.

1.5 Discussion

The three approaches presented above enable developers to decouple, to a significant extent, the service logic from the context management part of a service. On the other hand, they have quite a few differences; therefore, an important question from a developer's viewpoint is whether a specific approach is more suitable than others for a specific type of service or even a specific development methodology.

If a rigorous service development methodology is used, investing significantly in the design stage, then MDD approaches are probably more suitable for context-aware Web service development. One can choose from several

Web service and context models already available and impose modification to either model (Web service or context model) with minimal implications to the other. There is also the possibility of source code generation either in a popular programming language (usually Object-Oriented) or to an extended version of such language as the ones described in Section 1.2. Programming language extensions are obviously more suitable when an agile development method is employed involving frequent Web service releases keeping modeling efforts to a minimum. In the case where the ability to perform complex reasoning using context is a priority then an approach involving semantics is probably preferable. This of course comes at a certain cost in the sense that the Web service development team must feel at ease with ontology specification languages such as OWL and the use of rule/reasoning engines such as Jess. Of course nothing precludes the use of a hybrid approach combining two or more of the approaches presented above or even other approaches available for context-aware Web service development. This is the case with (33) and (22) where model-driven transformations are guided by respective ontologies. Also, as it has been mentioned above, Model-Driven Development blends well with programming language extensions at the source code generation stage.

The multiplicity of options faced by a potential developer indicate that we are still lacking a uniform approach regarding context-aware Web service development methodology that can be applied in conjunction with a variety of modeling techniques and programming language frameworks. In this respect, one very obvious issue for future research is the process of gathering and analyzing requirements for context-aware Web services. Another issue is whether effort should be placed on the specification of high-level programming languages especially for services (such as (17)) or even especially for context-aware Web services. Last but not least, handling of context-related information and exchange of such information between service nodes always raises issues concerning privacy and security. Indeed, context information related to sensitive user data, such as personal information, current location, ongoing and past transactions, etc., should not be revealed easily; significant research effort should be placed in this direction as well.

1.6 Summary

This book chapter was dedicated to the presentation of techniques exploited for the development of context-aware Web services. It has been concerned with three such directions: programming language extensions, model-driven development and semantic technologies, which have been proven quite popular taking into account the existing literature. For each approach it is shown how it facilitates context handling during the service design and development process. A brief evaluation of the three approaches is provided at the end along with ideas for future research directions. The main conclusion drawn is that we still have some distance to cover with respect to a software production methodology tailored to the specific requirements of context-aware Web services.

References

[1] Almeida, J. P. A., Iacob, M.-E., Jonkers, H. and Quartel, D. 2006. Model-Driven Development of Context-Aware Services. *Proceedings of the 6th IFIP International Conference Distributed Applications and Interoperable Systems*, 213-227.

[2] Ayed, D., Delanote, D. and Berbers, Y. 2007. MDD Approach for the Development of Context-Aware Applications. *Proceedings of the 6th International and Interdisciplinary Conference on Modeling and Using Context*, 15-28.

[3] Baader, F., Calvanese, D., McGuineness, D., Nardi, D. and Patel-Schneider, P. 2003. *The Description Logic Handbook.* Cambridge University Press.

[4] Budinsky, F., Steinberg, D., Merks, E., Ellersick, R. and Grose, T. J. 2003. *Eclipse Modeling Framework.* Addison Wesley Professional.

[5] Dantas, F., Batista, T. and Cacho, N. 2007. Towards Aspect-Oriented Programming for Context-Aware Systems: A Comparative Study. *Proceedings of the 1st International Workshop on Software Engineering for Pervasive Computing Applications, Systems, and Environments.*

[6] Goldman, N. 2003. Ontology-Oriented Programming: Static Typing for the Inconsistent Programmer. *Proceedings of the 2nd International Semantic Web Conference*, 850-865, Springer-Verlag.

[7] Gomez-Perez, A., Gonzalez-Cabero, R. and Lama, M. 2004. ODE SWS: A framework for designing and composing semantic Web services. *IEEE Intelligent Systems* 19(4):24-31.

[8] Grassi, V. and Sindico, A. 2007. Towards Model Driven Design of Service-Based Context-Aware Applications. *Proceedings of the International Workshop on Engineering of Software Services for Pervasive Environments: in conjunction with the 6th ESEC/FSE joint meeting*, 69-74.

[9] Gronmo, R., Skogan, D., Solheim, I., Oldevik and J. 2004. Model-driven web services development. *Proceedings of the IEEE International Conference on e-Technology, e-Commerce and e-Service (EEE'04)*, 42-45.

[10] Harel, D. and Marelly, R. 2003. *Come, Let's Play: Scenario-Based Programming Using LSCs and the Play-Engine.* Springer-Verlag.

[11] Hirschfeld, R., Costanza, P. and Nierstrasz, O. 2008. Context-oriented Programming. *Journal of Object Technology* 7(3):125-151.

[12] Kapitsaki, G. M., Kateros, D. A., Prezerakos, G. N. and Venieris, I. S. 2009. Model-driven Development of Composite Context-aware Web Applications. *Journal of Information and Software Technology* 51(8):1244-1260.

[13] Jin, Z. and Liu, L. 2006. Web Service Retrieval: An Approach Based on Context Ontology. *Proceedings of the 30th Annual International Computer Software and Applications Conference*, 513-520.

[14] Keays, R. and Rakotonirainy, A., 2003. Context-Oriented Programming. *Proceedings of the 3rd ACM International Workshop on Data Engineering for Wireless and Mobile Access*, 9-16.

[15] Knublauch, H. 2004. Ontology Driven Software Development in the Context of the Semantic Web: An Example, Scenario with Protege/OWL. *Proceedings of the 1st International Workshop on the Model-Driven Semantic Web Enabling Knowledge Representation and MDA Technologies to Work Together.*

[16] Knublauch, H., Fergerson, R., Noy, N. and Musen, M., 2004. The Protege OWL Plugin: An Open Development Environment for Semantic Web Applications. *Proceedings of the 3rd International Semantic Web Conference.*

[17] Labey, S. De, van Dooren, M. and Steegmans, E. 2007. ServiceJ A Java Extension for Programming Web Services Interactions. *Proceedings of the IEEE International Conference on Web Services*, 505-512.

[18] Maamar, Z., Narendra, N. and van den Heuvel, W. 2005. Towards an Ontology-based Approach for Specifying Contexts of Web Services. *Proceedings of the Montreal Conference on e-Technologies (MCETECH'2005).*

[19] Min, X., Jizhong, Z., Yong, Q., Hui, H., Ming, L. and Wei, W. 2007. Isotope Programming Model: a Kind of Program Model for Context-Aware Application. *Proceedings of the International Conference on Multimedia and Ubiquitous Engineering*, 597-602.

[20] OMG. 2003. Model-Driven Architecture v.1.0.1. http://www.omg.org/docs/omg/03-06-01.pdf (accessed March 29, 2009).

[21] OMG. 2006. Object Constraint Language Specification v.2.0. http://www.omg.org/docs/formal/06-05-01.pdf (accessed March 29, 2009).

[22] Ou S., Georgalas N., Azmoodeh M., Yang K. and Sun X. 2006. A Model Driven Integration Architecture for Ontology-Based Context Modelling and Context-Aware Application Development. *Proceedings of the Second European Conference on Model Driven Architecture Foundations and Applications*, 188-197, Springer-Verlag.

[23] Prezerakos, G. N., Tselikas, N. and Cortese, G. 2007. Model-driven Composition of Context-aware Web Services Using ContextUML and Aspects. *Proceedings of the IEEE International Conference on Web Services 2007*, 320-329.

[24] Rho, T. and Kniesel, G. 2004. Uniform Genericity for Aspect Languages. Technical report IAI-TR-2004-4. *CS Dept. III, University of Bonn.*

[25] Rho, T., Schmatz, M. and Cremers, A. B. 2006. Towards Context-Sensitive Service Aspects. *Proceedings of the European Conference on Object-Oriented Programming.*

[26] Sabou, M., Wroe, C., Goble, C. and Stuckenschmidt, H. 2005. Learning domain ontologies for semantic Web service descriptions. *Journal of Web Semantics* 3(4):340-365.

[27] Sheng, Q. Z. and Benatallah, B. 2005. ContextUML: A UML-Based Modeling Language for Model-Driven Development of Context-Aware Web Services. *Proceedings of the International Conference on Mobile Business*, 206-212.

[28] Subramanian, S., Narendra, N. and Maamar, Z. 2006. Ontologies for Specifying and Reconciling Contexts of Web Services. *Electronic Notes in Theoretical Computer Science* 146(1):43-57.

[29] Sun, J., Li, Y., Sun, J. and Wang, H. 2005. Visualizing and Simulating Semantics Web Services Ontologies. *Proceedings of the 7th International Conference on Formal Engineering Methods.*

[30] Tanter, E., Gybels, K., Denker, M. and Bergel, A. 2006. Context-Aware Aspects. *Proceedings of Software Composition 2006*, 227-242.

[31] Yang, S., Zhang, J. and Chen, I. 2008. A JESS-enabled context elicitation system for providing context-aware Web services. *Expert Systems with Applications* 34(4):2254-2266.

[32] Vale, S. and Hammoudi, S. 2008. Model Driven Development of Context-aware Service Oriented Architecture. *Proceedings of the 11th IEEE International Conference on Computational Science and Engineering Workshops*, 412-418.

[33] Vale, S. and Hammoudi, S. 2008. Context-aware Model Driven Development by Parameterized Transformation. *Proceedings of the Model Driven Interoperability for Sustainable Information Systems*, 121-133.

[34] World Wide Web Consortium. 2004. OWL Web Ontology Language Reference. http://www.w3.org/TR/owl-features/ (accessed April 10, 2009).

Chapter 2

Model-Driven Development of Context-Aware Web Services

Jian Yu, Quan Z. Sheng, Kewen Liao, and Hoi S. Wong

2.1 Introduction

Context awareness refers to the capability of an application or a service being aware of its physical environment or situation (i.e., context) and responding proactively and intelligently based on such awareness (1, 7, 11). With recent developments in computer hardware, software, networking, and sensor technologies, context awareness becomes one of the most exciting trends in computing today that holds the potential to make our daily lives more productive, convenient, and enjoyable. For example, a tour-guide service gives tourists suggestions on the attractions to visit by considering their current locations, preferences, and even the prevailing weather conditions.

In the last decade, Web services have become a major technology to implement loosely-coupled business processes and perform application integration. Through the use of context, a new generation of *smart* Web services are currently emerging as an important technology for building innovative context-aware applications. We call such category of Web services context-aware Web services, or CASs in short.

To date, CASs are still hard to build. One reason is that current Web services standards (e.g., UDDI, WSDL, SOAP) are not sufficient for describing and handling context information (13, 15, 22, 24). CAS developers must implement everything related to context management—including the collection, dissemination, and usage of context information—in an adhoc manner. Another reason is that, to the best of our knowledge, there is a lack of generic approaches for formalizing the development of CASs. As a consequence, developing CASs is a very cumbersome and time-consuming activity, especially when these CASs are complex.

Model Driven Architecture (MDA) (8) is an approach that supports system development by employing a model-centric and generative development process. MDA increases the quality of complex software systems based on creating high level system models and automatically generating system architectures from the models. It also eases system maintenance and evolution. Any changes can be easily made at the model level and propagated automatically to the implementation. Finally, MDA enhances the portability of service design due to technical independence of service models. The service models can be migrated to new technologies (e.g., new Web service languages or protocols) by simply developing new transformation rules.

This chapter presents ContextServ (21), a platform for rapid development of context-aware Web services. ContextServ uses a UML-based modelling language—ContextUML (20)—for formalizing the design and development of CASs. ContextUML provides constructs for i) generalizing context provisioning that includes context attributes specification and retrieval; and ii) formalizing context-awareness mechanisms and their usage in CASs. ContextServ supports the full lifecycle of developing CASs: it includes a visual ContextUML editor, a translator from ContextUML to WS-BPEL—the de facto Web services process language, and a WS-BPEL deployer working with the JBoss Application Server.

The rest of the chapter is organized as follows: Section 2.2 briefly overviews some basic concepts used in the chapter, and then gives an example CAS. Section 2.3 introduces the ContextUML language for modelling the concept of context and the context-aware mechanisms. Section 2.4 introduces the architecture and implementation of ContextServ. The main components of ContextServ, including the `Context Manager`, the `ContextUML Modeler`, and the `RubyMDA Transformer` will be discussed in detail. Section 2.5 showcases the real-life application of the ContextServ platform with a a context-aware Web application called *Smart Adelaide Guide*. Finally, Section 2.6 discusses related work and also compares the capability of several UML-based context modelling languages, and concludes the paper.

2.2 Background

In this section, we first briefly overview some basic relevant concepts, namely *model-driven development*, *UML*, *context*, and *context-aware Web service* (CAS).

2.2.1 Model-Driven Development

Model driven development (MDD) (8) is an approach that supports system development by employing a model-centric and generative development

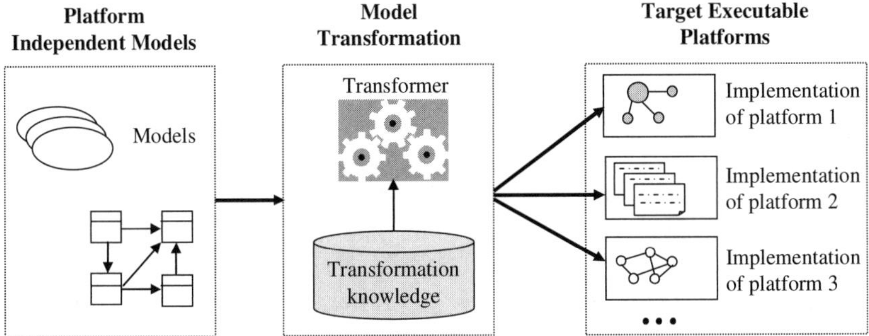

Figure 2.1: Model-driven development.

process. The basic idea of MDD is illustrated in Figure 2.1. Adopting a higher-level of abstraction, software systems can be specified in platform independent models (PIMs), which are then (semi)automatically transformed into platform specific models (PSMs) of target executable platforms using some transformation tools. The same PIM can be transformed into different executable platforms (i.e., multiple PSMs), thus considerably simplifying software development.

2.2.2 Unified Modelling Language

The Unified Modelling Language (UML)[1] is considered as the industry defacto standard for modelling software systems and plays a central role in Model Driven Architecture (MDA) (8). In UML, the structural aspects of software systems are defined as classes, each formalizing a set of objects with common services, properties, and behavior. Services are described by methods. Properties are described by attributes and associations. Object Constraint Language (OCL) can be used to express additional constraints.

UML can also serve as the foundation for building domain specific languages by specifying *stereotypes*, which introduce new language primitives by subtyping UML core types, and *tagged values*, which represent new properties of these primitives. Model elements are assigned to such types by labelling them with the corresponding stereotypes. In addition, UML can also be used as meta modelling language, where UML diagrams are used to formalize the abstract syntax of another modelling language.

[1]http://www.omg.org/technology/documents/formal/uml.htm.

2.2.3 Context

Dey and Abowd have defined context—which is widely used in the literature today—as "*any information that can be used to characterize the situation of an entity. An entity is a person, place, or object that is considered relevant to the interaction between a user and an application, including the user and applications themselves*" (6).

In a CAS environment, context contains any information that can be used by a Web service to adjust its execution and output. Examples of contexts are: i) contexts related to a service requester (mostly it is the client who invokes a service), including the requester's identification information, personal preferences, current situation (e.g., location), and other information (e.g., friends list, calendar); ii) contexts related to a Web service, such as service location, service status (e.g., available, busy), and its QoS attributes (e.g., price, reliability); and iii) other contexts like time and weather information. It should be noted that some contexts are application specific. Forecasted weather, for instance, could be a context in a vacation planning service, but not in a currency conversion service.

Some context information can be *sensed* directly (e.g., locations and temperatures using physical sensors), while others have to be *derived* from available context information. Contexts are provided by *context providers*. It is interesting to mention that more and more context providers advertise their services—called *context information services*[2]—over the Web that can be seamlessly integrated into CASs. Recently, quite a few research efforts on modelling the context provisioning services are proposed (18, 4, 10).

2.2.4 Context-Aware Web Service

A Web service is *context-aware* if it uses context information to provide relevant information and/or services to users, where relevancy depends on the users' task (6, 16, 19). A CAS can present relevant information or can be executed or adapted automatically, based on available context information. For example, a service can display restaurants that are around a user's current location, and if the weather is good, the service even suggests some restaurants that customers can sit outside.

To develop CASs, two important issues need to be considered. The first is the provisioning of context information. CAS developers have to identify what kind of context information will be used and how to derive it. Due to heterogeneity of context providers, sensor imperfection, quality of context information, and dynamic context environments, context provisioning is not trivial (18). In particular, various context providers may provide a same piece of context information (usually with different quality and data formats) and it is difficult to specify—at design stage—which context provider should be

[2]E.g., U.S. National Weather Service, http://www.nws.noaa.gov.

contacted for the provision of a particular context. Further, some context required by a CAS may not be able to find any context provider who can supply the context directly.

The second issue is the mechanisms that can be used by CASs to adapt their behaviors based on corresponding context information without explicit user intervention. In other words, it is the problem of how to use context information to achieve *context awareness* of services. In addition, the abstracted context awareness mechanisms should guarantee the efficiency of the development and maintenance of CASs. For instance, it should be possible to use legacy Web services to develop CASs without changing their implementation.

2.2.5 Attraction Searching Service

Suppose that there is a context-aware attraction searching service that is offered by a mobile network operator. Mobile users subscribed to the network operator can invoke this service using their mobile devices to get the recommended attractions when they visit new cities. The service works like the following:

- Users can subscribe their personal preferences to the service. For example, a user can specify what kinds of attractions (e.g., historical sites) she likes, and which language (e.g., Chinese) the description of attractions should be.
- The service recommends attractions according to a user's location (e.g., the city that the user is currently in).
- During the recommendation, the service also considers other contexts like weather and user preferences. If the weather is *harsh*, the service will only suggest indoor attractions (e.g., Adelaide Art Gallery). The definition of weather to be harsh depends on a couple of contexts like the temperature (e.g., above 30 degree Celsius) and the speed of wind (e.g., more than 50 km/h). The recommended attractions will also reflect the user's preferences, e.g., translating the attraction descriptions to the user's preferred language.

2.3 ContextUML

In this section, we introduce the ContextUML modelling language, a core component that underpins the model-driven development of CASs. ContextUML metamodel is shown in Figure 2.2, which can be roughly divided into two parts: *context modelling metamodel* and *context-awareness modelling metamodel*.

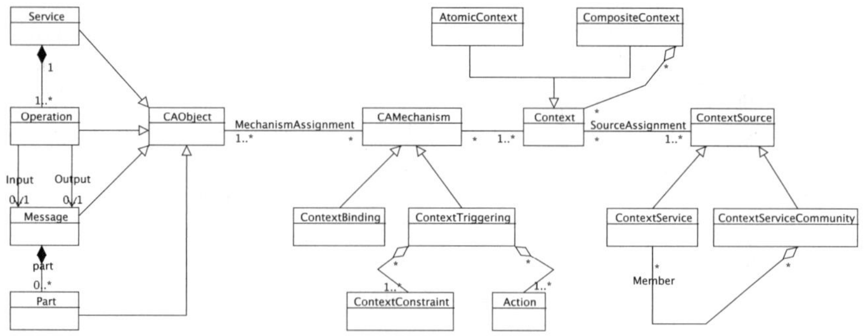

Figure 2.2: ContextUML metamodel.

2.3.1 Context Modelling

Context Type

A Context is a class that models the context information. In our design, the type Context is further distinguished into two categories that are formalized by the subtypes AtomicContext and CompositeContext. Atomic contexts are low-level contexts that do not rely on other contexts and can be provided directly by context sources (see Section 2.3.1). In contrast, composite contexts are high-level contexts that may not have direct counterparts on the context provision. A composite context aggregates multiple contexts, either atomic or composite. The concept of composite context can be used to provide a rich modelling vocabulary.

For instance, in the scenario of `attraction-search` service, `temperature` and `wind speed` are atomic contexts because they can be provided by e.g., `GlobalWeather`[3] Web service. Whereas, `harshWeather` is a composite context that aggregates the former two contexts.

Context Source

The type ContextSource models the resources from which contexts are retrieved. We abstract two categories of context sources, formalized by the context source subtypes ContextService and ContextServiceCommunity, respectively. A context service is provided by an autonomous organization (i.e., context provider), collecting, refining, and disseminating context information. To solve the challenges of heterogeneous and dynamic context information, we abstract the concept of context service community, which enables the dynamic provisioning of optimal contexts. The concept is evolved from *service community* we developed in (3) and the details will be given in Section 2.3.1.

[3]http://www.capescience.com/webservices/globalweather.

It should be noted that in ContextUML, we do not model the acquisition of context information, such as how to collect *raw* context information from sensors. Instead, context services that we abstract in ContextUML encapsulate sensor details and provide context information by interpreting and transforming the sensed information (i.e., raw context information). The concept of context service hides the complexity of context acquisition from CAS designers so that they can focus on the functionalities of CASs, rather than context sensing.

Context Service Community

A context service community aggregates multiple context services, offering with a unified interface. It is intended as a means to support the dynamic retrieval of context information. A community describes the capabilities of a desired service (e.g., providing user's location) without referring to any actual context service (e.g., `WhereAmI` service). When the operation of a community is invoked, the community is responsible for selecting the most appropriate context service that will provide the requested context information. Context services can join, leave communities at any time.

By abstracting ContextServiceCommunity as one of context sources, we can enable the dynamic context provisioning. In other words, CAS designers do not have to specify which context services are needed for context information retrieval at the design stage. The decision of which specific context service should be selected for the provisioning of a context is postponed until the invocation of CASs.

The selection can be based on a *multicriteria utility function* (23, 3) and the criteria used in the function can be a set of *Quality of Context* (QoC) parameters (5). The examples of QoC parameters are: i) `precision` indicating the accuracy of a context information; ii) `correctnessProbability` representing the probability of the correctness of a context information; and iii) `refreshRate` indicating the rate that a context is updated.

The quality of context is extremely important for CASs in the sense that context information is used to automatically adapt services or content they provide. The imperfection of context information may make CASs *misguide* their users. For example, if the weather information is outdated, our attractions searching service might suggest users to surf at the Bondi Beach although it is rainy and stormy. Via context service communities, the optimal context information is always selected, which in turn, ensures the quality of CASs.

2.3.2 Context Awareness Modelling

A CAMechanism is a class that formalizes the mechanisms for context awareness (CA for short). We differentiate between two categories of context awareness mechanisms by subtypes ContextBinding and ContextTriggering, which will be detailed in Section 2.3.2 and Section 2.3.2, respectively. Context awareness mechanisms are assigned to context-aware objects—modelled in the type CAObject—by the relation MechanismAssignment, indicating which objects have what kinds of context awareness mechanisms.

CAObject is a base class of all model elements in ContextUML that represent context-aware objects. There are four subtypes of CAObject: Service, Operation, Message, and Part. Each service offers one or more operations and each operation belongs to exactly one service. The relation is denoted by a composite aggregation (i.e., the association end with a filled diamond). Each operation may have one input and/or one output messages. Similarly, each message may have multiple parts (i.e., parameters). A context awareness mechanism can be assigned to either a service, an operation of a service, input/output messages of an operation, or even a particular part (i.e., parameter) of a message. It is worth mentioning that the four primitives are directly adopted from WSDL, which enables designers to build CASs on top of the previous implementation of Web services.

Context Binding

A ContextBinding is a subtype of CAMechanism that models the automatic binding of contexts to context-aware objects. By abstracting the concept of context binding, it is possible to automatically retrieve information for users based on available context information. For example, suppose that the operation of our example CAS has an input parameter `city`. Everyone who wants to invoke the service needs to supply a city name to search the attractions. Further suppose that we have a context `userLocation` that represents the city a user is currently in. A context binding can be built between `city` (input parameter of the service) and `userLocation` (context). The result is that whenever our CAS is invoked, it will automatically retrieve attractions in the city where the requester is currently located.

An automatic contextual reconfiguration (i.e., context binding) is actually a *mapping* between a context and a context-aware object (e.g., an input parameter of a service operation). The semantics is that the value of the object is supplied by the value of the context. Note that the value of a context-aware object could be derived from multiple contexts. For the sake of simplicity, we restrict our mapping cardinality as one to one. In fact, thanks to the introduction of the concept of composite context, we can always model an appropriate composite context for a context-aware object whose value needs to be derived from multiple contexts.

Context Triggering

The type ContextTriggering models the situation of contextual adaptation where services can be automatically executed or modified based on context information. A context triggering mechanism contains two parts: a set of *context constraints* and a set of *actions*, with the semantics that the actions must be executed if and only if all the context constraints are evaluated as true.

A context constraint specifies that a certain context must meet certain conditions in order to perform a particular operation. Formally, a context constraint is modelled as a predicate (i.e., a Boolean function) that consists of an operator and two or more operands. The first operand always represents a context, while the other operands may be either constant values or contexts. An operator can be either a prefix operator that accepts two or more input parameters or a binary infix operator (e.g., $=$, $\leq$) that compares two values. Examples of context constraints can be: i) `harshWeather`$= true$; ii) `windSpeed`≤ 50.

Considering our `attraction-search` service, we can have a context triggering mechanism assigned to its output message. The constraint part of the mechanism is `harshWeather`$= true$, and the action part is a transformation function `filter`$(\mathcal{M}, \mathcal{R})$, where $\mathcal{M}$ is the output message and $\mathcal{R}$ is a transformation rule (e.g., selecting only indoor attractions). Consequently, when a weather condition is not good, the output message will be automatically filtered (e.g., removing outdoor attractions) by the service.

2.4 ContextServ Platform

We have developed the ContextServ platform that provides an environment where a service developer specifies the required contexts and context-aware Web services using high-level and visual modelling languages. The service model is automatically transformed, using a set of transformation rules, to the executable specification of the target platform, which is then deployed to the corresponding execution engine. At this point, the service provider also needs to create a WSDL specification for the service and publish it (e.g., to UDDI registry) for free location and invocation.

The ContextServ architecture consists of three main components, namely the *Context Manager*, the *ContextUML Modeler*, and the *RubyMDA Transformer* (see Figure 2.3). All these components are implemented in Java. In the following subsections, we present the details of these three components.

2.4.1 Context Manager

Current implementation of the context manager supports atomic context, composite context, and context community.

Managing Atomic Context

As mentioned in Section 2.3.1, atomic contexts are low-level contexts that can be obtained directly from context sources. For the ContextServ platform to access context sources, *context providers* must be registered in the plat-

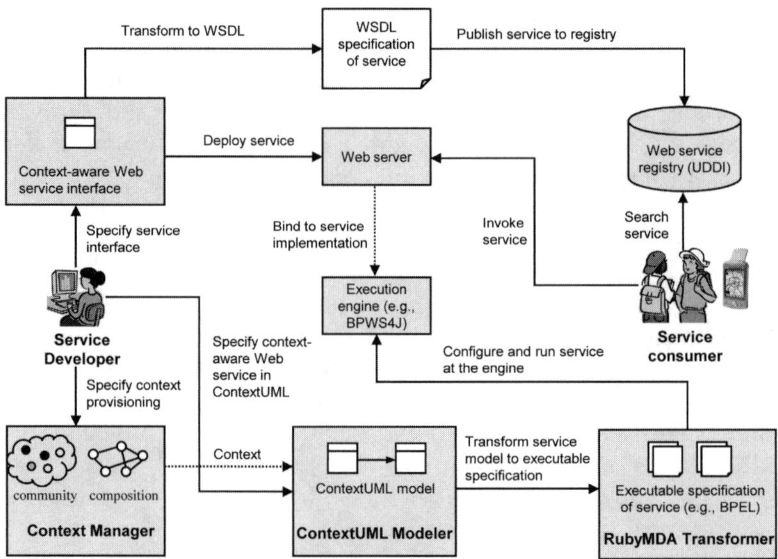

Figure 2.3: Architecture of ContextServ.

form. Currently the platform supports two types of context providers: local context providers and remote context providers. Local context providers are responsible for collecting local context information, such as device memory capacity, local temperature, etc. On the other hand, remote context providers gather context information from a remote sensor or device. Every context provider has at least one *agent*—a piece of program specifies the protocol of how to access the context information. For example, we use a Web service agent to access remote contexts and use a Java class to access local contexts. As illustrated in the atomic context GUI panel (Figure 2.4), on the left side we register a remote context provider containing a Web service agent to get the location context, and on the right side we define the collected context as an `location` atomic context with type `String`.

Managing Composite Context

As shown in Figure 2.5, composite contexts are modelled using state charts, a widely used formalism that is emerging as a standard for process modelling following its integration into UML. The state chart of a composite context is then exported into State Chart Extensible Markup Language (SCXML), an XML based language for describing generic state charts, and executed in a SCXML execution engine such as Commons SCXML[4].

[4]http://commons.apache.org/scxml.

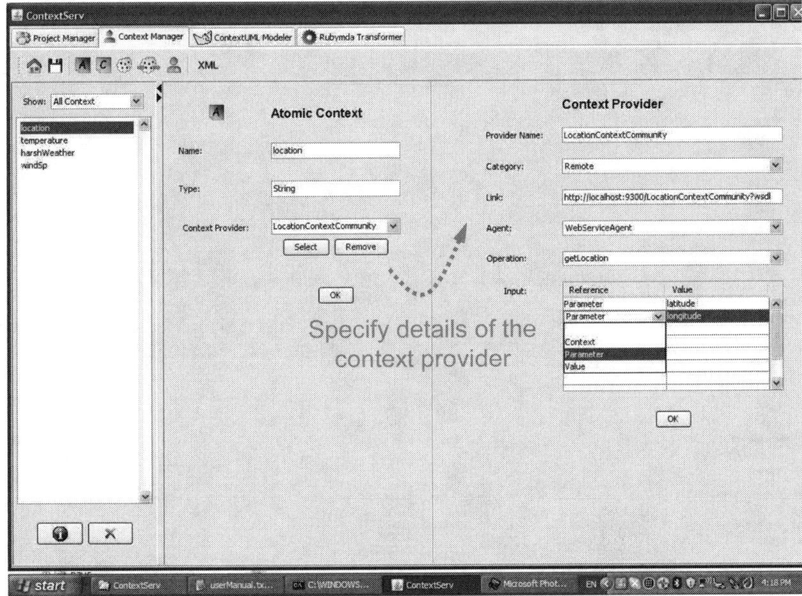

Figure 2.4: Panel for Defining Atomic Context

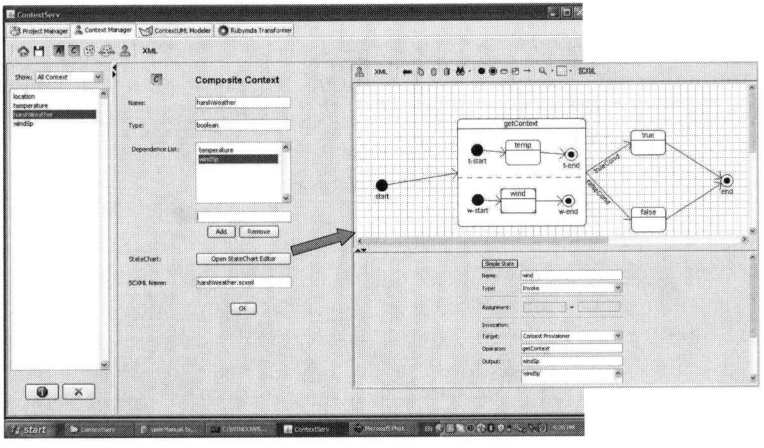

Figure 2.5: Specifying composite contexts.

Managing Context Community

A context community implements a common interface (`addContextSource()`, `removeContextSource()`, `selectContextSource()`) for context sources that provide same context information. The main purpose of a con-

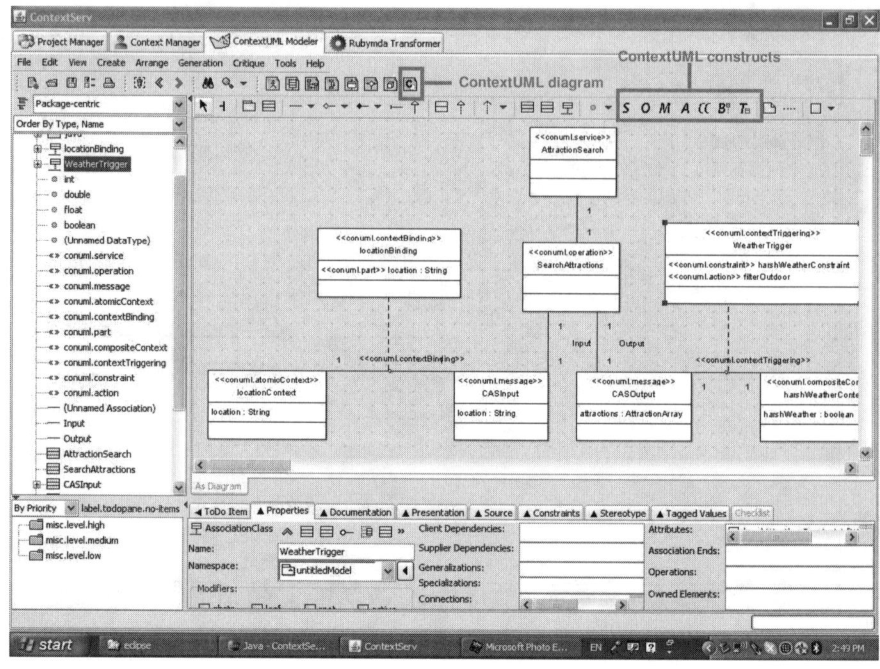

Figure 2.6: The ContextUML Modeler.

text community is to ensure robust and optimal provisioning of contexts to the consumer services so that on the one hand a candidate context source can take the place of an unavailable context source, on the other hand contexts having the best quality can be provisioned.

Let's assume that in a context community, a specific piece of context information can be provisioned by n context providers $\{CP_i\}_{1\leq i\leq n}$. The quality of a context provider $QoCP_i$ is modelled by a set of m quality attributes $\{A_j\}_{1\leq j\leq m}$ such as precision, trustworthiness, availability, response time, etc. And each quality attribute is assigned a weight W_j. So the quality of each context provider can be calculated using $\Sigma_{j=1}^{m}(A_j \times W_j)$. Details on how to model each quality attribute and how to implement the context community component can be found in (14).

2.4.2 ContextUML Modeler

Figure 2.6 shows the interface of the ContextUML modeler which is the visual environment for defining context-aware Web services using ContextUML. In the implementation, we extended ArgoUML, an existing UML editing tool[5],

[5]http://argouml.tigris.org.

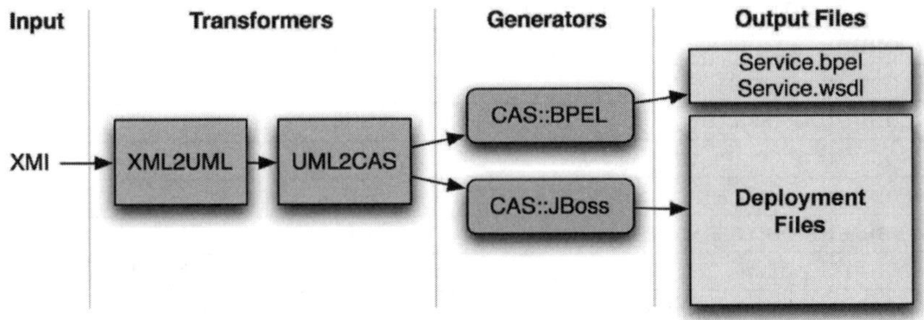

Figure 2.7: RubyMDA Data Flow.

by developing a new diagram type, ContextUML diagram, which implements all the abstract syntax of the ContextUML language.

2.4.3 RubyMDA Transformer

Services represented in ContextUML diagrams are exported as XMI files for subsequent processing by the RubyMDA transformer, which is responsible for transforming ContextUML diagrams into executable Web services, using RubyGems 1.0.1[6]. The ContextServ platform currently supports WS-BPEL, a de facto standard for specifying executable processes. Once the BPEL specification is generated, the model transformer deploys the BPEL process to a Web server and exposes it as a Web service. In the implementation, JBoss Application Server is used as the Web server since it is an open source and includes a BPEL execution engine, jBPM-BPEL 1.1. A set of mapping rules—from ContextUML diagram to BPEL and WSDL specifications—has been developed for the transformation purposes.

RubyMDA is developed based on the model transformation rules. The model transformation rules are mappings from ContextUML stereotypes to BPEL elements. Table 2.1 shows a summary of the model transformation rules of RubyMDA.

Figure 2.7 shows the data flow of RubyMDA model transformer. RubyMDA takes the XMI document as an input. The XMI document represents ContextUML diagram. RubyMDA reads the XMI document and constructs the UML model, which is a set of data structure representing the components in UML class diagram. After UML model is constructed, RubyMDA transforms the UML model into CAS model, which is a set of data structure representing the CAS described in ContextUML diagram. Finally, RubyMDA generates a BPEL process and WSDL document for a CAS. Moreover, it generates a

[6]http://rubyforge.org/projects/rubygems.

Map From: UML Stereotypes	To: BPEL Elements
«conuml.service»	<process>
«conuml.operation»	<invoke>
«conuml.message»	<variable>
«conuml.atomicContext»	<invoke>
«conuml.compositeContext»	<invoke>
«conuml.contextBinding»	<assign>
«conuml.part»	`part` attribute in <to>
«conuml.contextTriggering»	<switch> <invoke>

Table 2.1: RubyMDA's Model Transformation Rules

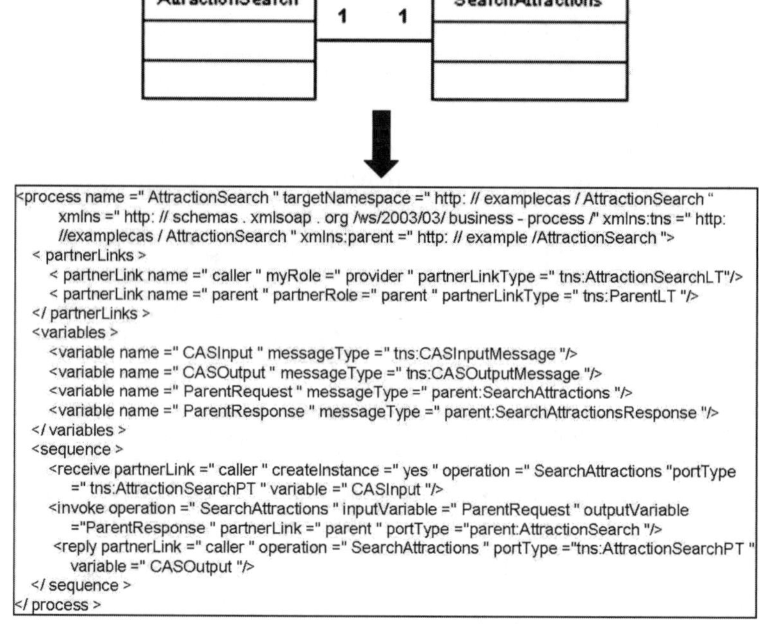

Figure 2.8: Generated BPEL Process.

set of deployment files needed to deploy CAS to a server. Figure 2.8 is an example of the generated BPEL process.

2.5 Applications

To showcase our ContextServ platform, we have implemented several context-aware Web applications. We present one application called *Smart Adelaide Guide* (SAG), which offers an interface helping tourists to find interesting places in Adelaide, the capital city of South Australia. SAG invokes the context-aware Web service depicted in Section 2.2.5, which has been developed by using our ContextServ platform.

SAG recommends attractions based on a user's current location, current weather condition in Adelaide[7], and a user's preferred proximity limitation (e.g., 500 meters) and language (e.g., Chinese). It should be noted that in this online demonstration, the location is simulated by a chosen point on the Google Map by clicking on the map.

SAG is hosted at our ContextServ project website[8]. To access SAG, launch any Web browser (e.g., Firefox or Internet Explorer) and then open the address: http://hs.cs.adelaide.edu.au/AttractionSearch.html. SAG application will appear in the browser, as shown in Figure 2.9.

To use SAG, click on any part of the map and a blue balloon will appear on the clicked point with its latitude and longitude. The next step is to choose a proximity and a preferred language for an attraction description from the drop-down lists next to the map. After clicking on the `Invoke` button, the context-aware attraction search Web service will be accessed and a list of attractions fitting the user's current context, including the location, the weather, and the proximity, will appear both as red balloons on the map and as a list of links to further information about the attractions next to the map. The names and descriptions of the attractions are automatically translated to the selected preferred language. A clear guideline on how to use this application is also given on the Web site.

What is happening at the back end is that the `location` is used as an atomic context source, and the temperature and wind speed information from the Weatherzone RSS feed[9] is used to derive the composite context `harshWeather`, which is defined as: `temperature` $> 30℃$ *or* `windSpeed` $>$ 20 km/h. The location context is bound to the input parameter of the Web service `attraction-search`, which returns a list of attractions upon the invocation. Then `harshWeather` and `proximity` contexts are used as triggers to filter the returned results of the Web service so that the attractions displayed to the user are within the specified proximity, and if weather is harsh (i.e., `harshWeather` is true), only indoor attractions are recommended.

[7] It is retrieved from a RSS (Really Simple Syndication) feed.

[8] http://www.cs.adelaide.edu.au/~contextserv

[9] http://rss.weather.com.au/sa/adelaide.

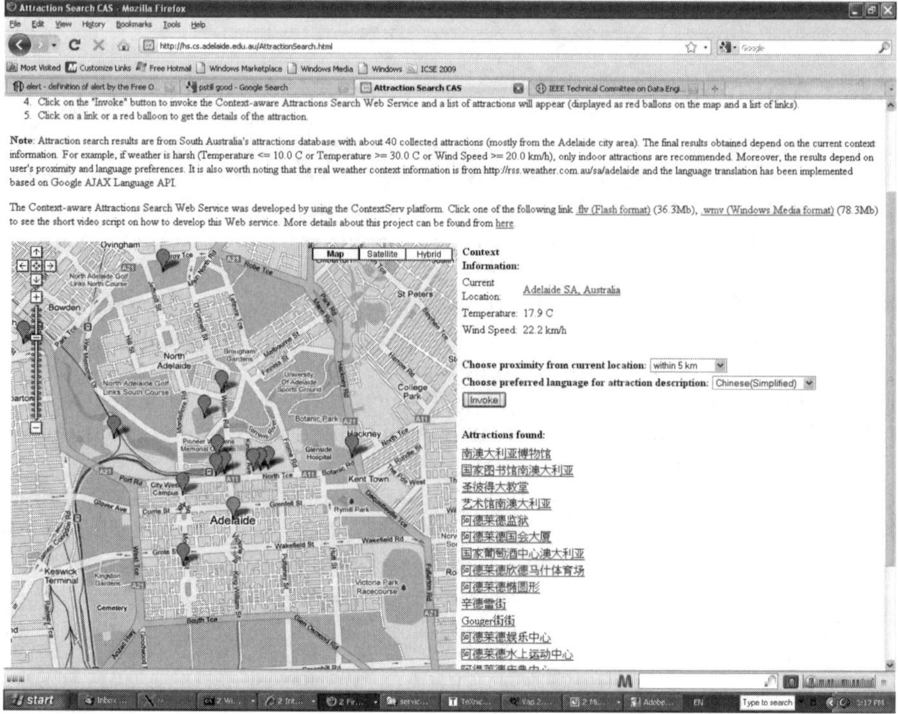

Figure 2.9: The Smart Adelaide Guide application.

2.6 Discussion and Conclusion

With the maturing and wide-adopting of Web service technology, research on providing engineering approaches to facilitating the development of context-aware services has gained significant momentum. Using a model-driven paradigm to develop CAS has proven to be a valuable and important strand in this research area considering the quality and efficiency it brings along. Apart from model-driven approaches, in the survey on context-aware service engineering (12), the authors propose 5 other categories of approach: *Middleware solutions and dedicated service platforms*, *Use of ontologies*, *Rule-based reasoning*, *Source code level programming/Language extensions*, and *Message interception*. In general, we agree with their viewpoint that any of the approaches has its pros and cons. For example, the source code level approach can give more freedom to developers to do all kinds of context-aware adaptation, but this approach does not separate apart the concerns on context-awareness and so suffers a significant maintenance cost. As to the model-driven approach, apart from its advantages, it requires to keep the consistency between high level models and low level executable code at all times, which brings extra complexity. We also agree that some approaches can be used at the same time to bring extra benefits. For example, we are planning to adopt ontologies in the context community to provide enhanced context organization and matching functionality.

In the literature on model-driven development of context-aware services, the following research work relates to ContextServ in particular. In (2), Ayed and Berbers propose a UML metamodel that supports context-aware adaptation of service design from structural, architectural, and behavioral perspectives. The structural adaptation can extend the service object's structure by adding or deleting its methods and attributes. The architectural adaptation can add and delete service objects to an application according to the context. Behavioral adaptation can adapt the behavior of the service object by extending its UML sequence diagram with optional context-related sequences. In (9), Grassi and Sindico propose a UML profile considering both model-driven development and aspect-oriented design paradigms so that the design of the application core can be decoupled from the design of the adaptation logic. In particular, this profile categorise context into `state-based` that characterizes the current situation of an entity and `event-based`, which represents changes in an entity's state. Accordingly, state constraints, which are defined by logical predicates on the value of the attributes of a state-based context, and event constraints, which are defined as patterns of event, are used to specify context-aware adaptation feature of the application.

		Context UML	**Ayed UML**	**Grassi UML**
Context Modelling	Atomic Context	+	+	+
	Composite Context	+	−	−
	Context Quality	+	+	−
	Context Collection	−	+	−
Service Modelling	Web Services	+	−	−
Context-Awareness Modelling	Context Binding	+	+	+
	Context Triggering	+	+	+
	Behavior Adaptation	+	−	−

Table 2.2: Language Capability Comparison

Table 2.2 gives a detailed language capability comparison between ContextUML and the above two UML metamodels from the perspective of context modelling, service modelling, and context-awareness modelling. As we can derive from the table, all languages support the modelling of atomic context, but only ContextUML supports composite context. As to service modelling, only ContextUML directly supports the structure of Web services, and the other two languages just use plain UML classes to represent services. Finally, ContextUML does not support behavior adaptation since it is impossible to change the internal logic of an encapsulated Web service. It is worth noting that among the three languages, ContextUML is the only one that has a comprehensive software tool, the ContextServ platform, to support its graphical modelling and automatic transformation. In (17), the authors extend ContextUML and implement an aspect-oriented transformation technique to transform UML models to AspectJ.

In this chapter, we have presented ContextServ, a comprehensive platform for simplifying the development of context-aware Web services. ContextServ adopts model-driven development where context-aware Web services are specified in a high-level modelling language and their executable implementations are automatically generated, thus contributing significantly to both design flexibility and cost savings. The platform has been validated by successfully creating a number of context-aware Web services. Currently, we are extending the platform to: i) support more context triggering mechanisms and ii) introduce semantic support of context provisioning.

References

[1] Gregory D. Abowd et al. Context-Aware Computing. *IEEE Pervasive Computing*, 1(3):22–23, 2002.

[2] Dhouha Ayed and Yolande Berbers. UML Profile for the Design of a Platform-independent Context-aware Applications. In *Proc. of the First Workshop on Model Driven Development for Middleware (MODDM'06), Melbourne, Australia*, pages 1–5, 2006.

[3] Boualem Benatallah, Marlon Dumas, and Quan Z. Sheng. Facilitating the Rapid Development and Scalable Orchestration of Composite Web Services. *Distributed and Parallel Databases, An International Journal*, 17(1):5–37, 2005.

[4] Thomas Buchholz, Michael Krause, Claudia Linnhoff-Popien, and Michael Schiffers. CoCo: Dynamic Composition of Context Information. In *Proc. of the First Annual International Conference on Mobile and Ubiquitous Systems: Networking and Services (MobiQuitous'04)*, Boston, Massachusets, USA, August 2004.

[5] Thomas Buchholz, Axel Küpper, and Michael Schiffers. Quality of Context: What It Is And Why We Need It. In *Proc. of the 10th Workshop of the OpenView Univeristy Association (OVUA'03)*, Geneva, Switzerland, July 2003.

[6] Anind K. Dey and Gregory D. Abowd. Towards a Better Understanding of Context and Context-Awareness. Technical Report GIT-GVU-99-22, GVU Center, Georgia Institute of Technology, June 1999.

[7] Anind K. Dey and Jennifer Mankoff. Designing Mediation for Context-aware Applications. *ACM Trans. on Computer-Human Interaction*, 12(1):53–80, 2005.

[8] David S. Frankel. *Model Driven Architecture^{TM}: Applying MDA^{TM} to Enterprise Computing*. John Wiley & Sons, 2003.

[9] Vincenzo Grassi and andrea Sindico. Towards Model Driven Design of Service-Based Context-Aware Applications. In *Proc. of International Workshop on Engineering of Software Services for Pervasive Environments, in Conjunction with the Sixth ESEC/FSE joint meeting, Dubrovnik, Croatia*, pages 69–74, 2007.

[10] Karen Henricksen and Jadwiga Indulska. A Software Engineering Framework for Context-Aware Pervasive Computing. In *Proc. of the Second*

IEEE Annual Conference on Pervasive Computing and Communications (PerCom'04), Orlando, Florida, USA, March 2004.

[11] Christine Julien and Gruia-Catalin Roman. EgoSpaces: Facilitating Rapid Development of Context-Aware Mobile Applications. *IEEE Trans. on Software Engineering*, 32(5):281–298, 2006.

[12] Georgia M. Kapitsaki et al. Context-aware Service Engineering: A Survey. *Journal of Systems and Software*, 82(8):1285–1297, 2009.

[13] Markus Keidl and Alfons Kemper. Towards Context-Aware Adaptable Web Services. In *Proc. of the 13th Intl. World Wide Web Conf. (WWW'04)*, New York, USA, May 2004.

[14] Kewen Liao. Optimal context provisioning in web service environments. Honors Thesis, School of Computer Science, The University of Adelaide, 2009.

[15] Wenjia Niu, Zhongzhi Shi, Changlin Wan, Liang Chang, and Hui Peng. A DDL-Based Model for Web Service Composition in Context-Aware Environment. In *Proc. of the IEEE International Conference on Web Services (ICWS'08)*, Beijing, China, September 2008.

[16] Jason Pascoe. Adding Generic Contextual Capabilities to Wearable Computers. In *Proc. of the 2nd International Symposium on Wearable Computers*, Pittsburgh, USA, October 1998.

[17] George N. Prezerakos, Nikolaos D. Tselikas, and Giovanni Cortese. Model-driven Composition of Context-aware Web Services Using ContextUML and Aspects. In *Proc. of the IEEE International Conference on Web Services (ICWS'07)*, pages 320–329, 2007.

[18] Daniel Salber, Anind K. Dey, and Gregory D. Abowd. The Context Toolkit: Aiding the Development of Context-Enabled Applications. In *Proc. of the Conference on Human Factors in Computing Systems (CHI'99)*, Pittsburgh, PA, USA, May 1999.

[19] Bill N. Schilit, Norman Adams, and Roy Want. Context-Aware Computing Applications. In *Proc. of the 1st International Workshop on Mobile Computing Systems and Applications*, Santa Cruz, CA, USA, December 1994.

[20] Quan Z. Sheng and Boualem Benatallah. ContextUML: A UML-Based Modeling Language for Model-Driven Context-Aware Web Service Development. In *Proc. of the 4th Intl. Conf. on Mobile Business (ICMB'05)*, Sydney, Australia, July 2005.

[21] Quan Z. Sheng, Sam Pohlenz, Jian Yu, Hoi S. Wong, Anne H.H. Ngu, and Zakaria Maamar. ContextServ: A Platform for Rapid and Flexible Development of Context-Aware Web Services. In *Proc. of the 31st International Conference on Software Engineering (ICSE'09)*, Vancouver, Canada, May 2009.

[22] Quan Z. Sheng et al. WS3 - International Workshop on Context-Enabled Source and Service Selection, Integration and Adaptation. In *Proc. of the 17th Intl. World Wide Web Conf. (WWW'08)*, Beijing, China, April 2008.

[23] Markus Stolze and Michael Ströbel. Utility-based Decision Tree Optimization: A Framework for Adaptive Interviewing. In *Proc. of the 8th International Conference on User Modeling (UM'01)*, Sonthofen, Germany, July 2001.

[24] Qi Yu, Xumin Liu, Athman Bouguettaya, and Brahim Medjahed. Deploying and Managing Web Services: Issues, Solutions, and Directions. *The VLDB Journal*, 17(3):537–572, 2008.

Chapter 3

Dynamic Software Product Lines for Context-Aware Web Services

Carlos Parra, Xavier Blanc, Laurence Duchien, Nicolas Pessemier, Rafael Leaño, Chantal Taconet, and Zakia Kazi-Aoul

Abstract As any software, Web Services have to face various execution environments. We then talk of context-aware Web Services. Moreover, Web Services could also be used in mobile environments. For such environments, it is necessary to have several products (e.g., different implementations of the same service), which may be deployed on various terminals. Furthermore, mobility involves the ability to dynamically change the systems functions at runtime in accordance with the environment variations. A Software Product Line (SPL) paradigm may be helpful to deal with the production of such a family of products. In this chapter, we introduce CAPucine, a Context-Aware Dynamic Service-Oriented Product Line for the production and execution of context-aware Web services. CAPucine proposes to express Web-Service variability through feature-diagrams and context-awareness models. CAPucine manages context-awareness (1) initially to derivate products using a Model Driven Engineering process and (2) with an iterative process to modify at runtime the product structure and behavior due to context changes using the SCA platform and the COSMOS context management framework through. CAPUcine presents the whole process, which includes context-aware Web service specification, context-aware platform code generation and context-awareness iterative loop. The whole process is demonstrated through a mobile-commerce scenario.

3.1 Introduction

Even if technologies used to develop Web services and mobile systems have completely different foundations such as communication mode or architectural pattern, they have to be conjointly used to develop nowadays applications on mobile phones.

On the one hand, SOA *Service-Oriented Architecture*, that groups all technologies used to develop Web services such as WSDL (38) or BPEL (25) among others, puts emphasis on loosely-coupled architectures and facility of reuse (20, 33). The main goal of SOA is to face challenges of integration and reuse of industrial scale services. On the other hand, technologies used to develop mobile applications put emphasis on dynamic adaptation and on context-awareness as they should be adapted to running platform having limited resources and should be aware of their environment and react properly to context evolutions (39, 27).

SOA and mobile systems technologies have to be conjointly used to develop nowadays mobile applications because: (1) mobile applications are now completely integrated into the Internet and are clients of existing Web services (last generation of mobile phones already support access to Web services like Google for instance), (2) mobile applications can be Web service providers as well (for peer-to-peer gaming for instance or when an update has to be dynamically installed), (3) even if mobile applications running platforms have more and more resources, their design still needs to be fitted, and (4) moreover, mobile applications continuously have to be adapted regarding their environment changes, e.g., they may become clients of new services depending on their location or their architecture may be automatically changed depending on the level of their battery.

Mobility also leads to the need for applications to be *aware* of the context and to react properly to its changes. Context information can be understood as a set of situations and events that occur on the environment and may have an impact on the behavior of the application. It is desirable to define a complete approach to build these mobile and context-aware applications. Standards such as the `Service Component Architecture` (SCA) (7) or `J2ME for Web services` (5) may be used to partially deal with this objective. Indeed, `SCA` defines how components can be used to realize Web services whereas `J2ME for Web services` defines how J2ME systems can interact with Web services. However, those standards only provide gateways between the Web services and the mobile systems world. They do not provide any support that goes from requirements to realization of such systems.

One possible solution involves SPLs that focus on variability management and aim at deriving different products from a same product family (8). In SPL, feature diagrams express the variability of a product family by defining its variants and its variation points (34, 37). Based on a feature diagram, one of the SPL's key principles is the *Product Derivation* (PD), which is the complete process of constructing a product from an SPL (13). The PD defines how assets are selected and combined in order to compose the final product. It should be noted that PD does not have to be confined to design or architectural phases. It can also be applied at runtime, so that, it is able to address the iteration phase of (13), making the SPL more dynamic (16). A Dynamic SPL (DSPL) then produces systems that can be adapted at runtime in order to dynamically fit new requirements or resource changes.

In this chapter, we propose a Context-Aware *Dynamic Software Product Line* (DSPL) named *CAPucine* for *C*ontext-*A*ware Service-Oriented *P*rod*uc*t L*ine*.[1] Our goal is to define at the same time a service-oriented and context-aware PD that monitors the context evolution in order to dynamically integrate the appropriate assets in a running system. Our approach is based on the two classical steps of product derivation (`initial step` and `iterative step`). The `initial step` aims at deriving an initial product from a product family whereas the `iterative step` aims at adapting dynamically this initial product in response to its environment changes. It should be noted that our target evaluation platform is based on the SCA standard that reconciliates SOA and *Component-Based Software Engineering* (CBSE) (7). Moreover, we use FraSCAti (26) (35), an SCA platform with dynamic properties enabling binding and unbinding of components at runtime. We also emphasize the use of sensed information from the environment, to dynamically realize the PD. We then use COSMOS (31), which is a context-aware framework connected to the environment by the use of probes and sensors. Thanks to COSMOS, the environment is abstracted by a set of software components—the so–called context nodes—that offer runtime operations reflecting the environment state.

The remainder of this chapter is organized as follows. In section 3.2, we present a motivating scenario that will be referred throughout the chapter to illustrate our approach. Section 3.3 introduces our proposal for a context-aware product derivation. Section 3.4 describes the product derivation technologies and implementation for the motivating scenario. In section 12.6, we briefly compare and position our work with other proposals found in the literature. Finally, in Section 3.6, we conclude and present some ideas for future work.

[1]This work is part of the `CAPPUCINO` project, funded by the *Conseil Régional Nord-Pas-de-Calais*, *Oseo/ANVAR*, and the *Fonds Unique Interministériel*

3.2 Motivating Scenario and Challenges

In order to motivate the challenges of context-aware Web service development, this section introduces a mobile commerce system as a scenario. This scenario highlights some of the new opportunities brought by this kind of system, such as the availability of context-aware Web services to *On-The-Go* users and context-aware social networks.

> *Mary uses a software system in her mobile phone to search and buy different items from an online catalog. In particular, this system can exploit Mary's profile to offer her personalized services. For example, when subscribing to the reward point program, Mary automatically receives special offers and prices. She can also register important events in her calendar like her best friends' birthdays. Using this information, customized notifications with a focused product offer may be pushed to her mobile phone. Mary can also find gift ideas by using a special service that connects to a social network, to retrieve a list of items selected by her friends as their favorite products. Additionally, the system can use Mary's current location to display the selected products available in the surrounding shops. If she does not have Internet access when she finally makes her choice, she can still order offline and the system automatically places the order, as an SMS message.*

3.2.1 Challenges

This scenario illustrates several challenges for the development of context-aware Web services systems. We further elaborate on the analysis and design of such systems making emphasis on three main aspects: variability, structure and behavior, and context-awareness.

Variability

Variability refers to the management of a product family that is defined by a set of reusable assets and a means to derive products from them. In context-aware mobile systems, we distinguish two types of variability: the variability that can be selected before the derivation of a product, and the variability that involves the modification of a product at runtime. Here we discuss the former type, and then, in section 3.2.1, we present the latter variability and analyze how it is related to context information. In SPL, one common way to express the variability before derivation is by using features and feature diagrams. With feature diagrams, it is possible to classify all the requirements that can be managed by a product family. As explained in (34), there are several variants for feature diagrams, which do not necessarily share the same semantics. In our case, we follow the notation introduced in (11). Figure 3.1 shows an example of this feature diagram for our scenario. With this notation, features are represented in a tree-like form. Dark circles indicate mandatory features, whereas white circles indicate optional features. Finally, inverted arcs describe a set of alternative features meaning that exactly one feature has to be chosen.

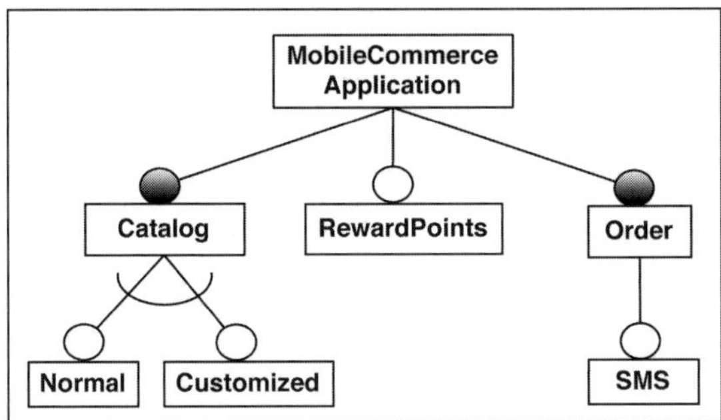

Figure 3.1: Feature diagram of the mobile commerce example.

In Figure 3.1, we see that the scenario system is characterized by two mandatory parts (`Catalog` and `Order`) and an optional one (`RewardPoints`). The `Catalog` may be in two different forms: the traditional form, which displays all the available items, and the customized form, that filters the items. The `Order` refers to the process of actually buying items in the catalog. It is usually performed online, which means that the client stays connected during the whole process. Optionally the system may support `SMS` orders, in which case, the order can be placed offline via `SMS`. With the optional reward points feature, it is possible to have a catalog that lists only low-priced products, and shows special discounts for a given order.

To derive a new product from a product family, it is necessary to first select the features to be supported by the product. Thus, a feature diagram defines the complexity of the product family since every optional feature increases the size of the product family. In our scenario, the feature diagram presents two optional features (SMS support, and reward points management) and one alternative feature (Catalog) with two different alternatives: normal and customized. In total, there are eight different combinations of features. Each combination represents a product as shown in Figure 3.2. For example, **Product 1** presents a customized catalog, with reward points to get special lists of items and discounts for every order, and also support for SMS orders.

The variability challenge, which is highlighted by our scenario, consists in dealing with such combinatorial complexity. The objective is to provide an

	Catalog		Reward Points	SMS
	Customized	Normal		
Product 1	Yes	No	Yes	Yes
Product 2	No	Yes	Yes	Yes
Product 3	Yes	No	Yes	No
Product 4	No	Yes	Yes	No
Product 5	Yes	No	No	Yes
Product 6	No	Yes	No	Yes
Product 7	Yes	No	No	No
Product 8	No	Yes	No	No

Figure 3.2: Product Family.

approach that supports (1) a definition of reusable assets and (2) a mechanism for combining them in order to derive a complete final product.

Structure and Behavior

By structure and behavior, we refer to the two classical kinds of software assets that constitute any software system. As it is highlighted by our scenario, each product of a product family has its own structure and behavior. However, as it is also highlighted by our scenario, all products of a product family share common elements in their structure and behavior. In fact, the structure and behavior of a product depend both on the features it has to support. For example, if the `RewardPoints` feature is selected, then, when placing an order, it is necessary to add the functionality to create a message with the order information and then, the normal execution of the system has to be modified to use this new functionality. The structure and behavior challenge consists in (1) modeling all those fragments of structure and behavior, (2) defining differences and compatibility between them, and (3) explaining how they can be combined when deriving the final product.

Context Information and Adaptation

The final challenge relates to the runtime variability introduced in section 3.2.1. In an SPL, every derived product has to support each of the selected features. However, in DSPL, depending on its context state, some of the product assets, used to support some features, will only be needed temporarily. Then, in order to fit the product's context, those assets should be deactivated when they are not needed. In other words, there is the notion of runtime variability, which indicates that the structure and behavior of the final product can be dynamically modified in order to fit the context state.

The scenario illustrates this fact with the `RewardPoints` and the `SMS` features of the product family. If those features are selected, which is the case for **Product 1**, then the product will contain assets that support them, but those assets will not always be needed. For instance, if the client is new and does not have a dedicated profile where the system stores information for providing her/him special offers, there is no need for the product to have special components that deal with this feature. In the same way, all the components that support the `SMS` feature do not have to be part of the architecture of the product when the Internet network is available.

The `Context information and Adaptation` challenge consists then in (1) defining such context-aware feature and (2) providing a mechanism to dynamically adapt the system according to its context state. It is important to notice though, that we do not extend the feature diagrams notation by introducing new concepts or restrictions to distinguish context-aware features. The way we define and deal with this type of variability is further explained in section 3.3.2.

3.2.2 Discussion

Summarizing the previous subsections, we identify several challenges for each mentioned aspect:

- *Variability:* It is clear that features help to understand the requirements of a software system. Nevertheless, as more features are identified, the scope of the family measured in terms of the number of different products that can be derived, grows exponentially because of the combinatorial explosion. As we have illustrated, even a small example, with two optional features and one alternative feature, produces eight different products.

- *Structure and Behavior:* Each feature is realized through assets that represent a part of the product. In this way, the feature is integrated into a software product, introducing the characteristics that make each product unique. Nevertheless, a feature may impact both the structure and the behavior of the product. This makes it complex to model the assets, and to define a process that composes them to derive a single and coherent product.

- *Context-aware adaptation:* Finally, including context awareness in SPL adds a new level of variability. Each product derived from the set of initial features can be configured in different ways to match specific environmental situations.

To face these challenges, we propose to implement a dynamic SPL. With SPLs, emphasis is put on identifying commonalities and variabilities of a product family in such a way that different products can be generated from a set of easily-composed assets. Identifying and building such assets is indeed a complex task. That is the reason why, using an SPL is initially more expensive than a direct development of one product. Even if there is no derived product, there is already an effort invested for producing the assets. The benefits of SPLs are perceived only after several product derivations using the same assets. In our case, the SPL is intended to build several mobile applications. To face the first two challenges, the SPL presents an initial phase where assets are associated with features. Thus, every asset represents several elements in the overall architecture of the products (structure and behavior), and its applicability depends on whether the associated feature has been selected or not. For the third challenge, we add an iterative phase to the SPL in order to dynamically derive the products. Thus, context situations trigger the derivation of different versions of the same product. In the remainder of this chapter, we present such dynamic SPL intended to build context-aware service-oriented systems using SOA.

3.3 CAPucine: Context-Aware Service-Oriented Product Line

Our claim is that using an SPL paradigm to build context-aware Web service systems permits a complete development from requirements to implementation, and the consideration of context throughout the software life cycle. Figure 3.3 presents the complete CAPucine approach that supports the two well-known phases of the SPL product derivation, i.e., the *initial phase* and the *iterative phase* (13). CAPucine supposes that the selection of features has already been made by the `application designer` who wants to derive a new product. The initial phase then aims at deriving an initial version of the product chosen by the application designer. To build this initial version, model assets that correspond to the selected features have to be identified and composed in order to make up the complete model of the product. In CAPucine, the model assets corresponding to the features have already been developed by the `product line designers`. The `application designer` just have to be composed in order to obtain the complete model of the final product. Once this complete model has been composed, it becomes the input of the automatic code generation step that aims at generating the source code of the product. At the end of this phase, the product is ready to be deployed. The iterative phase then starts with the first deployment of the initial version of the product. Once the product has been deployed, its environment is monitored in order to listen for context changes that may require an adaptation of the product.

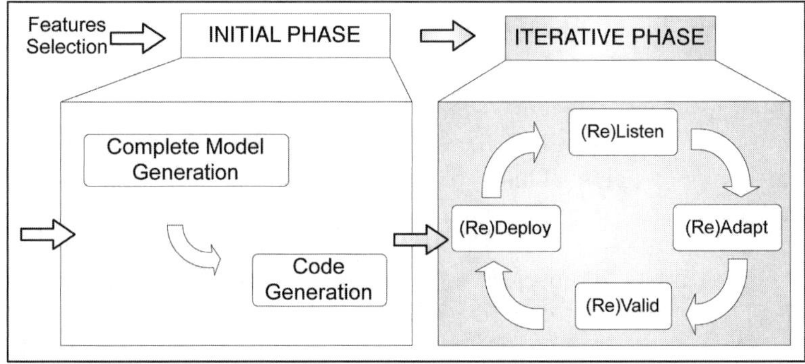

Figure 3.3: The CAPucine approach.

3.3.1 Initial Phase

The initial phase starts once the selection of the supported features has been made and ends when the first version of the final product is ready to be deployed. The process of deriving a product from a set of selected features involves several phases as described in Figure 3.4. First, features are associated with assets represented as partial models that use the *CAPucine* metamodel presented below. The assets are composed to obtain a complete model. Then, this model is transformed into platform (e.g., FraSCAti) and implementation specific models (e.g., Java). Finally, the product is built by generating the source code.

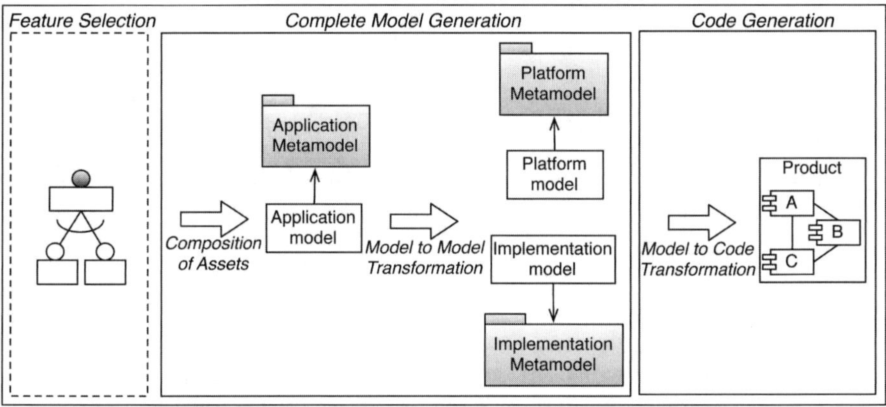

Figure 3.4: CAPucine's initial phase of product derivation.

Complete Model Generation

CAPucine's initial phase is based on a model-driven approach. For every selected feature, there is an associated asset which in our case, corresponds to a partial model of the product itself. This partial model contains both structure and behavior parts. This is similar to what has been proposed in (29), which advocates the use of *Aspect Oriented Modeling* (AOM) and proposes an automatic merge mechanism that integrates the models corresponding to the system features. In our case, we have specified the CAPucine metamodel, which is the target of the composition.

The CAPucine metamodel (Figure 3.5) describes from a high level of abstraction the structure and behavior of an SCA system. It is formed by three different parts: structure, behavior, and context-awareness. Since the context-aware part (left side of Figure 3.5) is directly related with the iterative phase, it is explained in section 3.3.2.

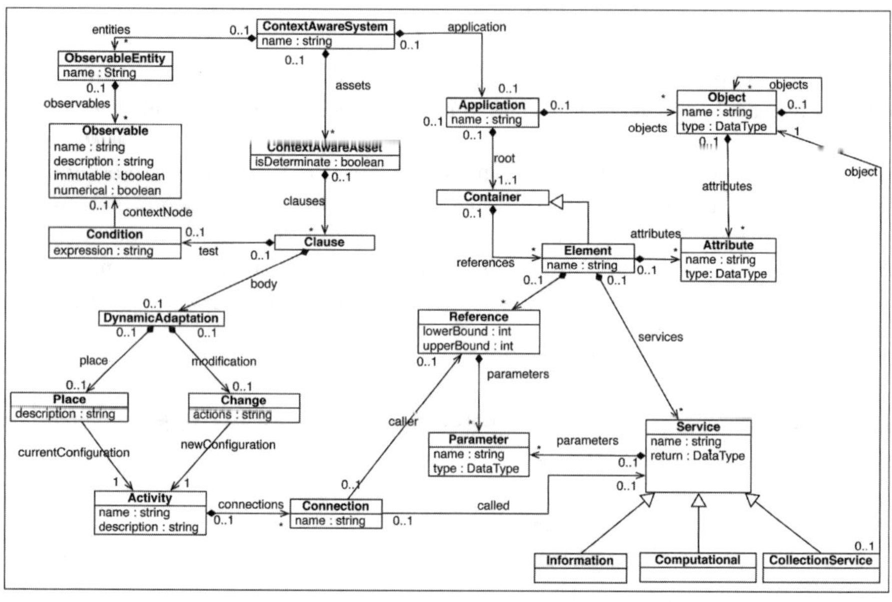

Figure 3.5: CAPucine Metamodel.

- *Structural and Business Modeling:* In the structural and business part, there are three main meta-classes: `Element`, `Service`, and `Reference`. They describe the generic structure of a service-oriented system. The information of these elements is used to build the SCA components and their implementations in Java. The `Element` is specialized by the

`Container`, which represents a composite of several elements. Additionally, by specializing the `Service` meta-class, we can specify different types of services for communication, collections, persistency, and user interface. This helps us to limit the scope of the line within a set of standard functionalities.

- *Behavior:* The behavior is represented by two meta-classes: `Activity` and `Connection`. An activity is composed of many connections. The general idea behind these elements is to represent a sequence of services communicating with each other. Every `Connection` links a reference of an element that makes the call with a service of an element that receives the call. An `Activity` is essentially, a basic set of links that represents a business process. The `Connection` meta-class might be further specialized in order to support different types of BPEL-like activities as: splits, forks, etc.

Code Generation

Once the model that corresponds to one product is obtained, it is transformed in order to add the platform, and implementation elements. This is done by performing a series of *model-to-model* transformations. Each transformation consists of a set of rules that maps the concepts of the CAPucine metamodel into the corresponding elements in Java, SCA, and COSMOS. For instance, an element will be mapped into (1) an SCA component, (2) a Java interface defining its services, and (3) a Java class implementing the Java interface and representing the SCA component. Finally, the product is built by generating the code from the target domain models.

3.3.2 Iterative Phase

In (13), the iterative phase is defined as the phase that starts as soon as the system has been deployed and ends at the end of the system's life cycle. During this phase, as the system environment changes continuously, its architecture may have to be adapted dynamically. The CAPucine approach proposes a dedicated runtime platform that is responsible for (1) listening to the system context, (2) deciding if the system has to be adapted once its context has been changed, (3) validating the adaptation, and (4) deploying the new version of the system.

In order to support those four steps (described in Figure 3.3), we distinguish the two main following challenges:

- *Context-aware assets:* The first challenge relates to clearly define how to introduce what we call `alternative architectures`. An `alternative architecture` is a variation of the system architecture that fits to a specific environment state. The problem of defining `alternative architecture` is then composed of two subproblems dealing with (1) how to formally represent them and their related context states as SPL runtime assets and (2) how to bind them with features that appear in a feature diagram.

- *Context-aware variability realization techniques:* The second challenge refers to the management and acquisition of context observations (listen), the decision of which adaptation has to be made (adapt), the validation of the adaptation (validate) and the realization of the dynamic adaptation of the system (deploy) CAPucine should then integrate a runtime platform that can monitor the context and, depending on changes, dynamically realize the adaptation of the running system.

Context-Aware Assets

We call *context-aware assets*, the assets that can be integrated at runtime, depending on the environment state. Context-aware assets own the different alternative architectures of a system and their conditions of existence that depend on the environment state.

The context-awareness part of the CAPucine meta-model defines how to model context-aware assets (see Figure 3.5). It defines that a `Context-aware Asset` is composed of several `Clauses`. Each clause is composed of one `Test` and one `Body`. The `Test` is a classical condition expressed on an `Observable` that is an object abstracting a state of the environment, i.e., the expression property of the meta-class `Condition` is used to specify the condition. The `Body` is a dynamic adaptation composed of two parts: a `Place` and a `Change`. The `Place` defines the place in the system where the dynamic adaptation will be realized. The `Change` defines the change actions that will be realized to adapt the system at runtime.

We propose to make explicit the fact that a context-aware asset may be determinate if at most one of its tests succeed whatever the state of the environment is. It should be noted that non determinate context-aware assets introduce non deterministic behavior because more than one dynamic adaptation can be realized for a particular environment state. Therefore they may cause conflicts or unexpected behavior depending on the order in which adaptations are realized.

Context-aware assets are, as defined by their name, dependent on the context. Our metamodel makes this dependency explicit thanks to the `Observable` concept. Our intention is to use this explicit dependency as a constraint for the software product line runtime platform. Indeed, the platform has to provide context information acquisition in order to support the realization of the assets.

It should be noted that context-aware assets are very different from classical assets. In particular, the decision of their integration into the system is undertaken dynamically and driven by context changes, their realization is done by adapting the behavior of the system at runtime, and finally, they depend on an existing system that is running and therefore, they are bound to classical assets.

Context-aware assets can be compared to Event-Condition-Action rules (21) as they express correspondences between changes in the environment and system adaptations. However, those two concepts are distinguishable as context-aware assets are coherent units that group together all adaptations and conditions that are related to a same conceptual context that is not the case of ECA where each rule is an independent unit. One context-aware asset can therefore be used to group opposite conditions and their corresponding adaptations in order to specify how the system should be adapted regarding alternative context states. Moreover, it should be noted that even if our current validation considers that the `Test` condition is a first order logical expression, we think about extending it in order to support other logics such as temporal logic in order to express constraints between orders of adaptations.

Context-Aware Asset Realization Techniques

In (37), Svahnberg et al. define that variability realization techniques are used to integrate assets while building the final products. Moreover, authors clearly identify *Component-Based Software Engineering* (CBSE) as one of the variability realization techniques that can be used at runtime. In this section, we propose a generic CBSE platform as a variability realization technique. This platform will support both acquisition of context information (listen), decision of which adaptation to realize (adapt), validation of adaptation (valid), and dynamic realization of the adaptation (deploy):

- *Acquisition of context information (listen) and decision of which adaptation to realize (adapt):* As it was explained in subsection 3.3.2, the context-aware assets include a definition of a context information that corresponds to the decision whether or not to adapt the system. Hence, the platform needs a context aggregation mechanism. Such a mechanism is in charge of getting the information from multiple sources and to provide a high level view of information, so that, it can be evaluated.

- *Validation of adaptation (valid) and dynamic realization of the adaptation (deploy):* Successful product derivation depends also on the ability of the platform to reconfigure the system. The platform has to be able to suspend and resume the execution of the system, modify its structure by performing different operations like deploy, add, bind, or delete components. This enables the platform to dynamically adapt to each context-aware asset.

The generic platform architecture is depicted in Figure 3.6. It starts with the *Context Manager*, which is in charge of *listening for changes* in the context information. It is composed of several nodes that collect data from different sources like a sensor layer that captures raw data from the environment, user preferences, and the *Runtime Platform* that provides information about current state and configuration of applications. The *Context Manager* aggregates this information and notify the *Decision Maker* whenever a change in the context occurs. The *Decision Maker* element *decides* about possible adaptations by evaluating the context using a repository of rules. Such rules represent the clauses of each context-aware asset. Finally, the *Decision Maker* uses the mechanisms offered by the *Runtime Platform*, where *Application Components* are executed and *reconfigured*, to adapt the application components by controlling their life cycle and modifying them at runtime if needed.

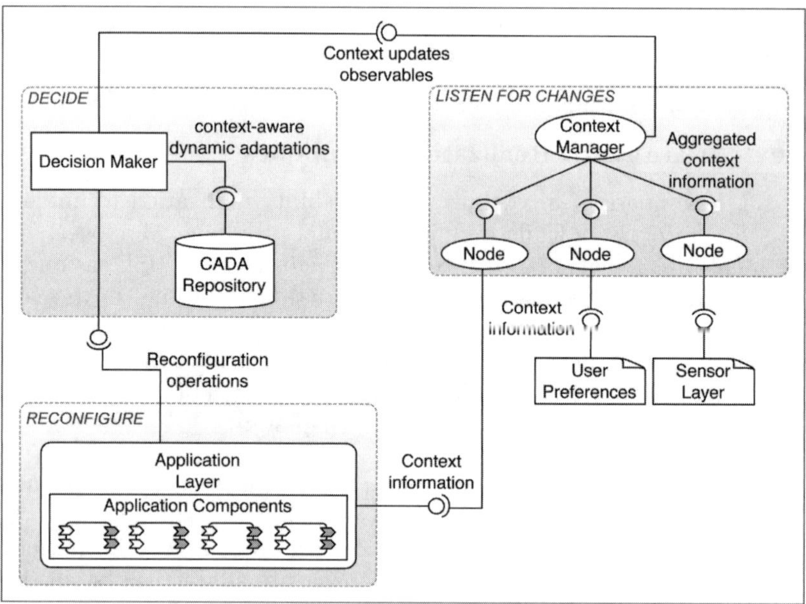

Figure 3.6: Generic Platform Architecture.

3.4 CAPucine Validation

In this section, we first explain the implementation of *CAPucine* based on the scenario presented in Section 3.2. The implementation is explained using the initial and iterative phases as in Section 3.3. Then we detail step by step the CAPucine process using this scenario. At the same time, we define each actor's role.

3.4.1 Initial Phase

Initial phase starts by associating an asset to each feature of the feature model and ends with the generation of the product's source code.

Complete Model Generation

The starting point of the initial phase is the feature model. Each feature is associated with an asset that has, at the same time, a structural and a behavioral part. The asset is a partial model that uses the elements defined in the CAPucine metamodel. Depending on the feature type, the associated asset either represent the basic elements that remain constant across the different variations in the product family (*core architecture*), or additional elements.

It is important to notice that the feature model is not meant to be modified frequently, besides there is no rule or pattern that indicates how the asset of each feature has to be built, or how it impacts the core architecture. We create these assets manually and afterwards we compose them according to each particular feature selection. In Figure 3.7, we use SCA to illustrate the result of this composition for the **Product 1** of Table 3.2. In SCA, components are represented as rounded-corner boxes. The dark chevrons in the left-side of components represent the offered services whereas white chevrons in the right-side represent the references (required services). In this case, there is a `GUI` component that offers a `run` service, and which is bound to the `CatalogManager` component through the `getCatalogList` reference to obtain the list of catalogs. The GUI also provides a `showNotification` service in order to display a message when new catalogs become available, like for example, when the user location changes. In turn, the `CatalogManager` chooses among different catalog providers like the `RewardPointManager` which offers a special type of catalog via the `getCatalogList` service. There is also an `Order` component that handles the process of buying an article. It can use a discount service to calculate the right price. Additionally, the order can be placed online through the `WS Order Handler`. Figure 3.7 also distinguishes the elements of the core architecture (single lines) from the additional components (dashed lines for optional features and double lines for alternative features). Finally, we use the stereotype notation (<<>>) to represent the *conforms to* relationship of every element of the model with the corresponding CAPucine meta-class.

In addition to the SCA architecture, the assets contain information about the behavior. Nevertheless, as explained in (15), SCA does not represent the order of the interactions between components of a given SCA system. This is the role of the business processes. In our case, we use the CAPucine meta-classes `Connection`, `Activity`, `Service`, and `Reference` to represent such processes. In Figure 3.8 we illustrate a simple example of business process to order an item using the **Product 1**. For a better understanding, we use the *Business Process Modeling Notation* (BPMN) (40) notation, which is a standard for modeling processes. In this case, there are three participants,

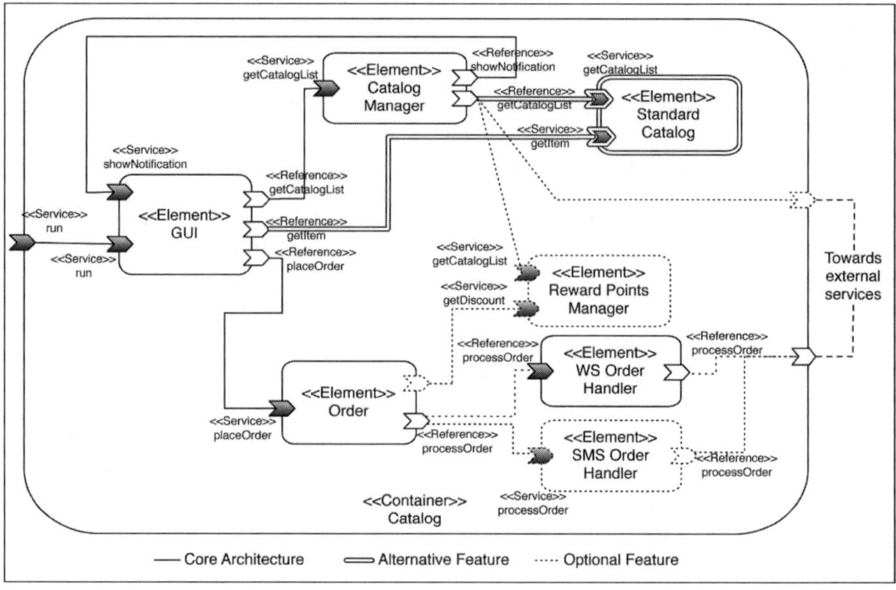

Figure 3.7: SCA architecture of **Product 1**.

the `Client`, who wants to place an order, the `Commerce Application` which corresponds to our **Product 1**, and the `External Services`. The client starts the process in order to obtain a list of items. Then the system fetches the catalog of items, depending on the available providers. Afterwards, the client picks an item to order. The order is placed according to the connectivity. Finally, the client receives a notification.

Code Generation

Figure 3.9 presents a conceptual view of how *model to model* transformations and code generation are performed. By first transforming CAPucine concepts into Java and SCA models, we can check the consistency among these new models. This would not be possible if we had directly generated the source code from the CAPucine metamodel. We have used *Spoon EMF* (23) as a Java metamodel, and the metamodel proposed by the *Open Service-Oriented Architecture* (OSOA) collaboration group (24) as SCA metamodel. Finally, code is generated from SCA and Java to obtain the composite descriptors and Java code using *Spoon* capabilities.

3.4.2 Iterative Phase

Dynamic product derivation is made by reconfiguring the system at runtime. This requires a platform with the characteristics described in Section 3.3. Here

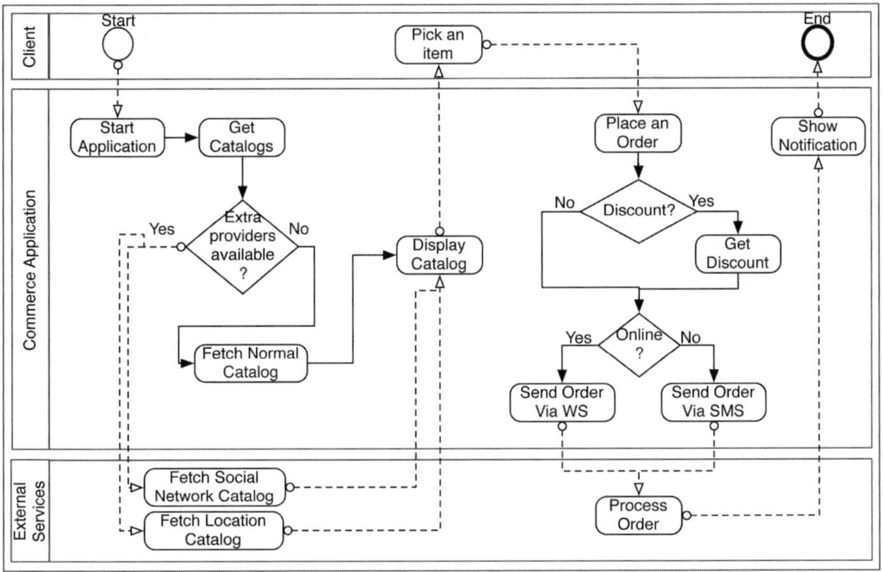

Figure 3.8: Business process to order an item with the **Product 1**.

we present a brief description of such a platform. For a detailed description, please refer to (30).

Runtime Platform

We use the FraSCAti platform (26). FraSCAti is a Fractal-based SCA implementation. SCA establishes that components are the basic building blocks. Each component requires and provides services. SCA supports several service description languages such as WSDL and Java interfaces, several programming languages such as Java, C++, and BPEL, several communication protocols between systems such as SOAP, CORBA, Java RMI, and JMS. On the other side, Fractal (6) is a hierarchical and reflective component model intended to implement, deploy, and manage complex software systems. Fractal offers several features like composite components (components containing subcomponents), sharing (multiple enclosing components for the same subcomponent), introspection, and re-configuration. A Fractal component can expose elements of its internal structure and offer introspection and intercession capabilities. Several controllers have been defined in the Fractal specification like the binding controller that allows the dynamic binding and unbinding of component interfaces, and the life cycle controller that makes possible to perform operations like stop and start the execution of a component. In FraSCAti, a Java-based SCA, components are simultaneously both SCA-compliant and Fractal-compliant. The main benefit of this particular property is that all the

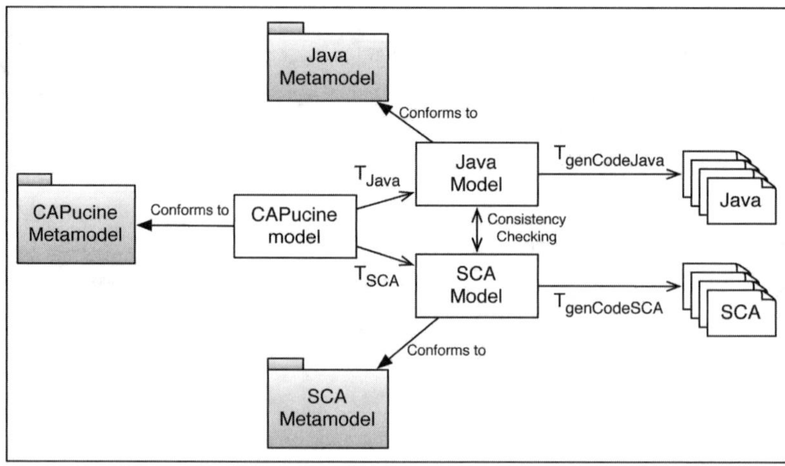

Figure 3.9: Model to model transformation and code generation.

components can be dynamically reconfigured at runtime. In addition to the runtime platform, we use the following technologies for context management, decision making and dynamic adaptation.

- Context manager: As shown in Figure 3.6, our platform needs a context manager to gather high level context information. Several context managers may be used as soon as model to model and model to code transformations are provided (see Figure 3.4). The context manager used for validation purpose is COSMOS (31). It is a component-based framework for gathering and processing context information. It obtains the context information from different sources like sensors, network probes, or systems, and processes it according to defined policies. These policies are described as hierarchies of context nodes using a dedicated composition language. Each COSMOS node in the top of the hierarchy is connected to a single *observable* of the context aware asset.

- Decision Making: The Decision Making component is attached to the COSMOS notification service. We configure COSMOS nodes to send notifications on some events. The notification is linked with an observable in order to enable the decision making to filter the context-aware assets that are impacted by the change. The decision algorithm is quite straightforward. It only evaluates the clauses linked to an observable for which a change notification has been received. Then, for the ones that have been evaluated to true, it realizes their body.

- Dynamic Adaptation: The adaptation takes place at runtime. Two activities have to be performed to adapt a running system. First, the *places* where the changes have to be realized should be identified. Second, components of the system should be adapted safely regarding their states. We use FPath and FScript (12) for these two activities. FScript and FPath notations are two domain-specific languages to code dynamic adaptation of Fractal-based systems. FPath eases the navigation inside a Fractal architecture with simple and readable queries. FScript, makes use of FPath, to enable the application designer to define adaptation scripts to modify the architecture of a Fractal system. One of the key advantages of FScript is that it provides a transactional support for architectural reconfigurations and then guarantees that the system remains consistent even if the reconfiguration fails at a given point.

3.4.3 Adapting the Mobile Commerce Example

In Figure 5.7, we present an extended SCA assembly of the example of Figure 3.7 with additional components. The basic architecture has been dynamically adapted in two different ways: (1) a geolocation context node has detected the location of Mary, triggering the deployment of a new catalog proposing regional products (see a new connection coming from the `CatalogManager` component to a new component `Geolocation-Based Catalog`); (2) during the payment operation, wifi connection with the server is lost, triggering other adaptation of the architecture. The old connection using the Web Service is replaced by a new one passing through an SMS component which process the order without an Internet connection. Based on this example, we describe in Figure 3.11 the actions that adapt the system. The first part of the code is the `place` part of the adaptation. It is written in FPath and consists in identifying the three subcomponents that will be adapted. The second part of the code is the `body` part of the adaptation. It is written in FScript and consists in (1) stopping the identified subcomponent, (2) changing their bindings and (3) re-starting them. If one of these FPath or FScript instruction fails, then the complete action is rolled-back, keeping the system in a consistent state.

3.4.4 CAPucine Process Step by Step

The CAPucine process is partially automatic. Firstly, the product line designer defines the feature diagram of the mobile commerce application from the customers' requirements and, then, she determines the product family represented by the set of variants (see Figures 3.1 and 3.2 in Section 3.2). In a second step, she prepares the associated assets. These assets are partial models that conform to the CAPucine metamodel for representing the structure and behavior. This part of the process is manual and depends on the application requirements and their variability.

After these steps, the CAPucine product line for this product family is ready

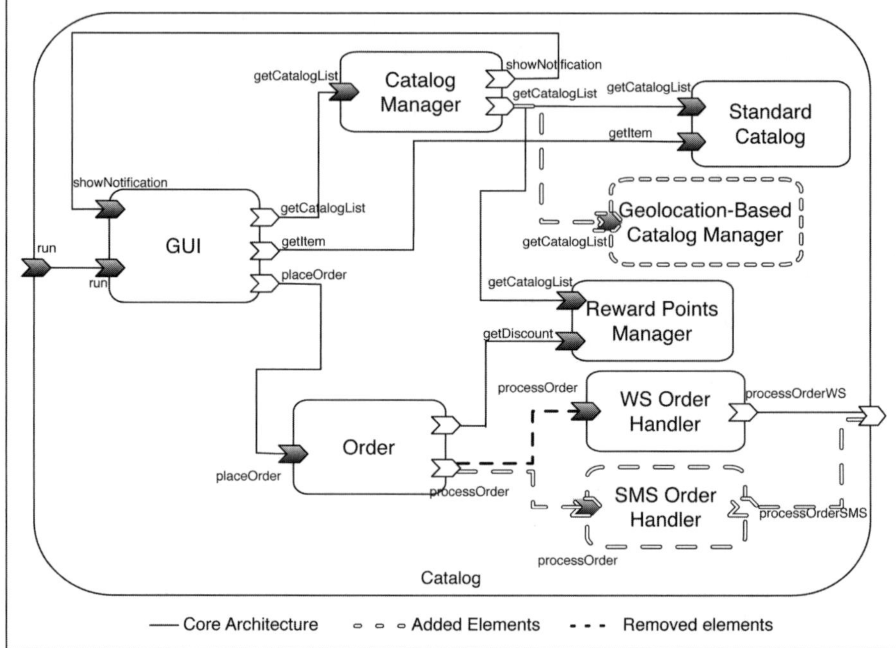

Figure 3.10: FraSCAti system adaptation example.

to be used by an application designer. This kind of user selects a product by selecting a given set of features in the feature diagram. The associated assets are manually composed in order to have a model in conformance with the CAPucine metamodel. Then, the model to model transformation and code generation can be applied. Finally, Java classes and SCA composite descriptors are ready to be deployed and processed on the runtime platform.

In the runtime platform, the context nodes represent the observables of the context-aware assets. The Decision Making component evaluates the observables changes and applies the context-aware dynamic adaptations. These ones have to be written in the *FPath* and *FScript* languages manually by the application designer and are deployed at the same time as the SCA components. We plan to automate the generation of this scripts in the next CAPucine version in the same way as we did for the Java classes and the SCA composite descriptors.

```
action adapt() {
 --Place
 catalog = $explorer/descendant::CatalogManager;
 order = $explorer/descendant::Order;
 ws=$explorer/descendant::WSOrderHandler;
 geoloc = $explorer/descendant::GeolocationBasedCatalogManager;
 sms = $explorer/descendant::SMSHandler;
 --Body
 -- First reconfiguration (Geolocation-Based Catalog)
 stop($catalog);
 stop($geoloc);
 bind($catalog/interface::getCatalogList,
      $geoloc/interface::getCatalogList);
 start($catalog);
 start($geoloc);
  -- Second reconfiguration (SMS Handler component)
 stop($order);
 stop($ws);
 unbind($order/interface::processOrder);
 bind($order/interface::processOrder,
      $sms/interface::processOrder);
 bind($sms/interface::processOrderSMS,
      $catalog/internal-interface::processOrderSMS);
 start($order);
 start($sms);
}
```

Figure 3.11: FPath and FScript implementation of the adaptation.

3.5 Related Work

In (36), Sheng and Benatallah present one of the first work concerning a specification language to define context-aware Web services. They propose a UML language to define filters to be applied on Web service parameters according to constraints on context data. Our work also concerns context-aware Web services. We propose more than a language for defining context-aware Web services, we propose a whole framework for product line definition, context-awareness specification, code generation, and context awareness at runtime. CAPucine handles adaptations on the structure of a complex Web service and the orchestration of its services. Our work is based on DSPLs. DSPLs are Software Product Lines where variation points can be bound not only during the design and deployment steps but also at runtime (16). Optionally, DSPLs can be context-aware when the decision of binding a variation point comes from a change within their environment. As we already explained, our proposal is a context-aware DSPL since we propose to develop a complete SPL approach for SOA and to make the context explicit within the variant specification. DSPLs need to collect high-level observations and to identify situations under which they need adaptations. High-level observations may be computed from different distributed sources such as operating systems, user profiles, and environment sensors. Context managers are services in charge of computing those high-level observations (2, 10, 14, 28). Few DSPLs introduce the notion of context manager. In our approach, we have validated the DSPL with a context manager called COSMOS (31).

Few successful approaches refer to the definition of SPL for SOA. Most of these approaches are in a preliminary state or propose some ways for a reconciliation between SOA and SPL (41), (1), (9), (19). In (3), Bastida et al. develop dynamic self-reconfiguring and context-aware compositions by applying a multi step methodology based on product-line engineering notions of variability management. They base their service composition infrastructure on Event-Condition-Action Rules. Their approach is mainly based on a composition service at design-time. At runtime, a BPEL engine and a rule engine compose the middleware part. The context part is not explicitly defined in the complete approach.

In a different way, work on adaptive systems and context awareness in SPL is prolific. In (4), Bencomo et al. propose software product lines for adaptive systems. In their approach, a complete specification of the context and supported changes has to be provided thanks to a state machine. Each state then represents a particular variant of the system and transitions between states define dynamic adaptation that are triggered by events corresponding to context changes. The main limitation of this approach is that the state machine has to be provided during the design step and cannot be extended, making the system quite static as it can only be adapted for a fixed set of change events. Nevertheless, they manage to introduce optimizations for the selected transitions.

In (22), Morin et al. propose an approach that deals with dynamic variability in software product lines and that encompasses the limit of (4). Their approach relies on AOM for specifying variants and for realizing the binding of variation points. This use of AOM replaces the use of the state machine of the (4) approach and then makes the adaptation more dynamic. Our approach is similar to their approach for this particular point. Authors also claim to propose a context-aware adaptation model that is in charge of selecting adequate variants depending on the context. Unfortunately, no detail is presented in the paper in order to understand how the context is specified and monitored. Our approach can then be considered as an extension of their work.

In (29), Perrouin et al. propose a SPL based on AOM. In their proposal, variants are specified thanks to model fragments and the product derivation is done automatically by merging them together. The initial phase of our approach is based on the same principle but it extends the product derivation with context-based adaptations.

In (17), Hallsteinsen et al. use product lines techniques for building adaptive systems. In their proposal, adaptive systems are implemented using component-based architecture and variability modeling and they delegate the adaptation complexity to a reusable adaptation platform. Our approach is similar. It is based on the same kind of platform but we integrate at the same time the notion of context in the product line and in the runtime platform. In (32), Salifu et al. propose an approach to variability for software product-families that deal with context-awareness. They link requirements to software architecture using product-family paradigms, but they do not support the context integration in the runtime platform.

Finally, in (18), Hartmann and Trew propose the concept of a context variability model that contains the primary drivers for variation (e.g., different geographic regions). This model constrains the feature model in order to choose one dimension in the context space. This context variability model remains as a static way to manage a product derivation in accordance with an orthogonal variation description.

3.6 Conclusion

In this chapter, we identified and faced three challenges for the development of mobile service-oriented and context-aware systems. To face the first one related to variability, we proposed a software product line named CAPucine that permits the selection of features in accordance with the variability of the user requirements. For the second one, which considers the structure and the behavior of the system, we translate the set of selected features to (1) a set of assets corresponding to an architecture of SCA components and (2) the associated business processes. These transformations are made in the initial phase by a product derivation based on composition and transformation of models. Finally, context information and adaptability are considered as first-class entities at runtime. They allow the platform to change the structure and the behavior of the product during the iterative phase. Our approach is homogeneous as we propose two different processes for the initial and iterative phases of product derivation. The advantages of CAPucine are twofold. First, we propose a context-aware asset that introduces alternatives in the software product line and that is considered at runtime. The context-aware asset defines the information of an adaptation when the context changes. Second, we are based on a set of realization techniques for context-aware variability by the way of context-aware tools, such as COSMOS and dynamic reconfiguration tools such as FraSCAti. As a future work, we plan, in the short term, to study the reduction of non deterministic behaviors when we introduce non determinate context-aware assets. Indeed, we consider an open environment where more than one dynamic adaptations can be realized for a particular environment state. Therefore, they may cause conflicts or unexpected behavior depending on the order in which adaptations are realized. We plan to reduce these conflicts by a set of strong preconditions that will clarify the condition for introducing new behaviors.

References

[1] Sven Apel, Christian Kaestner, and Christian Lengauer. Research challenges in the tension between features and services. In *SDSOA '08: Proceedings of the 2nd International Workshop on Systems Development in SOA Environments*, pages 53–58, New York, NY, USA, 2008. ACM.

[2] Matthias Baldauf, Schahram. Dustdar, and Florian. Rosenberg. A survey on context-aware systems. *International Journal of Ad Hoc and Ubiquitous Computing*, 2(4):263–277, 2007.

[3] Leire Bastida, Francisco Javier Nieto, and Roberto Tola. Context-aware service composition: A methodology and a case study. In *SDSOA '08: Proceedings of the 2nd International Workshop on Systems Development in SOA Environments*, pages 19–24, New York, NY, USA, 2008. ACM.

[4] Nelly Bencomo, Pete Sawyer, Gordon Blair, and Paul Grace. Dynamically adaptive systems are product lines too: Using model-driven techniques to capture dynamic variability of adaptive systems. In *2nd International Workshop on Dynamic Software Product Lines (DSPL 2008)*, Limerick, Ireland, 2008.

[5] Jean-Yves Bitterlich and al. JSR 172: J2ME Web Services Specification, Java community process, 2004. http://jcp.org/en/jsr/detail?id=172.

[6] Eric Bruneton, Thierry Coupaye, Matthieu Leclercq, Vivien Quéma, and Jean-Bernard Stefani. The FRACTAL component model and its support in Java: Experiences with Auto-adaptive and Reconfigurable Systems. *Softw. Pract. Exper.*, 36(11-12):1257–1284, 2006.

[7] David Chappell. Introducing SCA. white paper, Chappell & Associates, July 2007.

[8] Paul Clements and Linda Northrop. *Software Product Lines: Practices and Patterns.* Addison-Wesley Professional, August 2001.

[9] Sholom Cohen and Robert Krut, editors. *Proceedings of the First Workshop on Service-Oriented Architectures and Software Product Lines (CMU/SEI-2008-SR-006).* Software Engineering Institute, Carnegie Mellon University, 2008. http://www.sei.cmu.edu/publications/documents/08.reports/08sr006.html.

[10] Joëlle Coutaz, James L. Crowley, Simon Dobson, and David Garlan. Context is key. *CACM*, 48(3):49–53, March 2005.

[11] Krzysztof Czarnecki and Ulrich Eisenecker. *Generative Programming: Methods, Tools, and Applications*. Addison-Wesley Professional, June 2000.

[12] Pierre-Charles David, Thomas Ledoux, Marc Léger, and Thierry Coupaye. FPath and FScript: Language support for navigation and reliable reconfiguration of Fractal architectures. *Annals of Telecommunications*, 64(1-2):45–63, February 2009.

[13] Sybren Deelstra, Marco Sinnema, and Jan Bosch. Experiences in software product families: Problems and issues during product derivation. In Robert L. Nord, editor, *SPLC*, volume 3154 of *Lecture Notes in Computer Science*, pages 165–182. Springer, 2004.

[14] Anind K. Dey, Gregory D. Abowd, and Daniel Salber. A conceptual framework and a toolkit for supporting the rapid prototyping of context-aware applications. *Special issue on context-aware computing in the Human-Computer Interaction Journal*, 16(2–4):97–166, 2001.

[15] Mike Edwards. Relationship between SCA and BPEL. Technical report, Open Service Oriented Architecture Collaboration (OSOA), March 2007.

[16] Svein Hallsteinsen, Mike Hinchey, Sooyong Park, and Klaus Schmid. Dynamic software product lines. *Computer*, 41(4):93–95, 2008.

[17] Svein Hallsteinsen, Erlend Stav, Arnor Solberg, and Jacqueline Floch. Using product line techniques to build adaptive systems. In *SPLC '06: Proceedings of the 10th International on Software Product Line Conference*, pages 141–150, Washington, DC, USA, 2006. IEEE Computer Society.

[18] Herman Hartmann and Tim Trew. Using feature diagrams with context variability to model multiple product lines for software supply chains. In *SPLC '08: Proceedings of the 2008 12th International Software Product Line Conference*, pages 12–21, Washington, DC, USA, 2008. IEEE Computer Society.

[19] Andreas Helferich, Georg Herzwurm, Stefan Jesse, and Martin Mikusz. Software product lines, service-oriented architecture and frameworks: Worlds apart or ideal partners? In *Trends in Enterprise Application Architecture*, volume 4473/2007, pages 187–201. Lecture Notes in Computer Science, Springer, 2007.

[20] Nicolai Josuttis. *SOA in Practice, The Art of Distributed System Design.* O'Reuilly, August 2007.

[21] Dennis McCarthy and Umeshwar Dayal. The architecture of an active database management system. *SIGMOD Rec.*, 18(2):215–224, 1989.

[22] Brice Morin, Franck Fleurey, Nelly Bencomo, Jean-Marc Jézéquel, Arnor Solberg, Vegard Dehlen, and Gordon S. Blair. An aspect-oriented and model-driven approach for managing dynamic variability. In Krzysztof Czarnecki, Ileana Ober, Jean-Michel Bruel, Axel Uhl, and Markus Völter, editors, *Model Driven Engineering Languages and Systems, 11th International Conference, MoDELS 2008, Toulouse, France*, volume 5301 of *Lecture Notes in Computer Science*, pages 782–796. Springer, September 2008.

[23] Carlos Noguera and Laurence Duchien. Annotation framework validation using domain models. In *Fourth European Conference on Model Driven Architecture Foundations and Applications*, pages 48–62, Berlin, Germany, June 2008.

[24] Open Service Oriented Architecture Collaboration (OSOA). *Service Component Architecture*, November 2007.

[25] Organization for the Advancement of Structured Information Standards (OASIS). *Web Services Business Process Execution Language (WS-BPEL) Version 2.0*, April 2007.

[26] OW2 consortium. FraSCAti project. `http://frascati.ow2.org`.

[27] Nearchos Paspallis, Frank Eliassen, Svein Hallsteinsen, and George A. Papadopoulos. *Developing Self-Adaptive Mobile Applications and Services with Separation of Concerns*, chapter 6, pages 129–158. MIT Press, 2009.

[28] Nearchos Paspallis, Romain Rouvoy, Paolo Barone, George A. Papadopoulos, Frank Eliassen, and Allessandro Mamelli. A Pluggable and Reconfigurable Architecture for a Context-aware Enabling Middleware System. In *Proc. 10th of DOA*, volume 5331 of *lncs*, pages 553–570, Monterrey, Mexico, November 2008. Springer.

[29] Gilles Perrouin, Jacques Klein, Nicolas Guelfi, and Jean-Marc Jézéquel. Reconciling automation and flexibility in product derivation. In *12th International Software Product Line Conference (SPLC 2008)*, pages 339–348, Limerick, Ireland, September 2008. IEEE Computer Society.

[30] Daniel Romero, Romain Rouvoy, Sophie Chabridon, Denis Conan, Nicolas Pessemier, and Lionel Seinturier. *A Middleware Approach for Ubiquitous Environments*, chapter X. Chapman and Hall/CRC, 2009. Submitted.

[31] Romain Rouvoy, Denis Conan, and Lionel Seinturier. Software architecture patterns for a context-processing middleware framework. *IEEE Distributed Systems Online*, 9(6), 2008.

[32] Mohammed Salifu, Bashar Nuseibeh, and Lucia Rapanotti. Towards context-aware product-family architectures. In *IWSPM '06: Proceedings of the International Workshop on Software Product Management*, pages 38–43, Washington, DC, USA, 2006. IEEE Computer Society.

[33] Anne-Marie Sassen and Charles Macmillan. The service engineering area: An overview of its current state and a vision of its future. Technical report, European Commission, Network and Communication Technologies, Software Technologies, 2005.

[34] Pierre-Yves Schobbens, Patrick Heymans, and Jean-Christophe Trigaux. Feature diagrams: A survey and a formal semantics. In *Requirements Engineering Conference, 2006. RE 2006. 14th IEEE International*, pages 139–148, Minneapolis/St. Paul, Minnesota, USA Minneapolis/St. Paul, Minnesota, USA Minneapolis/St. Paul, Minessota, USA, September 2006.

[35] Lionel Seinturier, Philippe Merle, Damien Fournier, Nicolas Dolet, Valerio Schiavoni, and Jean-Bernard Stefani. Reconfigurable SCA Applications with the FraSCAti Platform. In *6th IEEE International Conference on Service Computing (SCC'09)*, Bangalore Inde, 2009. IEEE. IST FP7 IP SOA4All.

[36] Quan Z. Sheng and Boualem Benatallah. ContextUML: A UML-Based Modeling Language for Model-Driven Development of Context-Aware Web Services. In *The 4th International Conference on Mobile Business (ICMB'05), IEEE Computer Society. Sydney, Australia.*, pages 206–212, July 11–13 2005.

[37] Mikael Svahnberg, Jilles van Gurp, and Jan Bosch. A taxonomy of variability realization techniques. *Softw., Pract. Exper.*, 35(8):705–754, 2005.

[38] W3C. Web Services Description Language (WSDL) 1.1, 2001 March. `http://www.w3.org/TR/wsdl`.

[39] Jules White, Douglas C. Schmidt, Egon Wuchner, and Andrey Nechypurenko. Automatically composing reusable software components for mobile devices. *Journal of the Brazilian Computer Society Special Issue on Software Reuse SciELO Brasil*, 14(1):25–44, March 2008.

[40] Stephen White. Business process modeling notation specification. Technical report, Object Modeling Group, February 2006. `http://www.bpmn.org`.

[41] Christoph Wienands. Studying the common problems with service-oriented architecture and software product lines. In *Service Oriented Architecture (SOA) & Web Services Conference*, October 2006.

Chapter 4

Context Constraint Integration and Validation in Dynamic Web Service Compositions

Claus Pahl, Kosala Yapa Bandara, and MingXue Wang

Abstract System architectures that cross organisational boundaries are usually implemented based on Web service technologies due to their inherent interoperability benefits. With increasing flexibility requirements, such as on-demand service provision, a dynamic approach to service architecture focussing on composition at runtime is needed. The possibility of technical faults, but also violations of functional and semantic constraints require a comprehensive notion of context that captures composition-relevant aspects. Context-aware techniques are consequently required to support constraint validation for dynamic service composition. We present techniques to respond to problems occurring during the execution of dynamically composed Web services implemented in WS-BPEL. A notion of context–covering physical and contractual faults and violations–is used to safeguard composed service executions dynamically. Our aim is to present an architectural framework from an application-oriented perspective, addressing practical considerations of a technical framework.

4.1 Introduction

System architectures that cross organisational boundaries are usually implemented based on Web service technologies due to their inherent interoperability benefits. With increasing flexibility requirements, such as an on-demand service provision, a dynamic approach to service architecture focusing on composition at runtime is needed. The possibility of technical faults, but also violations of functional and semantic constraints require a comprehensive notion of context that captures composition-relevant aspects. Contractual definitions reflect the needs of partners–flexibility to deal with these constraints at the level of contracts and service-level agreements is essential in dynamic, on-demand applications. The physical environment needs to be monitored as faults can be caused by device, platform, and other technical factors. Context-aware techniques are required to support constraint validation and fault management for dynamic service composition. A flexible solution is sought that guarantees that these constraints are satisfied.

We present techniques to respond proactively to problems occurring during the execution of dynamically composed Web services implemented in WS-BPEL (or BPEL for short) (7). A notion of context–covering physical and contractual faults and violations–is used to safeguard composed service executions dynamically. In particular, we introduce the following aspects:

- context ontology: an ontology-based context model that integrates business and technology context aspects,
- context constraint integration: we add monitoring instrumentation to BPEL using a context model-driven constraint validation,
- constraint monitoring and fault handling: use of BPEL fault handling to capture context constraint violations and faults at runtime,
- constraint violation analysis: implementation of recovery and remedial strategies using intelligent mapping of faults to context aspects.

The techniques can realize central functions of a middleware platform for dynamic and context-adaptive Web services. Our aim is to present an architectural framework from an application-oriented perspective, addressing practical considerations of a technical framework. We illustrate the benefits of the proposed technologies through an electronic payments application.

We discuss principles of service composition and specifically dynamic composition and process execution in Section 13.2. Then, the core techniques are introduced: a context ontology in Section 4.3, the constraint integration technique in Section 4.4, an introduction of fault tolerance in Section 4.5, a monitoring and fault handling technique in Section 4.6, and a fault analysis approach in Section 4.7. An evaluation of the technical framework in terms of reliability and performance is provided in Section 4.8. Finally, we discuss trends and current issues before ending with some conclusions.

4.2 Dynamic Service Composition

We discuss different aspects and stages in the autonomic composition of Web services. These aspects and stages are supported by the individual components of the overall architecture, see Figure 4.1.

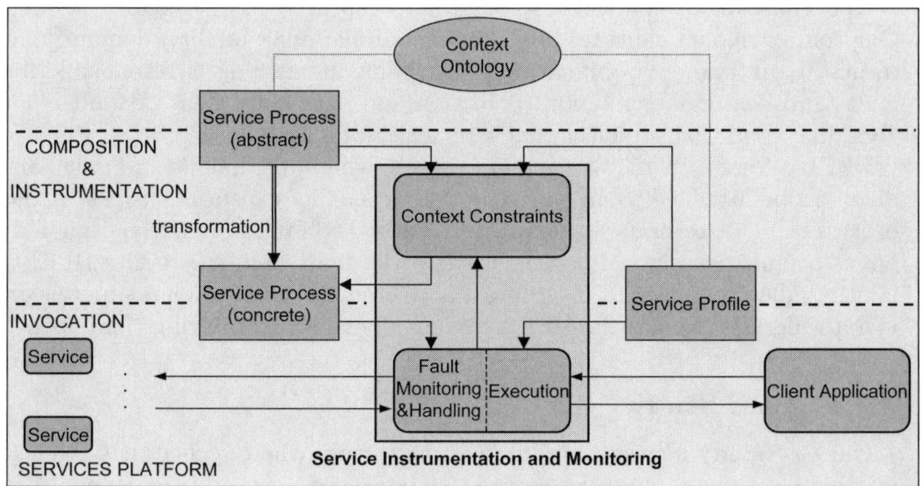

Figure 4.1: System architecture.

4.2.1 Context and Constraints

The autonomic composition of Web services usually starts with a planning process based on an abstract goal (13). This approach allows a planner to consider a set of loosely coupled goals as a planning problem. An abstract composition plan is produced from the goals, from which an executable service process can be derived. Plan generation forms the starting point for the integration of context constraints. Context is the sum of all factors that can influence dynamic composition. While functional aspects are often considered during planning and composition, some aspects such as quality can only be determined and validated during execution. We call these dynamically validated aspects context constraints, or constraints for short.

We present a context ontology capturing a wide range of aspects of a service in relation to its environment–which we define as its context. Context constraints to be validated are generated (possibly based on a contract or service-level agreement between user and provider) and linked to constraint checkers, which validate the constraint during service process execution. Data collectors are used to determine actual context attributes dynamically. Constraint validation is woven into the BPEL application process.

4.2.2 Fault Monitoring and Fault Handling

Faults can occur during the execution of service processes–as a consequence of technical runtime problems such as the unavailability of external services or the violation of contractual constraints captured in the context model. Our context model captures and integrates functional, quality, domain, and technical runtime environment aspects. Fault monitoring is responsible for fault (and constraint violation) detection and data collection. A fault is an abnormal condition or defect that may lead to failure.

BPEL allows to catch and manage faults using fault handlers. Fault handlers can be attached to an entire process or smaller execution scopes. If the process or scope terminates normally, the attached fault handlers get ignored, but if a fault occurs, it is propagated to the fault handler. Using BPEL's fault handlers for monitoring can avoid overheads on additional supervision monitoring process and the BPEL engine-dependent monitoring component.

4.2.3 Case Study

Our case study focuses on a broker architecture where a client can request utility bills from a range of devices. The service broker (e.g., a bank) is responsible for providing the requested utility bill in the requested currency to the requested device. We assume that both user and service provider have registered with the service broker. This example illustrates the effect of context on local and external services in a composed service process.

An initial user request is a goal that results in a dynamic generation of a service process, composing the application Web services and weaving in context-dependent constraint validation services. Constraints and their validation support are generated based on context information of the service involved. The user calls the `UserBillRequest` at the service broker. This process is composed of `ProviderBillRequest`, `ProviderBillResponse` and `UserBillResponse` application services. Initially, the request is internally analyzed, then each provider of services is contacted (invocation of external services to proved bill responses), and finally, the response is adapted to the needs of the end user.

All related context constraints are integrated into the Web service process as preconditions or postconditions, i.e., all context constraints are grouped under these two categories by a context constraint generator. The bill format may vary depending on the destination context. The user might expect a bill on her/his mobile device (e.g., user-friendly format in appropriate resolution) whereas the service broker expects it in machine-processable format (e.g., XML). Fault monitoring–based on BPEL's fault handlers that capture constraint violations–is the start of an analysis process that determines a remedial strategy.

4.3 Context Ontology for Service Composition

4.3.1 Context Model

Context has recently been explored in many projects to facilitate the development and deployment of context-aware and adaptable Web services (19). Wang et al. propose CONON for modelling context in pervasive computing environments, identifying location, user, activity, and computational entities as fundamental context categories (24). Doulkeridis et al. define two types of context called service context and user context for mobile services (10). The service context implies the location of the service, its version, the provider's identity, the type of the returned results, and its cost of use. The user context characterizes the users' current situation including location, time, temporal constraints, device capabilities, and user preferences. Hong et al. propose a context-aware learning architecture ontology for ubiquitous learning environments defining context in four top-level classes as Person, Place, Activity, and Computational Entities (11). Often, location is the central context concern.

In order to determine a context model for autonomic services composition, we followed an empirical approach by looking at a number of case studies. Three case study scenarios have been defined in three domains illustrating the needs of a complete and flexible context model and applicable context-determined services. The scenarios–a traditional financial service (billing and payment), an e-learning application, and convenience services–were analyzed. In these case studies, context stems from different domains, ranging from classical business scenarios to modern services and convenience infrastructures. The important context aspects often vary significantly from application to application and some context aspects emerge more frequently such as device type or security. We organize these context concerns in a comprehensive and extensible model to capture context for Web services composition.

We derived a context model ontology. Four major context categories are identified in the proposed context ontology as Functional Context, which is useful in autonomic services composition in general; Quality of Service Context, which is useful in achieving dynamic composition; Domain Context, which is useful in achieving autonomic composition in different organizations;

and Platform Context, which captures the technical environment. Together, the four categories capture all aspects (knowledge) of potential relevance for dynamic composition. The syntactical aspects of the service interface are part of this knowledge, as is device or domain-specific knowledge.

Medjahed and Atif have proposed a context categorization and context matching approach for Web services (14). However, a lack of integrative context models that can be used in autonomic service composition led us to develop our context model for changing service environments.

4.3.2 The Service Composition Context Ontology

Based on our empirical observations, we define context as any static or dynamic–client, provider or service related information, which enables or enhances efficient integration of clients, providers, and services. Services need to be aware of their context if they were to automatically adapt to changing circumstances. A number of individual context categories are defined in the context ontology. These are grouped into four top-level context areas: functional, quality, domain, and platform. Our aim is to be comprehensive, i.e., to embrace the functional focus of planning and composition, but also the device and location focus of many current context notions.

Functional Context: This describes the operational features of services. The notion of functional context in Web services is subgrouped:

- Syntax: includes input/output parameters that define operations, messages, data types of the parameters for invoking a service.
- Effect: includes the preconditions and postconditions, i.e., the operational effect of an operation execution.
- Protocol: refers to a consistent exchange of messages among services involved in services composition to achieve their goals. It includes context on conversation rules and data flow.

Quality of Service Context (QoS): Qualitative properties can be organized into four groups (15) of quantifiable attributes based on the type of measurement performed by each attribute (21).

- Runtime Attributes: relates to the execution of a service. Performance; the measurement of the time behaviour of services in terms of response time, throughput etc. Reliability; the ability of a service to be executed within the maximum expected time frame. Availability; the probability that the service is accessible.
- Business Attributes: assess a service from a business perspective. Cost; the price for execution. Reputation; measures the service's trustworthiness. Regulatory; a measure of how well a service is aligned with government or organizational regulations.

- Security Attributes: describe whether the service is compliant with security requirements. Integrity; protecting information from being deleted or altered. Authentication; ensure that both consumers and providers are identified and verified. Nonrepudiation; ability of the receiver to prove to a third party that the sender did send a message. Confidentiality; protecting information from being read by anyone not authorized.

- Trust Attributes: refer to establishment of trust relationships between client and providers–a combination of technical assertions (measurable and verifiable quality) and relationship-based factors (reputation, history of cooperation).

Other quality of service attributes or groups can be added to these four fundamental groups.

Domain Context: Each application domain may need its own context (locale) for interacting with services:

- Semantic: refers to semantic framework (i.e., concepts and their properties) in terms of vocabularies, taxonomies, or ontologies.

- Linguistic: language used to express queries, functionality, and responses.

- Measures and Standards: refers to locally used standards for measurements, currencies, etc.

Platform Context: The technical environment a service is executed in.

- Device: refers to the computer/hardware platform on which the service is provided.

- Connectivity: refers to the network infrastructure used by the service to communicate.

4.3.3 Context Ontology Construction

The starting point for the ontology construction are existing service specifications. WSDL provides the syntactic input; semantic service ontologies like OWL-S or WSMO provide further functional and nonfunctional aspects. Often, formally or informally described service-level agreements provide other details such as domain and platform-specific aspects.

Our context ontology is a knowledge framework to capture composition-relevant information. It is not meant to replace existing descriptions. Ontology mappings can be defined between service and context ontologies. Context model information comes from very different sources. The functional and quality contexts are derived from service descriptions; platform contexts are captured based on system and platform data. Domain context is based on external information sources such as domain models or external settings (like languages or units). This diversity requires an integrating framework, which we provide in the form of a context model ontology. These context instances are modelled into a single context ontology based on OWL (8). The context model ontology is an OWL-DL ontology that, at its core, captures the context model categories in the format of a taxonomy (concept level of the ontology) (22). We only illustrate a few excerpts of the context ontology with Manchester OWL syntax. For instance, we define:

```
Class:         FunctionalContext              Class:         Syntax
SubClassOf:    Context                        SubClassOf:    FunctionalContext
DisjointWith:  QoS or Domain or Platform      DisjointWith:  Effect or Protocol
```

Specific links–e.g., between Trust and Security–can be formalized by using `SubClassOf` instead of `DisjointWith` to define trust as a specific computer security aspect. Specific properties can be formulated, for instance `Syntax hasInterface MIN 1 and hasInterface SOME string`, which requires a syntax element to have at least one interface of type string associated to it.

4.3.4 Context Constraints

Context model instances express concrete requirements, for instance, concrete values for expected response times. Context reasoning for generation is used in two forms: checking the consistency of context and deducing context constraints based on defined context properties. For instance, when a client makes a request about her gas bill from her mobile device, an abstract process (plan) is generated to fulfil her request. A context constraints list is generated based on the agreements made between the parties participating in fulfilling the client's request. This list contains context instances of context types used in agreements and also instances of deduced contexts types–e.g., Mobile Phone is a context instance of device type and Bill Format is a deduced context type for the device type.

We capture concrete constraints as context model instances. We illustrate this using service `UserBillRequest` with parameters `UserID`, `UserName`, `UserAddress`, `UtilityType`, `BillRequestDevice`, and `BillRequestCurrency`. Each parameter has a data type and the Web service has functionality, both specified as context information. For an Interface element, we can express `hasUserID VALUE 123 and hasUserName VALUE ŚJohnŠ and hasUserAddress VALUE ŚDublinŠ`. `UserAddress` and `UtilityType` are the other syntax context elements of this service. `BillRequestCurrency` is a domain context element (measures and standards). Parameter `BillRequestDevice` is a device context element, part of the platform context.

4.4 Constraint Integration

The ontology-based context instances, which define and describe a concrete situation, are converted into context constraints. Constraint validation services (or short constraint services as opposed to application services) validate these constraints. At composition time, context constraint validation is integrated with the application Web service process, see Figure 4.2.

Each context aspect is validated by a constraint service. Constraint services use data collectors to support the validation of context constraints, e.g., when a client is using a mobile phone, the mobile phone becomes a context instance (of the device context aspect) and setting the bill format to mobile phone display becomes a required context constraint. Data collectors are used to collect device settings and constraint checkers validate the settings with the given device. Data collectors are also needed for performance constraints to determine for instance response-time behavior. All constraints become pre- or postconditions of the service within the integrated, composed Web service process. For response time, we determine a concrete required value for the quality attribute from the respective service profile. This is then converted into two data collector calls (creating begin and end time stamps) and a validation constraint, which checks whether the required value is achieved. A fault exception is raised if not.

4.4.1 Constraint Generation

The constraint service invocations are generated based on constraint templates for the specific context aspect. A context template has link, condition type, and expression elements. The link is used to support service binding. Path expressions (XPath) are used to specify the location of the constraint checker. The condition type explains both the type of the constraint validation (precondition or postcondition–based validation) and the order of execution. The expression specifies the constraint to be checked.

A key observation here is that the implementation of constraint validation is context category-dependent:

- The Functional context details, e.g., parameters and protocol aspects. Pre/postcondition validation is used, but no data collectors.
- The Quality of Service constraints usually require data collectors to monitor variable quality properties before validating constraints.
- The Domain and Platform constraints refer to data collectors to determine environment conditions. In contrast to the quality monitors, these are static properties (such as language or device) that need to be queried, but not measured.

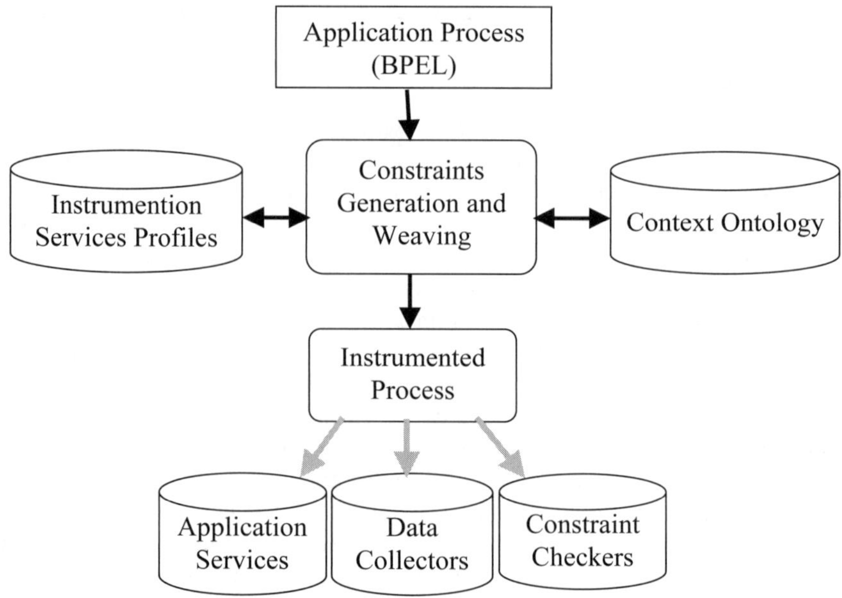

Figure 4.2: Constraint generation architecture.

This category-dependency allows for uniform constraint monitoring within the categories, which is an advantage for efficient constraint integration.

4.4.2 Constraint Language

The Java Modelling Language (JML) is our context constraint language. JML is a behavioural interface specification language. JML is suitable as it supports pre- and postconditions–the format in which we express validation constraints. The keyword `requires` is used in specifying preconditions. A precondition is a condition that must be satisfied before calling a service. The `ensures` keyword prefixed a postcondition that must be established. The `UserBillRequest` service proceeds further only if the user is verified (authenticated). For the `userVerification` with parameter UserID, we define:

```
<Link: path expression to the service process/>
<Condition>
  <Type> post-condition </Type>
  <Order> 1 </Order>
</Condition>
<Expression>
   @ensures returnBoolean( Context:userVerification(UserID) ) == True;
</Expression>
```

The @ensures expression requires the return value of the `userVerification` context service (located at `Context` with parameter is `UserID`) to be true. Similarly, for the `UserBillResponse` service, the constraint depends on the user device and device type (platform context):

```
<Expression>
   @ensures returnBoolean(
        Context:compareBillFormat(),
        DeviceType,
        Context:setBillFormat( Context:getBillFormat(DeviceType) ) ) == True;
</Expression>
```

This postcondition ensures that `compareBillFormat` returns true. `Set-BillFormat` calls `getBillFormat`, located at `Context`, with the parameter `DeviceType` to set the bill format. The parameters are the Context-based `CompareBillFormat` service, the `DeviceType` parameter and the currently set bill format (which is set by calling `getBillFormat` in `setBillFormat`). Then `compareBillFormat` compares the device type with the set bill format and only returns true if the bill format matches the device type.

4.4.3 Instrumentation and Weaving

We implement dynamic constraint integration and monitoring as an instrumentation of the application process, achieved through weaving (Figure 4.3) (3). For our case study, user and service broker agree on using different devices and different currency types in the utility bill process. The user request generates both an abstract business processes and a context constraints list, which is based on the agreements made between the parties. The constraints list contains the context instances and constraint validation service bindings.

At the center of the validation instrumentation is a mapping:

- Context model attributes are connected to concrete values at instance level to form abstract constraints. Thus, attributes like `UserID` or `BillRequestDevice` are extracted from the ontology.
- A preparation step for the final mapping is the determination of data collectors (e.g., `getBillFormat`) and data initializers (e.g., `SetBillFormat`) that support the constraint condition (e.g. `compareBillFormat`). These can be application-specific (however, `getBillFormat` might depend on a generic device context attribute). Other constraints are directly based on generic data collectors (e.g., performance monitoring services).
- The abstract constraints are mapped to JML pre- or postcondition constraints. Constraint service calls for constraint checking are generated based on context ontology instances. These service invocations are based on information given in the constraint templates, i.e., path expression

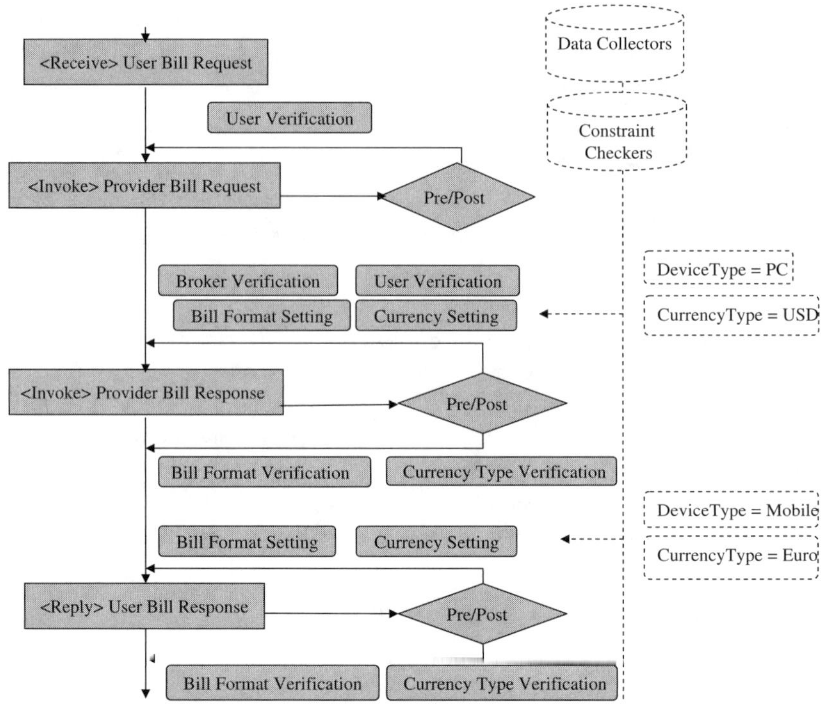

Figure 4.3: Constraint validation instrumentation.

(link), context constraints (condition type), constraint services, and context constraint language (expression).

Constraint services encapsulate the constraint checker. A service-related context specification, the context ontology, and constraint service invocation shells can be precomputed at development time of the main services. However, the weaving process needs to be executed in parallel with Web service process planner and constraints generator.

```
<bpel:invoke name="UserVerificationInvoke" ... </bpel:invoke>
<bpel:if name="userVerificationConstraint">
  <bpel:condition>
    <![CDATA[\$UserVerificationResponse.parameters/return ='true']]>
  </bpel:condition>
```

The execution of the application process only proceeds if the respective constraint is not violated. The monitoring, analysis, and handling of possible violations is integrated using BPEL's fault handling mechanisms, which will be described in the subsequent sections.

4.5 Fault Tolerance and Remedial Strategies

Any of the context constraints might be violated at runtime. A violation causes a fault. A fault analysis takes fault data as input and is responsible for choosing a suitable remedial strategy for that fault from predefined fault remedial knowledge. The remedial strategy is applied to the faulty process execution. There are three steps for defining the fault remedial knowledge: defining a fault taxonomy, defining remedial strategies, and matching each fault category with remedial strategies.

Our fault monitoring and analysis covers application-level violations (e.g., functional of quality attributes) as well as technical, environment-specific faults (domain and platform attributes). As a consequence, our fault taxonomy is based on the context model. Thus, the root fault categories are the context categories Functional, Quality of Service, Domain, and Platform.

4.5.1 Defining Remedial Strategies

Some common strategies such as retry or replace have been introduced (12, 9, 2, 18). In dynamic service composition, remedies are selected and applied dynamically. We categorize remedial strategies into goal-preserving and nongoal-preserving strategies. Goal-preserving strategies aim to recover from faults; the business goal of a process would be completed through a continued process execution after fault recovery. Nongoal preserving strategies do not attempt recovery. They assist possible future recovery.

4.5.2 Goal-Preserving Strategies

In BPEL, an invocation calls a business activity performed by a Web service. This can be monitored with a fault handler using the scope attachment. We identify four goal-preserving strategies for fault handling as follows:

- Ignore is a simple strategy, which does not take any action on a fault. The objective of fault-tolerance is that assuring the business goal is achieved by service processes in fault situations, rather than recovery of all faults. We can ignore faults that do not affect a business goal.

- Retry is suitable for communication faults. For example, an invoked service is temporarily unavailable; messages are lost during network transmission or replies are missing. In this case, this strategy suspends the execution of the process and retries the invocation of the fault services. Maximum retry number and interval before each retry can be defined.

- Replace is similar to retry, but with an alternative service that has the same abilities as the original faulty service. Since we cannot alter remote services, replacing faulty services can be effective. However, this strategy is limited to stateful Web service composition. The client is required to keep an instance or session data to support business requirements, such as conversational message exchange patterns. If the session data is nonretainable in the ongoing service, the service is tightly coupled to the process workflow. Thus, the replace strategy is unworkable.

- Recompose: Ignore, Retry, and Replace are inner process-level remedial strategies try to recover faults within the current process. Recompose is different in that it discards the fault process and re-establishes an alternative process, which has same business goal. As a consequence, recomposition is suitable for all categories of faults.

Ignore and Retry are lower-level recoveries that keep the original process workflow. Applying them requires less time resources. In higher-level recovery (Replace, Recompose), an additional component is needed for discovering alternatives, which requires more time and computation resources. Lower-level goal preserving strategies should be applied first, as they require less time resources with less impact on processes. The following example allows one Retry opportunity before applying Ignore:

```
<sequence>
   <retry><max>1</max><waitingTime>P0Y0M0DT0H0M1.0S</waitingTime></retry>
   <ignore><value>true</value> <log>level_1</log></ignore>
</sequence>
```

There are two ways to provide alternative replacements. First, alternative services are pre-assigned to remedial strategies. Replace can be applied instantly. Second, alternative services are dynamically discovered based on functional and nonfunctional properties. Recompose is different, as in dynamic composition, we presume service processes are only discovered at runtime. However, depending on business goal and size of the registry, Recompose can be time consuming. Hence, we have also developed a selective process repository to minimize time (17). The process repository saves composed services and processes with a categorized fault ratio. Alternative processes can be retrieved and selected from the process repository.

Replace is a passive technique; the backup is only called after a primary service fault. (9) introduces a parallel strategy. Several alternative services are invoked in parallel for one invocation. The first response received is chosen for ongoing process execution. A disadvantage in dynamic composition is that all alternative services need to be discovered dynamically at composition time. It also causes overheads on computation and network resources to execute alternative services. Moreover, it could cause business goal violations on state update, e.g., a bill is paid twice. The advantage is that only the best performing service is picked, and does not need to be replaced. We get similar results and avoid some disadvantages by selecting alternative services for replacement dynamically. Alternative services' fault ratio and response times in a process repository are used to determine the most suitable one.

Replace and Recompose might call for compensation or rollback. Compensation would be a precondition of these remedial strategies in many cases. Deploying an alternative process, the system needs to clear up partially executed faulty processes (rollback), i.e., the process execution needs transactional behavior. However, this is difficult as no common protocol exists for Web services (16, 4). BPEL's compensation handler enables one to define an activity at the scope or process level whose execution reverses some previously executed application logic. However, there is no automatic restoration of data during compensation. The application might define its own compensation behavior. We assume for state-updating services that there is at least one service that can rollback its effect and does not depend on any state for execution. For Replace, compensation may also be required for postcondition faults before an alternative service is retried.

4.5.3 Nongoal Preserving Strategies

Nongoal preserving strategies do not impact on process execution. They can be combined with other strategies including goal-preserving strategies. We define three nongoal preserving strategies. **Log** records the captured fault. It could be applied at different levels, e.g., Level-1 logs fault source and fault message. Level-2 logs data transmission of fault sources as well. This data is saved in a fault log database. **Alert** notifies relevant stakeholders. **Suspend** suspends the faulty service or process until future investigation, if the fault element exceeds an acceptable fault ratio. The purpose is to isolate the fault elements to avoid possible repeat faults.

4.5.4 Fault Categories and Remedial Strategies

Matching fault categories with remedial strategies needs to consider different levels of data. From low to high, there are default remedial data, services, and process-specific remedial data and application-specific remedial data.

Default remedial data comes from an analysis of fault categories. It is the proposed solution for all fault categories (see Table below). Retry is suitable for most remote faults from remote services where postcondition constraints are violated. For instance, a `missingOutput` fault might result from a temporary unavailable service. Retry is not suitable for precondition constraint violations. Replace and Recompose are suitable for all fault categories. Recompose would be last option as it is the most resource consuming.

	Precond. constraint violation	Postcond. constraint violation
Ignore	All fault categories	All fault categories
Retry	Not suitable	Functional context fault; Platform context fault
Replace	All fault categories	All fault categories
Recompose	All fault categories	All fault categories

The following XML code is a precondition remedial strategy for securityFaults. The system needs to assign a remedial strategy for each context aspect that is validated at runtime. The strategy definition is used when a fault in the respect category has been identified by the constraint service.

```
<securityFault>
   <preConditionViolationRemedy>
      <sequence>
         <ignore><value>false</value><ignore>
         <retry><max>0</max><waitingTime>P0Y0M0DT0H0M0.0S</waitingTime></retry>
         <replace><value>any</value></replace>
         <recompose><value>true</value><log>level_1</log></recompose>
      </sequence>
   </preConditionViolationRemedy>
   <postConditionViolationRemedy>...</postConditionViolationRemedy>
</securityFault>
```

Service-specific remedial data is defined according to service descriptions for specific services only. State-updating services need compensation. Services can have fault and compensation handlers associated to it.

```
<service>
   <serviceReference>
      <endpointUrl>http://localhost:8080/.../BankPaymentService</endpointUrl>
      <operation>BankPayment</operation>
   </serviceReference>
   <faults>
      <securityFault>...</securityFault>
      ...
   </faults>
   <compensation>
      <serviceReference>
         <endpointUrl>http://localhost:8080/.../BankRefundService</endpointUrl>
         <operation>BankRefund</operation>
      </serviceReference>
   </compensation>
</service>
```

Process-specific remedial data is defined according to business goals and application domains. It needs to comply with application requirements and organizational policies. In processes involving financially sensitive data, security-level mismatch faults are not acceptable; some processes would mark minor security faults as ignorable. Organizations might define their own trusted alternative service as a Replace remedy.

```
<process>
   <processReference>
```

```
      <onDemandRequest>GasBillPayment</onDemandRequest>
   </processReference>
   <services>
      <service>
         <serviceReference>...</serviceReference>
         <faults>...</faults>
      </service>
      ...
    </services>
    <faults>...</faults>
</process>
```

For a fault instance, the system searches for remedial strategies from high to low levels. Higher levels are customizations of lower level data.

4.6 Architecture and Core Components

We divide our architecture into three layers (Figure 4.4): process execution layer, composition and fault-tolerance layer, and database layer. A BPEL engine is responsible for the process execution layer. The three databases of the database layer have been discussed in the previous section. The four composition components in the fault-tolerance layer form the architectural core. They directly interact with the instrumented BPEL process, thus our approach is BPEL engine independent. We discuss the core components now.

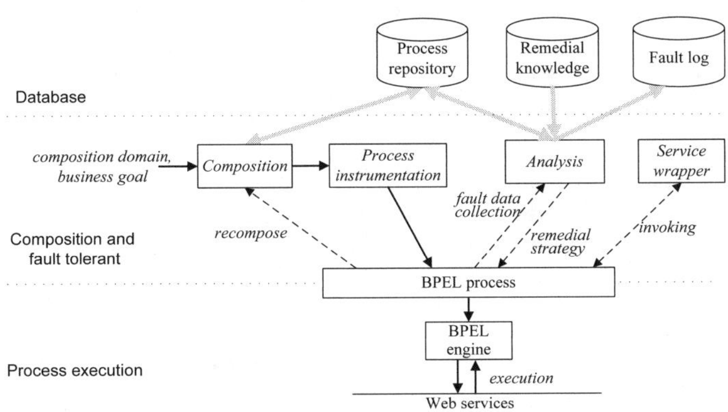

Figure 4.4: Three layered system architecture.

The **composition component** composes services to service processes based on user requirements–a classical AI planner can be used for this purpose (13, 25). This requires a semantic service description to define the composition domain. Service processes are saved in an indexed process repository for possible future reuse, e.g., during recomposition. The repository supplies a suspended-list of services and processes for composition to filter invalid processes. It also contains a list of suitable replacement services for each application service (dynamic discovery and matching is beyond the scope of this investigation). To enable recomposition, the composition component is exposed as a Web service `recompose` with a `processReference` as input. `ProcessReference` contains the process name as business goal and the process index, which differentiates multiple processes for the same business goal.

The **process instrumentation component** converts application processes to instrumented processes, which includes the fault monitoring and handling mechanisms within the BPEL process.

The **analysis component** utilizes remedial knowledge to provide remedial strategies for a fault instance. It also updates the fault ratio of services and processes in the process repository and updates the fault log if the Log strategy is required. All nongoal preserving remedial strategies can be implemented by the analysis component. Its Web service interface `analyse` has five inputs. `faultData` is a fault variable or constraint violation collected by the BPEL fault handlers. `processReference` denotes the current BPEL process. `invokingServiceReference` is an instance of `ServiceReference` the identifies a Web service. A `ServiceReference` contains an endpointUrl and an invoking operation. `RequestData` and `responseData` record fault service data transmission for the Log strategy.

The **service wrapper component** is a dynamic service invoker. It wraps actual application services into a unified service interface `genericOperation`. The `genericOperation` has two input parts. `requestData` is input of the service; `invokingServcieReference` is the identity of the service. `responseData` returned by `genericOperation` is output of the service. The purpose is to provide a dynamic binding partner link. In BPEL, partner links define how a process interacts with other processes and services. Dynamic binding partner links allow that application service endpoints are selected and assigned to the partner link through configuration or runtime input. The limitation is these services must have the same interface. Our wrapper component achieves dynamic binding without this limitation.

4.7 Instrumentation Template for Violation Handling

The purpose of the process instrumentation is to add fault monitoring and handling capability to BPEL business processes, i.e., dynamically selected remedies are able to act on process executions when faults occur. The instrumentation is based on a designed instrumentation template for violations and fault handling for each service-invoking activity (Figure 4.5). Since we use constraint services for pre- and postactivity validation, two constraint service invocations may be bound to each invoking service.

There are two types of fault handler activities in a fault handler construct: a set of custom catch activities and at most one at catchAll. A catch fault handler, which only catches a specified fault that has an optional fault name and/or fault variable. The fault data can be forwarded to the analysis component as needed. A catchAll fault handler executes if a fault is not caught by existing catch fault handlers. The catchAll ensures that no fault is ignored.

We describe the violation and fault handling template in two parts. The first part in Section 4.7.1 is the <`repeatUntil`> container in the top half of Figure 4.5. It supports the Ignore, Retry, and Replace remedial strategies. The second part in Section 4.7.2 is for the Recompose strategy; see bottom half of Figure 4.5. Since nongoal preserving strategies do not impact the processes, they will not be considered by the violation handling template. Again, the bill payment service process acts as an example to illustrate the application of the template to instrument the application process with fault violation handling. For instance, `billPay` is a sample service invocation that is used. Pre- and post-condition violation handling is described.

4.7.1 Ignore, Retry, and Replace

Five variables for each invoking activity define the violation handling context and determine the handler execution: `invokingServiceReference` provides the current invoking activity service reference, e.g., `billPay`. Variable `composition` denotes if compensation of the current invocation is needed for the Recompose strategy. It has the default value true. `compensationServiceReference` names the compensation service of the current invocation–the initial value is empty. `waitingTime` defines the waiting duration for the Retry strategy with an initial value 0. The execution `path` is by default initialized as the uninstrumented original execution path: precondition constraint service, invoking service, and postcondition constraint service.

In addition to context constraints, paths are the second key concept in the template. We create a conditional service composition with all possible paths, which for decisions of the analysis component include the necessary recovery as well. In a fault-free scenario, only the default path is executed. The other paths will only be executed based on a corresponding selection of `analyse`. Otherwise, the fault is caught by attached fault handlers and the `analyse` service inside the handlers determines the following actions, including selection of a new execution path. In the template, the <`repeatUntil`> container is important. It only ends when a path is executed successfully (`path=0`) or `analyse` decides to recompose the current process (`path=-1`). In the following, we distinguish pre- and postcondition based constraint violations.

For faults caused by **precondition constraint violations**, the fault handler passes the fault variable thrown by a constraint service (`constraint-`

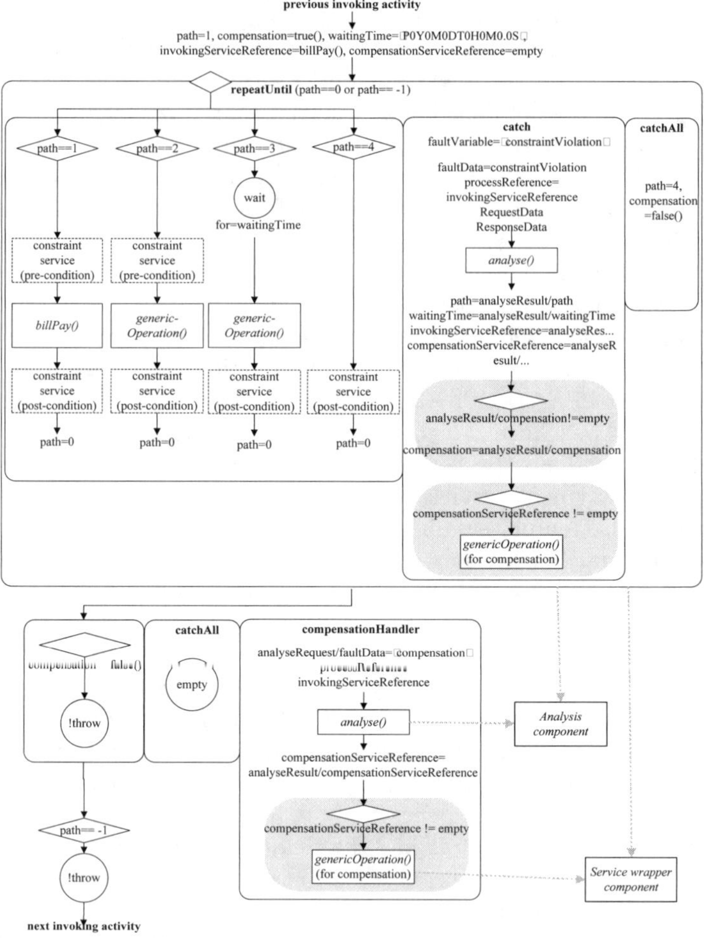

Figure 4.5: Violation and fault handling template

Violation) to analyse. analyse uses processReference, invokingServiceReference and other additional variables.

1. If the remedial knowledge suggests Ignore, analyse returns path = 3, compensation = true, compensationServiceReference = empty and waitingTime = 0 and keeps invokingServiceReference = billPay. The <repeatUntil> forces the process to execute path 3. The billPay is executed through the wrapper component genericOperation and the postcondition is validated.

2. If the Replace strategy is applied, analyse sets path = 2, compensation = true, compensationServiceReference = empty,

waitingTime = 0 and assigns the invokingServiceReference to an alternative service. The second path is similar to the first, except genericOperation replaces the original application service billPay. This allows the alternative service to be executed through the wrapper component.

3. If analyse suggests recomposition, it keeps invokingServiceReference, sets compensation = true, compensationServiceReference = empty, waitingTime = 0, and path = -1 to end <repeatUntil>. We discuss this in the next subsection.

For faults coming from the **invoking service activity**, i.e., either billPay or genericOperation, nonconstraint faults are caught by the catchAll handler. The fault handler assigns path=4, which means the postcondition constraint service deals with the fault and throws a constraint violation fault for analyse, i.e., a syntax constraint violation is expected thrown from the constraint service for faultData. Variable compensation is also set to false, as no compensation is required for this invocation during recomposition.

For faults caused by **postcondition constraint violation** we distinguish the four strategy cases:

1. For Ignore, analyse keeps invokingServiceReference, sets compensation = empty, compensationServiceReference = empty, waitingTime = 0, and path = 0 to end <repeatUntil>. An empty compensation variable means to keep its previous value, i.e., the fault comes from the invoking service.

2. For the Retry strategy, analyse sets path = 3, compensation = true, compensationServiceReference = empty and keeps the last invokingServiceReference.

3. For Replace, analyse sets path = 2, compensation = true, waitingTime = 0 and an alternative service for invokingServiceReference. Replace for postcondition faults also needs to check if compensation is required. If analyse returns a compensationServiceReference, genericOperation within the fault handler executes the compensation service.

4. For Recompose, path = -1, compensation = empty, compensationServiceReference = empty, waitingTime = 0 is set and invokingServiceReference is kept.

In case of faults with alternative replacement services, the same strategy as above is applied again.

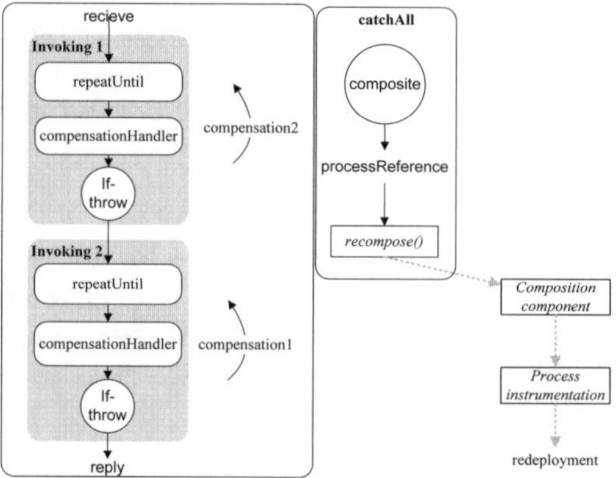

Figure 4.6: The structure of a process scope

4.7.2 Recomposition

We continue with the second `scope` (with `compensationHandler` attachment) of Figure 4.5. The second scope is responsible for compensating of an entire process, i.e., Recompose is applied. If `compensation==false`, a defined fault is thrown. An attached <`catchAll`> handler catches the fault and does nothing. The purpose is to mark this scope as faulty. The BPEL compensation handler can only be triggered by a successful scope for process compensation. In that case, such as Ignore with a postcondition fault, a compensation handler attached to the scope within <`repeatUnit`> will never be triggered, as a fault occurred. We create this scope for invocation activity compensation, and a `compensation` variable decides whether to trigger it.

If `analyse` decides to recompose (`path=-1`), a fault is thrown. The process scope catches this fault and starts scope composition before calling `recompose` (Figure 4.6). All fault-free compensation scopes are executed in backward order. If any invocation activity requires compensation, `analyse` provides a rollback service, which is executed through `genericOperation`.

4.8 Evaluation

We have implemented a service process monitoring architecture. Prototypes for the constraint generation and the fault handling (the two main components) exist and are the basis of the evaluation. The BPEL engine (Active BPEL) running the process integrates the constraint validation checker (JML checker).

Performance is central for runtime composition and monitoring. Two aspects emerge that we have investigated using our prototype:

- The constraint processing and weaving is time consuming. Our strategy favors early precomputation of constraint templates as instrumentation profiles (from the ontology as soon as changes to the ontology are known if the application services are determined).

- Our experiments with different variants of the architecture in terms of the constraint weaving into the BPEL service process–with respect to how constraint violations are handled–shows an acceptable overhead.

Process instrumentation makes the instrumented process more complex than the original one. For constraint violation handling, the total overhead depends on the plan execution time, the number of constraint violations, and the constraint engine performance, etc. In the case study scenarios we investigated, there is only a violation handling overhead of in average 10.05% in a total plan execution time of in average 17,110 milliseconds (not including replanning). In case no faults occur, a central benefit of our solution is that there is no overhead. Since all Web services were hosted locally, we expect overheads in networked environments to be much lower.

With in total 35 test cases executed and successful, our approach provides a reliable violation and fault handling for dynamic service composition. Scalability is another central issue. In general, more complex service processes do not impact the performance and reliability results as long as the degree of concurrency does not increase significantly.

However, the overall success also depends on alternative services and processes being supplied for the replacement remedies. Dynamic replanning is another aspect that can be affected negatively with increased complexity (which is not part of this investigation). We discuss this further in the next section.

4.9 Discussion–Related Work, Trends, and Challenges

Related work in this area covers context modelling and constraints used in dynamic service architectures. Broens proposes a context binding infrastructure called the Context-Aware Component Infrastructure (CACI) (5). This realizes context-binding transparency and is composed of a context binding mechanism and a context discovery interoperability mechanism. However, this approach is not specific to service process composition with its binding and fault handling mechanisms. Hong and Cho present a context-aware manager architecture to support user-centric ubiquitous learning services and describe an ontology-based context model for intelligent school spaces (11), however, do not address constraint integration in service compositions. Medjahed and

Atif propose a context-based matching approach for Web service composition (14). This approach introduces policy-based context binding, but does not cover violations and fault handling.

The METEOR-S project focuses on constraint-driven Web services composition (1). It distinguishes data, function, and quality of service semantics. The use of SWRL and OWL to provide descriptive rules for specifying constraints is planned, but not realized yet. Chen et al. propose a semantic matchmaker architecture that consists of a service planner for capability matching and a context reasoner for context matching (6).

In (25, 23) solutions using various planning techniques are provided–which have received significant attention–for dynamic service composition, but they lack fault-tolerance. In (4), a constraint language is proposed for the Dynamo monitoring platform. We use a more simple and more efficient standard BPEL fault handling without requiring additional execution monitoring subsystems.

The solution that we have implemented demonstrates that fault handling mechanisms can support high-performance constraint violation handling based on standard BPEL engines. While constraint integration solves some problems, some challenges have also arisen. To provide flexibility in application processes, various types of constraints are required. Constraints need to be defined at a context model level to capture business and technical aspects and need to be integrated dynamically. Our prototype is able to integrate context constraints into application services and weave these into BPEL processes for violation handling using the instrumentation template.

Our infrastructure can support context-dependent fault handling in a composition approach. Other challenges, however, remain. A composition planner often has incomplete information initially (13). Planners need observers or gather information. As a solution to this problem, an interleaved approach integrates service execution with sensing services as part of a planning and composition process. The composition tool dynamically queries the environment for context conditions rather than searching for all possibilities in a tree-like conditional plan. Our solution can support this through data collectors and constraint checkers. Semantic enhancements of WS-BPEL would also bring closer composition and the process execution (20).

A signalled fault does not unambiguously identify the reason. While in the context of our framework it is sufficient if a suitable remedy can be put in place (like a replacement service), the possibility to optimize the remedial strategy arises. Our aim was only the immediate dynamic reaction. An offline analysis could probe deeper in order to improve the remedial strategy definition.

4.10 Conclusions

Context-based composition is an essential ingredient for autonomic Web service composition. We have introduced techniques for contractual constraint violation and runtime fault handling for dynamic service composition.

An ontology-based context constraint definition and validation framework is the conceptual core. Constraint validation is woven into application service processes. We have defined the notion of context for Web services and formalized an approach to specify semantic context information in a comprehensive context ontology. We have illustrated an intelligent remedial knowledge base for dynamic remedial strategy selection. We have provided a BPEL constraint violation and runtime fault handling template to enable a lightweight engine-independent implementation. We have also evaluated our implementation in terms of violation and fault handling ability and overhead and performance aspects on instrumented process execution.

These techniques are stepping stones towards implementing a general approach for autonomic services composition. On-demand service provision is an example of changing approaches to software provision and utilization. These new forms, however, depend heavily on dependable dynamic composition and execution.

Acknowledgment

The authors would like to thank the Science Foundation Ireland for its support of this work through the RFP project CASCAR.

References

[1] R. Aggarwal, K. Verma, J. Miller, and W. Milnor. Constraint driven web service composition in METEOR-S. In *Proceedings of the 2004 IEEE International Conference on Services Computing*, 2004.

[2] D. Ardagna, C. Cappiello, M.G. Fugini, E. Mussi, B. Pernici, and P. Plebani. Faults and recovery actions for self-healing Web services. In *15th international World Wide Web conference*, 2006.

[3] L. Baresi and S. Guinea. *Towards Dynamic Monitoring of WS-BPEL Processes*, volume 3826 of *Lecture Notes in Computer Science*, pages 269–282. Springer Berlin/Heidelberg, 2005.

[4] L. Baresi and S. Guinea. A dynamic and reactive approach to the supervision of BPEL processes. In *1st India Software Eng. Conf.*, 2008.

[5] T.H.F Broens. *Dynamic context bindings–Infrastructural support for context-aware applications.* PhD thesis, Univ. of Twente, 2008.

[6] I.Y.L. Chen, S.J.H. Yang, and J. Zhang. Ubiquitous provision of context aware Web services. In *IEEE International Conference on Services Computing (SCC'06)*, 2006.

[7] The WS-BPEL Coalition. WS-BPEL business process execution language for Web services–specification version 1.1. In *http://www-106.ibm.com/developerworks/webservices/library/ws-bpel*, 2004.

[8] The World-Wide Web Consortium. Owl web ontology language. In *http://www.w3.org/TR/owl-features/*, 2004.

[9] G. Dobson. Using WS-BPEL to implement software fault tolerance for Web services. In *32nd EUROMICRO Conference on Software Engineering and Advanced Applications*, 2006.

[10] C. Doulkeridis, N. Loutas, and M. Vazirgiannis. A system architecture for context-aware service discovery. *Electronic Notes in Theoretical Computer Science*, 146:101–116, January 2006.

[11] M. Hong and D. Cho. Ontology context model for context-aware learning service in ubiquitous learning environments. *International Journal of Computers*, 2, July 2008.

[12] J. Lau, L.C. Lung, J.D.S. Fraga, and G.S. Veronese. Designing fault tolerant Web services using BPEL. In *7th IEEE/ACIS International Conference on Computer and Information Science*, 2008.

[13] S. McIlraith and T.C. Son. Adapting GOLOG for composition of semantic Web services. In *8th International Conference on Principles of Knowledge Representation and Reasoning*, 2002.

[14] B. Medjahed and Y. Atif. Context-based matching for Web service composition. *Distributed and Parallel Databases*, 21:5–37, 2007.

[15] B. Medjahed and A. Bouguettaya. A dynamic foundational architecture for semantic Web services. *Distrib. Parallel Databases*, 17:179–206, 2005.

[16] T. Mikalsen, S. Tai, and I. Rouvellou. Transactional attitudes: Reliable composition of autonomous Web services. In *Workshop on Dependable Middleware-based Systems*, 2002.

[17] C. Moore, M.X. Wang, and C. Pahl. An architecture for autonomic Web servive process planning. In *3rd Workshop on Emerging Web Services Technology*, 2008.

[18] O. Moser, F. Rosenberg, and S. Dustdar. Non-intrusive monitoring and service adaptation for WS-BPEL. In *17th World Wide Web Conf.*, 2008.

[19] M. Mrissa, C. Ghedira, D. Benslimane, and Z. Maamar. *Context and Semantic Composition of Web Services*, volume 4080 of *Lecture Notes in Computer Science*, pages 266–275. Springer, 2006.

[20] J. Nitzsche, T. van Lessen, D. Karastoyanova, and F. Leymann. BPEL for semantic Web services (BPEL4SWS). In *On the Move to Meaningful Internet Systems 2007: OTM 2007 Workshops*, pages 179–188, 2007.

[21] J. O'Sullivan, D. Edmond, and H. M. Arthur. Formal description of non-functional service properties. Technical report, Queensland University of Technology, Centre for Information Technology Innovation, 2005.

[22] C. Pahl. An ontology for software component matching. *Int. J. Softw. Tools Technol. Transf.*, 9(2):169–178, 2007.

[23] M. Pistore, F. Barbon, P. Bertoli, D. Shaparau, and P. Traverso. Planning and monitoring Web service composition. In *Workshop on Planning and Scheduling for Web and Grid Services*, 2004.

[24] X.H. Wang, D.Q. Zhang, T. Gu, and H.K. Pung. Ontology based context modeling and reasoning using OWL. In *Proceedings of the Second IEEE Annual Conference on Pervasive Computing and Communications Workshops*. IEEE, 2004.

[25] D. Wu, E. Sirin, J. Hendler, and D. Nau. Automatic Web services composition using shop2. *Workshop on Planning for Web Services*, 2003.

Part II

Architecture

Chapter 5

Enabling Context-Aware Web Services: A Middleware Approach for Ubiquitous Environments

Daniel Romero, Romain Rouvoy, Sophie Chabridon, Denis Conan, Nicolas Pessemier, and Lionel Seinturier

Abstract In ubiquitous environments, mobile applications should sense and react to environmental changes to provide a better user experience. In order to deal with these concerns, *Service-Oriented Architectures* (SOA) provide a solution allowing applications to interact with the services available in their surroundings. In particular, context-aware Web Services can adapt their behavior considering the user context. However, the limited resources of mobile devices restrict the adaptation degree. Furthermore, the diverse nature of context information makes difficult its retrieval, processing and distribution. To tackle these challenges, we present the CAPPUCINO platform for executing context-aware *Web Services* in ubiquitous environments. In particular, in this chapter we focus on the middleware part that is built as an autonomic control loop that deals with dynamic adaptation. In this autonomic loop we use FRASCATI, an implementation of the *Service Component Architecture* (SCA) specification, as the execution kernel for *Web Services*. The context distribution is achieved with SPACES, a flexible solution based on REST (*REpresentational State Transfer*) principles and benefiting from the COSMOS (*COntext entitieS coMpositiOn and Sharing*) context management framework. The application of our platform is illustrated with a mobile commerce application scenario that combines context-aware *Web Services* and social networks.

5.1 Introduction

Context awareness enables nowadays systems to sense and react to changes observed in their physical environment (e.g., location, connectivity, brightness, resource availability). This capability is particularly critical in ubiquitous environments, where context is the central element of mobile applications. In order to cope with these concerns, *Service-Oriented Architectures* (SOA) have provided an attractive solution to build distributed applications, which can communicate with services discovered in their surroundings. In particular, *Web Services* (2, 8) enable mobile devices to interact spontaneously with services available on the Internet (weather forecast, news, etc.) in order to improve the *Quality of Service* (QoS) offered to end-users. Context-aware Web Services can therefore dynamically tune their behavior in order to better integrate the user's context. Nevertheless, on a mobile device, the degree of adaptation of these applications is often restricted by the device capacities. In order to meet this challenge and provide an uniform approach to context-aware adaptation, we propose to exploit the *Service Component Architecture* (SCA) technology (5) for enabling end-to-end adaptive applications. We believe that SCA provides an elegant solution to build self-adaptive ubiquitous applications based on Web Services.

This chapter introduces CAPPUCINO, [1] which is a SCA-based middleware platform for executing context-aware Web Services in ubiquitous environments. The CAPPUCINO platform is divided into two technical parts: context-aware software product line [2] and a middleware platform for context-aware Web Services. The architecture of our Web Services platform is built as an *autonomic control loop* (16). The principle of the control loop from the autonomic computing domain brings several characteristics, such as self-configuration, self-optimization, self-healing, and self-reparation (28, 17). We rely on several of these characteristics to build our platform architecture, which is structured around the following threefold contribution:

The *autonomic control loop* is built of three parts: i) A "knowledge" part for abstracting modelling of autonomous systems; ii) an "autonomic manager" part for setting adaptation rules; and iii) a "managed resources" composed of context management, deployment, and reconfiguration services.

The *execution kernel* is based on FraSCAti, our implementation of the SCA specification and its FraSCAme version for Java ME-based mobile

[1]The development of this platform is supported by the project of the same name from the research agency PICOM, Nord-Pas-de-Calais, France.

[2]The context-aware software product line is published as a separate chapter to this book (22).

devices.

The *context-awareness* of Web Services is dealt with by a *context mediation service*: SPACES. This service applies the principles of *REpresentational State Transfer* (REST) in order to distribute COSMOS context policies in ubiquitous environments.

The remainder of this chapter is organized as follows. Section 5.2 motivates the use of context-awareness in mobile commerce applications. Section 5.3 introduces the principles and the background of the contribution. Then, the threefold contribution, that is the realization of an autonomous SOA system with the control loop, the SCA runtime, and the distributed RESTful context manager, is described in Section 5.4 and then illustrated in a mobile commerce scenario in Section 5.5. Finally, Sections 5.6 and 5.7 discuss the contribution with regard to related work and conclude the chapter, respectively.

5.2 Motivating Scenario

In order to motivate the challenges of ubiquitous applications development, this section introduces a mobile commerce application scenario. This scenario takes advantage of some of the new opportunities brought by ubiquitous systems, such as the availability of context-aware Web Services to *On-The-Go* users and context-aware social networks.

> *Mary uses an application on her mobile phone to search and buy different items from an online catalog. This application exploits Mary's profile to offer her personalized services. For example, when subscribing to the reward point program, Mary automatically receives special offers and prices. She can also register important events in her calendar like her best friends' birthdays. Using this information, customized notifications with a focused product offer are pushed to her mobile phone. Mary can also find gift ideas by using a special service that connects to a social network to retrieve a list of items selected by her friends as their favorite products. Additionally, according to her location, Mary is informed of Flash Sale offers currently running in the shops nearby. If she does not have Internet access when she finally makes her choice, she can still complete her order off-line. The application registers the order in memory, as an SMS message, and sends it automatically once the network coverage becomes sufficient again.*

Challenges This scenario highlights some of the challenges of the development of context-aware Web services systems concerning the capacity to detect context changes in the user environment and to dynamically adapt the provided services. We now discuss how the CAPPUCINO platform can fulfill these challenges.

Dynamic adaptation: The scenario requires that applicative services are context-aware and can be dynamically reconfigured after a context change. The customized catalog depends on the user's context, including both profile and location information. When a special event registered in Mary's profile (a birthday approaching) occurs, the CAPPUCINO platform prepares a new customized catalog. In order to do that, the platform considers the account preferences and product suggestions coming from several sources (e.g., a social network). A more challenging situation results from combining Mary's location and the current date and time. When Mary comes close to a shop, where there is a Flash Sale offer running, her mobile device is detected and a new version of the customized catalog is proposed to her. The Flash Sale offer is only available at a given place and during a precise time period. This situation demonstrates the capacity of the CAPPUCINO platform to deal with time and space constraints. Furthermore, the purchase is usually performed online. This means that the client remains connected during the whole purchase process. However, the network connection is not necessarily guaranteed all the time. For this reason, the CAPPUCINO platform provides service continuity in case of a network disconnection. To do that, the order can be continued off-line and registered as an SMS message. This message will be sent automatically when the network coverage becomes sufficient again.

Device limitations: As mobile devices vary in terms of hardware and software capabilities, tailoring software components to run on multiple devices is a difficult and time-consuming task that must be dealt with carefully. Therefore, the deployment of software components on mobile devices and the dynamic reconfiguration of the application depending on context variation represents a key challenge of the CAPPUCINO approach.

Context mediation: The context information manipulated in this scenario is associated to the user profile, location and network connectivity. Mary's profile aggregates a very rich set of information including her favorite products (published in a social network), the characteristics of her mobile device, and her current subscriptions to the reward point program and to the gift idea service. These pieces of information are not all registered at a single place but are rather distributed on different devices. Location information can be determined using a GPS-like system in outdoor conditions. However, other geo-location mechanisms are necessary for indoor environments involving sensors dispersed in the monitored area. Network connectivity information is critical for *On-The-Go* users. Therefore, it is required to monitor different network interfaces, such as WiFi, Bluetooth, or long-distance access and to anticipate potential disconnections.

5.3 Principles and Background

This section presents the foundations of our contributions, namely the MAPE-K control loop (cf. Section 5.3.1), the SCA component model (cf. Section 5.3.3) and the COSMOS context framework (cf. Section 5.3.2). The combination of these three building blocks represents the foundations of the CAPPUCINO platform.

5.3.1 MAPE-K: Control Loop for Autonomic Computing

The CAPPUCINO runtime follows the principle of autonomic computing, which is a computing environment able to manage itself and dynamically adapt to change according to business policies and objectives. Autonomic systems are characterized by their ability to devise and apply counter measures when necessary, including the ability to detect these adaptation situations. Self-managing environments can perform such activities based on situations they observe or sense in the IT environment rather than requiring IT professionals to initiate the task. The following properties are the ones that we elect in the design of the CAPPUCINO runtime (taken from (13, 16, 17)). *Self-configuration* stipulates that the system shall be capable of adapting its behavior to the execution context. For instance, it shall be possible to add or remove some functionalities (e.g., a business component providing a service) without a complete interruption of all the services. Moreover, system parts not directly impacted by the changes shall be able to progressively adapt themselves to these changes. *Self-optimization* states that the system shall control and monitor the system resources it consumes. It shall then detect a degradation of service and cater with the necessary reconfigurations to improve their own performance and efficiency. *Self-healing* establishes that the detection, analysis, prevention, and resolution of damages shall be managed by the system itself. The system shall be capable to survive hardware and software failures. *Self-protecting* aims at anticipating, detecting, identifying, and protecting against threats. The system automatically defends against malicious attacks or cascading failures. It uses early warning to anticipate and prevent system wide failures.

These properties are generally reached in autonomic computing by applying the principles of the MAPE-K control loop (upper side of Figure 5.1). The Knowledge is the standard data shared among the Monitor, Analyze, Plan, and Execute functions of an *Autonomic Manager*, such as symptoms and policies. This runtime knowledge must be complete—i.e., including the whole aspects influencing adaptation decisions—, modifiable—i.e, following the application changes—, and at a high-level of abstraction—i.e., comprising only relevant information. The Monitor part provides the mechanisms that collect, aggregate, filter, and report details (such as metrics and topologies) collected from

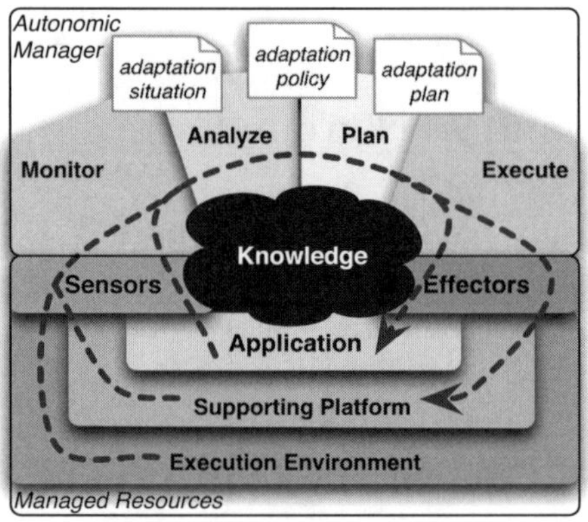

Figure 5.1: Overview of the MAPE-K autonomic control loop.

a managed resource. The Analyze part contains the mechanisms that correlate and model complex *adaptation situations*. The Plan function encloses the mechanisms that construct the actions needed to achieve goals and objectives. The planning mechanism uses *adaptation policies* information to guide its work. The Execute function groups the mechanisms that control the execution of an *adaptation plan* with considerations for dynamic updates. The *Managed Resources* are entities that exist in the runtime environment of an IT system and that can be managed. This encloses the Execution Environment, Supporting Platform, and Application entities. All these managed resources provide Sensors to introspect the entity states, and Effectors to reconfigure them dynamically.

5.3.2 COSMOS: Collecting and Composing Context Data

COntext entitieS coMpositiOn and Sharing (COSMOS) is a component-based framework for managing context information in ubiquitous environments for context-aware applications (6, 26). In particular, COSMOS identifies the contextual situations to which a context-aware application is expected to react. These situations are modeled as *context policies* that are hierarchically decomposed into fine-grained units called *context nodes* (cf. Figure 5.2).

A Context Node represents context information modeled by a software component. The relationships between context nodes are *sharing* and *encapsulation*. The sharing of a context node—and, by implication, of a partial or complete hierarchy—corresponds to the sharing of a part of or a whole context policy. Context nodes leaves (the bottom-most elements, with no descendants) encapsulate raw context data obtained from collectors, such as operating system probes, sensors in the vicinity of the user terminal, user preferences in profiles, and remote devices. Context nodes should provide all the inputs necessary for reasoning about the execution context. For this reason

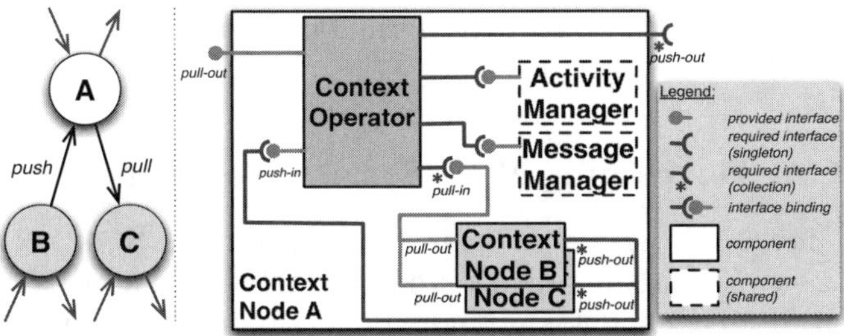

Figure 5.2: COSMOS context node and context policy.

COSMOS considers user preferences as context information. Thus, the role of a context node is to isolate the inference of high-level context information from lower architectural layers responsible for collecting context information.

Context nodes also possess *properties* that define their behavior with respect to the context policy: (i) **active/passive** An active node is associated with a thread of control (attached to the Activity Manager). Typical examples of active nodes include a node in charge of the centralization of several types of context information, a node responsible for the periodic computation of higher-level context information, and a node to provide the latter information to upper nodes. A passive node gets information upon demand; (ii) **observation/notification** Communication into a context node's hierarchy can be top-down or bottom-up. The former—implemented by the interface `Pull`—corresponds to observations that a parent node triggers, while the latter—realized by the interface `Push`—corresponds to notifications that context nodes send to their parents; (iii) **pass-through/blocking** Pass-through nodes propagate observations and notifications while blocking nodes stop the traversal. In other words, for observations, a blocking node transmits the most up-to-date context information without polling child nodes, and for notifications, a blocking node uses context data to update the node's state, but it does not notify parent nodes.

COSMOS provides the developer with predefined generic context operators, organized following a typology: elementary operators for collecting raw data; memory operators, such as averaging and translation operators; data mergers; and abstract or inference operators, such as adders or thresholds operators. The only programming is in the Context Operators. If a sufficiently large library of context operators is available and well targeted to a developer's business, no programming should be necessary—only declarative composition of context nodes.

The context nodes exchange *context reports* (messages), which are created and manipulated by the Message Manager. The context nodes send and receive these context reports through the `Pull` and `Push` interfaces. Each context report is composed of a set of `chunks` and encloses a set of sub messages, while each chunk reflects a context information as a 3-tuple (`name, value, content type`).

5.3.3 SCA: Developing Web Service Components

Initiated in 2005, *Service Component Architecture* (SCA) (21) is a set of specifications for building distributed applications using the principles of SOA and CBSE. The model is promoted by a group of companies, including BEA, IBM, IONA, Oracle, SAP, Sun, and TIBCO. The specifications are now defined and hosted by the *Open Service Oriented Architecture* (OSOA) collaboration [3] and promoted in the Open CSA section of the OASIS consortium. [4]

SOA, e.g., when based on Web Services, provides a way for exposing coarse-grained and loosely-coupled services, which can be remotely accessed. Yet, the SOA approach seldom addresses the issue of the way these services should be implemented. SCA fills this gap by defining a distributed component model. SCA entities are software *components*, which may provide interfaces (called *services*), require interfaces (called *references*), and expose *properties*. References and services are connected through *wires*. The model is hierarchical with components being implemented either by primitive language entities or by subcomponents (the component is then said to be a *composite*). Figure 5.3, which is taken from the SCA specifications (21), provides the graphical notation for these concepts. An XML-based assembly language is available to configure and assemble components.

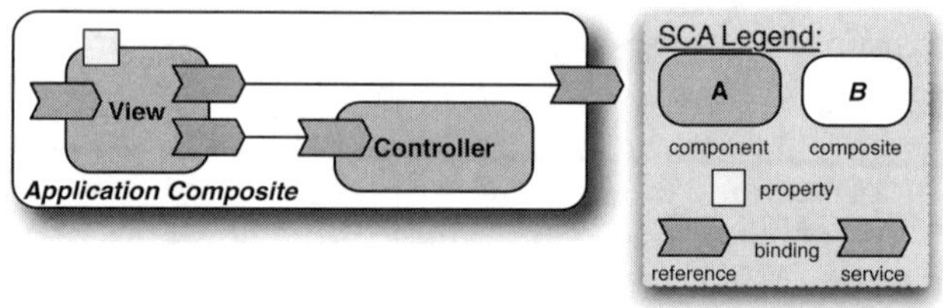

Figure 5.3: SCA component architecture.

In order to cover a wide range of SOA applications, the SCA specifications are not bound to a particular middleware technology, nor a programming language. They have been kept as open as possible thanks to four main principles: Independence from *programming languages*, from *interface definition*

[3] OSOA collaboration: `http://www.osoa.org`

[4] OASIS consortium: `http://www.oasis-opencsa.org`

languages, from *communication protocols*, and from *non functional properties.* First, SCA does not assume that components will be implemented with a unique programming language. Rather, several language mappings are supported and allow programming SCA components in Java, C++, PHP, BPEL, or COBOL.

Second, SCA components provide (*resp.* require) functionalities through precisely defined interfaces. SCA does not assume that a single *Interface Definition Language* (IDL) will fit all needs. Rather, several IDL are supported, such as WSDL and Java interfaces.

Third, although Web Services are the preferred communication mode for SCA components, this solution may not fit all needs. In some cases, protocols with different semantics and properties may be needed. For that, SCA provides the notion of a *binding*: a service or a reference will be bound to a particular communication protocol, such as SOAP (3, 8) for Web Services, Java RMI, Sun JMS, or REST (10).

Last, concerning the non functional properties, they may be associated to an SCA component with the notion of a *policy set* (this notion is also referred to with the term *intent*). The idea is to let a component declare the set of policies (non functional services) that it depends upon. The platform is then in charge of guaranteeing that these policies are enforced. So far, security and transactions have been included in the SCA specifications. Yet, developers may need other types of non functional services. For that, the body of supported policy sets may be extended with user-specified values.

Therefore, these principles offer a broad scope of solutions for implementing SCA-based applications. Developers may think of incorporating new forms of language mappings (e.g., SCA components programmed with EJB or OSGi, which are two solutions currently investigated by the community), IDLs (e.g., CORBA IDL), communication bindings (e.g., *Java Business Integration* [JBI] bindings), and non functional properties (e.g., persistence, logging).

5.4 CAPPUCINO: Enabling Context-Aware Adaptive Services

The CAPPUCINO runtime combines the strengths of both the SCA and COSMOS models to provide a versatile infrastructure for supporting context-aware adaptations in SOA environments. After an introduction of the overall infrastructure (cf. Section 5.4.1), this section focuses on the reconfiguration capabilities offered by the FraSCAti platform (cf. Section 5.4.2) as well as on the distribution of context policies using SPACES (cf. Section 5.4.3).

5.4.1 Overview of the CAPPUCINO Runtime

The novelty of CAPPUCINO lies in the exploitation of SCA as an unified model to build both applications and their associated control loops. Each control loop is in charge of monitoring the execution of an SCA application and of adapting it with regards to the evolutions of the ambient context. Figure 5.4 illustrates our contributions and introduces the CAPPUCINO distributed runtime as well as its core constituting components. This illustration considers a deployment on three distinct nodes: the *adaptation server*, a *mobile client*, and an *application server*. The adaptation server hosts the control loop dedicated to an application. This control loop is deployed as a SCA system using the FraSCAti platform. The control loop interacts with the mobile client(s)—i.e., hand-held device(s) hosting a lightweight version of FraSCAti (so called FraSCAme)—to deploy and reconfigure dynamically the client-side application. The server-side application can also be deployed and reconfigured dynamically using the FraSCAti platform available on the application servers.

Specifically, these control loops are implemented as distributed COSMOS context policies using SPACES in order to describe end-to-end adaptations of the system. In particular, such a distributed context policy correlates the ambient context information to infer the reconfiguration actions to execute for adapting the SCA application. The reconfiguration scripts are automatically produced depending on changes observed in the surrounding environment of a mobile device. As the adaptation decision is taken remotely, the context information sensed by the mobile device requires therefore to be propagated to the remote server. Thus, the combination of the SCA and COSMOS concepts enables CAPPUCINO to i) build context-aware SCA applications and ii) distribute the adaptation policies across a set of distributed nodes.

Inspired by the MUSIC definition (25), an *adaptation domain* is a partition of CAPPUCINO runtime instances executing a distributed application under the control of one adaptation server. This node acts as the nucleus around which the adaptation domain organizes itself as satellite nodes, such as mobile devices and application servers, enter and leave the domain. The movement of satellite nodes or changes in the network topology also cause the dynamic configuration of an adaptation domain. Adaptation domains may overlap in the sense that satellite nodes may be members of multiple adaptation domains. This adds to the dynamics and increases the complexity since the amount of resources the satellite nodes are willing to provide to a given adaptation domain may vary depending on the needs of other served domains. The user of a mobile device may start and stop applications or shared services. The set of running SCA components is therefore safely adapted by the Reconfiguration Engine according to the end-user actions and context changes, while taking into account the resource constraints.

This Reconfiguration Engine pilots the SCA Platform to deploy the SCA application components, as further described in Section 5.4.2. Once deployed,

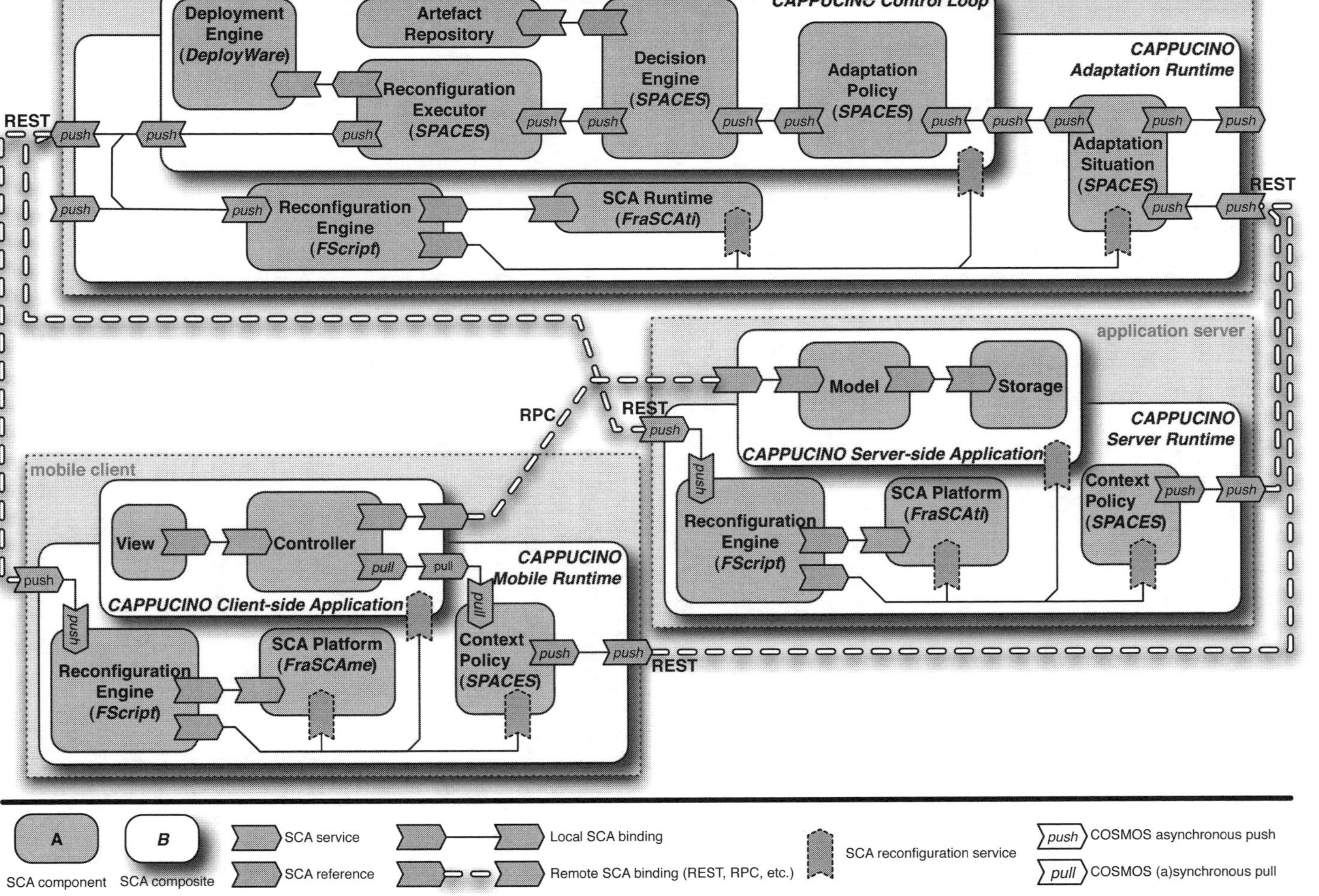

Figure 5.4: Overview of the CAPPUCINO distributed runtime.

these components can be reconfigured dynamically via a SCA reconfiguration service, provided by the SCA component, in order to reflect changes in the computing environment. Reconfigurations in this application layer include the deployment of new SCA components as well as their bindings to available services. This capability is an original SCA feature enabled by the integration of the FSCRIPT reconfiguration language (9) within the CAPPUCINO runtime. Besides, thanks to the adoption of a uniform component model for all the parts of the control loop (including the execution kernel), the Reconfiguration Engine can also exploit the SCA reconfiguration service provided by the SCA Platform and the Context Policy to dynamically reconfigure the CAPPUCINO runtime. Possible reconfigurations in this infrastructure layer include the deployment of new communication protocols (e.g., SOAP) and new context collectors (e.g., GPS sensor).

The Context Policy is responsible for i) providing contextual information and ii) detecting changes in the device's context. The former supports the realization of context-aware applications, while the latter initiates the adaptation process by sending the reified context information to the context policy hosted on the adaptation server. The distribution of a context policy among hosts is based on the REST architecture style as described in Section 5.4.3. In particular, SPACES enriches COSMOS with a flexible RESTful architecture for supporting lightweight communication protocols and alternative resource representations. Once received by the Adaptation Situation, the context information—in addition to application variation points and device characteristics—is processed by the Adaptation Policy to filter out the input data relevant for the Decision Engine. Note that for the sake of clarity, Figure 5.4 does not depict observation interactions between context policies performed using the `pull` calls of COSMOS; the notification mode—i.e., COSMOS `push` calls—suffices to demonstrate the principles of the control loop.

The Decision Engine applies an application-specific reasoning algorithm (e.g., situation-action rules or constraint satisfaction problems) to produce the *adaptation plans* (e.g., the reconfiguration script) to execute. The application artefacts as well as the adaptation knowledge are provided by the Artefact Repository. If a reconfiguration is requested, the Reconfiguration Executor can deploy the adaptation artefacts using the Deployment Engine, which is based on the DEPLOYWARE framework (11), and then trigger a reconfiguration by sending the generated reconfiguration script(s) to the Reconfiguration Engine(s) available in the adaptation domain. This typically means that the result of an adaptation may impact not only the mobile device(s), but also the application server(s) as well as the adaptation server. For instance, adaptations performed in the application server can balance the load and handle potential failures.

5.4.2 FraSCAti: Deploying Adaptive Distributed Systems

The FraSCAti platform (27) enables the execution of SOA applications with advanced properties, such as the ability to reconfigure at runtime existing assemblies of SCA components (cf. Section 5.3.3). Other available features include a binding factory supporting various communication protocols, a transaction service, a semantic trader service, a service for deploying autonomic SCA-based applications, and a graphical administration and management console. As illustrated in Figure 5.5, the FraSCAti platform is composed of the following elements. The SCA Meta Model defines all the concepts available in the SCA specifications. The meta-model is used by the FraSCAti Assembly Factory to load component assembly descriptors. The Tinfi Runtime Kernel is in charge of hosting the SCA components. This kernel implements the execution semantics defined by the SCA specifications and provides some additional properties, such as dynamic reconfiguration. This kernel is based on generative programming techniques where most of the code for managing components is generated on-demand at runtime. The Binding Factory is in charge of establishing communication channels between SCA components. This factory supports various protocols, such as SOAP (3, 8), REST (10), and RMI. The Trading Service provides functionalities to perform semantic trading on components published in a registry. This service associates to each external element of a component (service, reference, property, implementation) a concept defined in a semantic model. This service allows, for example, the replacement of components which would have become inaccessible, by some semantically equivalent ones. The Assembly Factory is in charge of analyzing, validating, and instantiating assembly descriptors. The analysis step uses the SCA metamodel to load and validate descriptors. The instantiation step uses the runtime kernel and the assembly factory for installing and connecting components.

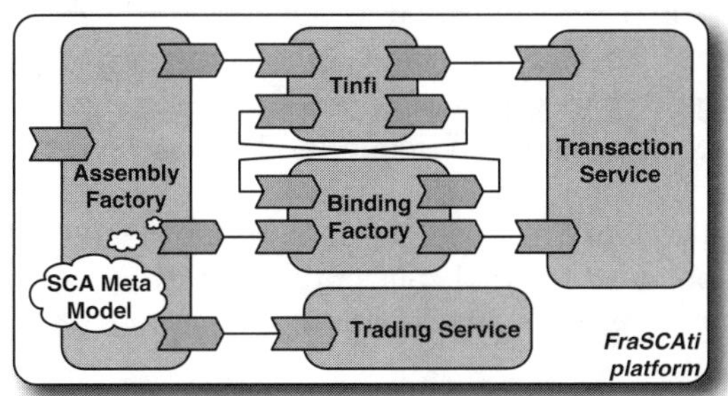

Figure 5.5: Architecture of the FraSCAti execution kernel.

Besides deploying and instantiating an SCA application, the FRASCATI platform provides some runtime services for dynamic reconfiguration. These features are of particular importance for adapting applications to new and unforeseen requirements. The runtime adaptation actions, which are enabled by FRASCATI, are described below. They concern the architecture of an SCA application. They can take place at any time during the execution of the application.

First, FRASCATI enables instantiating new components. For example, these new components can provide additional functionalities or provide alternate implementations for existing ones. Second, new wires can be created and existing wires can be removed. Third, component hierarchies can be queried and modified. The query functionality enables traversing an existing application much like what is available with component browsers. The modification of the hierarchy, in conjunction with the creation of new wires, enables creating different component compositions. Finally, components can be selectively started and stopped. This enables performing maintenance operations in a safe manner.

5.4.3 SPACES: Deploying Distributed Context Policies

This section introduces SPACES, a *distributed processing architecture integrating context as RESTful services.* SPACES is a lightweight middleware solution enabling the versatile and efficient mediation of context information. Context policies have been introduced by the COSMOS context framework as a scalable model for processing context information (cf. Section 5.3.2). Therefore, we propose to extend COSMOS with the principles of *REpresentational State Transfer* (REST) (10) in order to distribute context policies in ubiquitous environments. These RESTful context policies leverage the provision of *Context as a Service* and enable the efficient mediation of context information among heterogeneous devices.

Figure 5.6 illustrates the integration of SPACES into a COSMOS context policy. Specifically, SPACES acts as a mediation middleware, which is responsible for disseminating the context information between two physical entities. This distribution is implemented as a connector based on the *HyperText Transfer Protocol* (HTTP). This connector is composed of i) an HTTP server publishing part of the context nodes as REST resources and ii) an HTTP client sending context requests as REST requests. By considering the context mediation as a software connector, the realization of the context policy is not impacted by the distribution concerns. In addition to that, the use of a software connector opens up for multiple implementations of the dissemination mechanism, including *message-oriented* (19) and *peer-to-peer* (14) middleware approaches.

In this section, we report the design of SPACES using HTTP to facilitate the integration of adaptation policies with legacy systems. In particular, we can monitor the context surrounding a mobile device from a traditional Internet

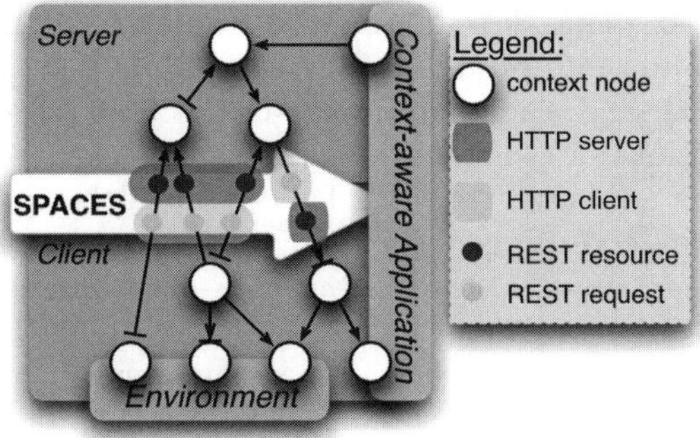

Figure 5.6: Principles of the SPACES approach.

browser and access this information from any systems supporting HTTP. This integration is facilitated by the support for multiple representation of a given resource. This means that depending on the client requirements, a REST resource can be retrieved as an HTML document (when accessed from an Internet browser) or as an XML document (when accessed from a legacy system). Therefore, the integration of the REST principles within the context policies requires to map the REST triangle of nouns, verbs, and content types to COSMOS context nodes, which are the first class entities of CAPPUCINO adaptation policies.

SPACES Nouns. The advertisement of context nodes as REST resources requires the definition of *context identifiers.* These context identifiers are unique nouns described using the *Uniform Resource Identifier* (URI) format (1). Therefore, context identifiers include a communication scheme, a server address, a context path, and a sequence of request parameters: `scheme://context-server/context-path?request-parameters`. The *communication scheme*, such as HTTP, FTP (*File Transfer Protocol*), URN (*Uniform Resource Name*), RTSP (*Real Time Streaming Protocol*), or file, describes the communication protocol used to transfer the resource representation between the hosting server and the requesting client. Then, the *server address* description is specific to a given scheme. For example, an HTTP server address can be specified using the syntax `user:password@host:port` to describe the Web server host and port as well as the credentials for accessing the context information. The *context path* points to context node name under which the context information is published. This context path can be hierarchically organized in context domains using the syntax `parent-domain/child-domain/context`. Finally, the *request parameters* are used to specify the representation of the requested (or submitted) context information using the syntax `format=application/xml`.

SPACES Verbs. The dissemination of context information is based on *observation* and *notification primitives.* COSMOS implements these primitives as `Pull` and `Push` interfaces (cf. Section 5.3.2). In SPACES, resource observations are implemented as side-effect free `GET` requests, while notifications are based on `POST`. For example, the context information *geolocation-dps* can be retrieved from a mobile device using the HTTP request `GET http://device.inria.fr:8080/geolocation-dps?format=application/xml`, which returns the requested context information as an XML document. Context information can also be pushed into a remote entity by using an HTTP `POST` request. In this case, the submitted request points to the server-side resource associated to the context policy. For example, the HTTP request `POST http://server.inria.fr/adjusted-bit-rate?format=application/json` notifies the server-side context policy that the uploaded JSON document refers to a change of the context information *GPS geolocation.*

SPACES Content Types. Finally, SPACES supports various types of content for representing the context information. This representation multiplicity benefits from the structure of COSMOS messages used to encode the context information (cf. Section 5.3.2). Therefore, the content types supported by SPACES are based on the MIME media types classification (15). In particular, SPACES promotes the *Java object serialization* as the default resource representation (`application/octet-stream`) for performance concerns. Nevertheless, SPACES provides also representations of resources as XML (`application/xml`) and JSON (7) (`application/json`) documents.

5.5 Illustrating Dynamic Context-aware Web Services with a Mobile Commerce Scenario

We present in this section the way the CAPPUCINO platform is able to realize the mobile commerce scenario presented in Section 5.2. Figure 5.7 shows the corresponding component architecture obtained using CAPucine (*Context-Aware Service-Oriented Product Line*), a Context-aware *Dynamic Service-Oriented Product Line* (DSOPL) approach described in (22). We describe in Section 5.5.1 how deployment takes place on a mobile device. We then detail in Section 5.5.2 a reconfiguration scenario taking place on the mobile device after the detection of an adaptation situation. We focus on the deployment and reconfiguration on the mobile device side for more concision. Nonetheless, the application server side follows the same principles.

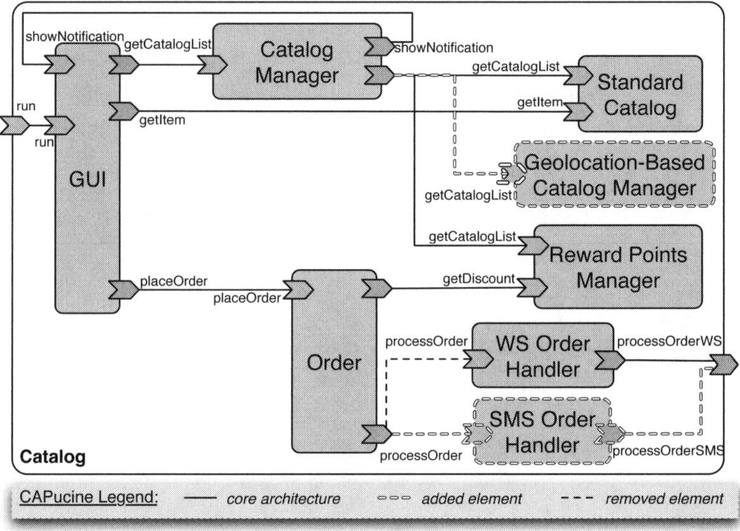

Figure 5.7: Mobile commerce application architecture using CAPucine.

5.5.1 On-demand Deployment on a Mobile Device

This section presents the dynamic aspects of the CAPPUCINO architecture as UML activity diagrams. We consider that the deployment of the CAPPUCINO adaptation server followed by the deployment and the start of the application server have already taken place. The deployment on a mobile device can then be made on-demand. For legal issues, the user intervention is required to request and then validate the deployment of software applications on their device. The artefacts deployed on a mobile device are lightweight versions of the ones deployed on an application server, e.g., the lightweight version of FraSCAti called FraSCAme.

The deployment of the CAPPUCINO runtime on a mobile device is detailed in Figure 5.8. (1) We suppose that the service provider has preliminarily installed access points able to detect mobile devices in their vicinity. We do not detail here the functioning of these access points. We only assume that the user either declares their presence by passing their mobile device in front of a bluetooth antenna or has agreed beforehand to get automatically detected by providing the physical characteristics of their mobile device. Consequently, the interactive point notifies the adaptation server of the arrival of a mobile device. (2) The adaptation server sends to this mobile device an electronic message inviting the detected user to use available services. This message may be a SMS or a MMS message proposing to download the CAPPUCINO runtime to enable the access to personalized services. (3) The user gives an explicit authorization to download and deploy the CAPPUCINO runtime. (4)

The adaptation server analyzes the type of the mobile device, prepares the adequate deployment plan and uploads the required software artefacts. (5) The adaptation server then transmits the CAPPUCINO runtime artefacts to the mobile device. (6) A volunteer action is required here from the end user: The end user must explicitly accept the effective installation of the artefacts on their mobile device. (7) Following the installation of the artefacts, the CAPPUCINO runtime is automatically started. This step concludes the first phase of the deployment on a mobile device.

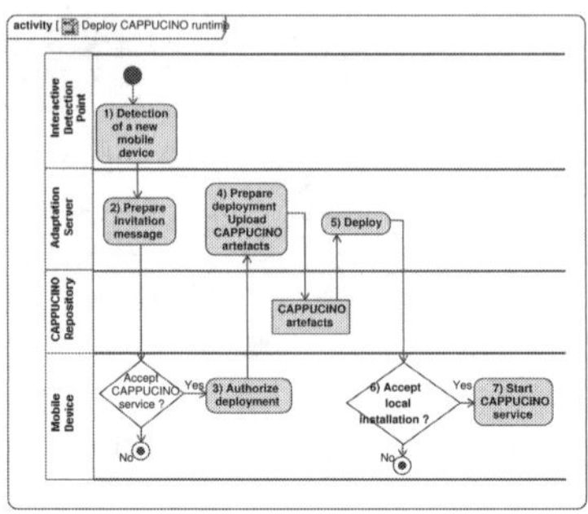

Figure 5.8: Deploying the CAPPUCINO runtime on a mobile device.

The second phase for deploying on a mobile device corresponds to the deployment of an application component. We show as an example the deployment of the `Order` component on Mary's mobile device (cf. Figure 5.7). The `Order` component will allow Mary to buy an item chosen in a catalog. The different deployment steps are depicted in Figure 5.9. (1) At the end of the first phase of the deployment of the CAPPUCINO runtime, this runtime informs the adaptation server that the deployment is completed. (2) The adaptation server then builds the list of the relevant Web Services to be proposed to Mary according to the type of her mobile device (e.g., video feature available or not), her user profile (e.g., recently accessed services, preferred services) and her user network connectivity (e.g., Internet access and high rate network connection). (3) The adaptation server transmits the list of the available services to Mary. (4) Mary chooses the `Order` service in the list. (5) The adaptation server prepares the `Order` service deployment plan. It includes to upload the `WS Order Handler` and the `SMS Order Handler` components on the mobile device. The former allows an order to be handled directly by the remote web server when the network coverage is sufficient. The latter provides service continuity in case of network disconnection registering the order as a low-memory consuming SMS message and placing it as soon as the network is available again. Both handlers are deployed but only `WS Order Handler` is initially bound to the `Order` service. (6) The deployment on the client side may require to preliminarily deploy a server part on the application server. (7) Moreover, the deployment on the server side may require to upload non CAPPUCINO artefacts from the repository of the business information system. For instance, Mary is registered as a new potential mobile clients and the payment service is activated. (8) Following the deployment of the required applicative services on the server side, the adaptation server sends the `Order` artefacts together with the `WS Order Handler` and the `SMS Order Handler` artefacts and drives their deployment on Mary's mobile device. This includes initially binding the `Order` service with the `WS Order Handler`. (9) Finally, Mary is notified of the availability of the `Order` service and can use it.

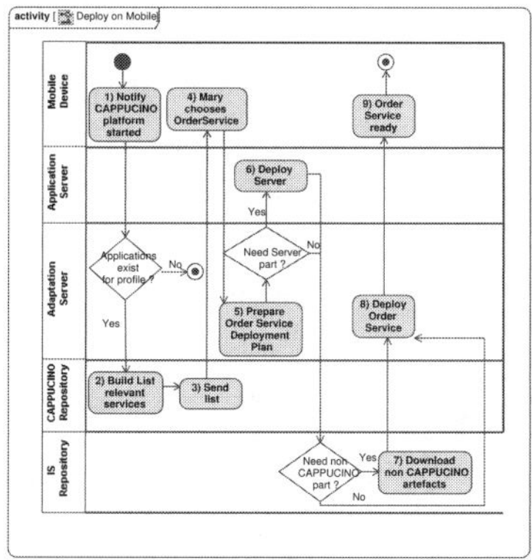

Figure 5.9: Deployment of an applicative service on a mobile device.

5.5.2 Dynamic Reconfiguration on a Mobile Device

We present in this section how reconfiguration is enabled at runtime illustrating the reaction path from the detection of an adaptation situation to the execution of the reconfiguration actions on the mobile device. We describe two reconfiguration cases corresponding to different runtime reconfiguration services of the FraSCAti platform (see Section 5.4.2) namely the possibility to modify the wires between existing components and the possibility to instantiate new components.

Case 1: Change connections between existing components. As shown in Section 5.5.1, the `Order` service has been deployed when network connectivity was high, implying that the `Order` component has been bound to the `WS Order Handler`. However, the `SMS Order Handler` has also been deployed in order to tolerate any network disconnection and to provide service continuity transparently to Mary. When the network connectivity becomes lower than a threshold indicating there is a risk of disconnection, the adaptation server detects an adaptation situation and establishes the reconfiguration plan. The reconfiguration actions are performed on Mary's device corresponding to unbinding the `WS Order Handler` and binding instead to the `SMS Order Handler`.

Case 2: Deploy a new component and create connections. Consider Mary arriving nearby a shop where a Flash Sale offer is currently running. Mary has accepted to periodically notify the adaptation server with her geographical position. When the adaptation server detects that Mary can benefit from the nearby Flash Sale, it pushes the special offer by presenting a customized catalog. This new customized catalog is specifically prepared according to Mary's preferences as mentioned in her profile and requires the reconfiguration of the application components present on her mobile device. The `Geolocation-Based Catalog` component shown on Figure 5.7 is instantiated. It is then bound to the `Catalog Manager` component. These reconfiguration steps taking place on the mobile device are shown on Figure 5.10.

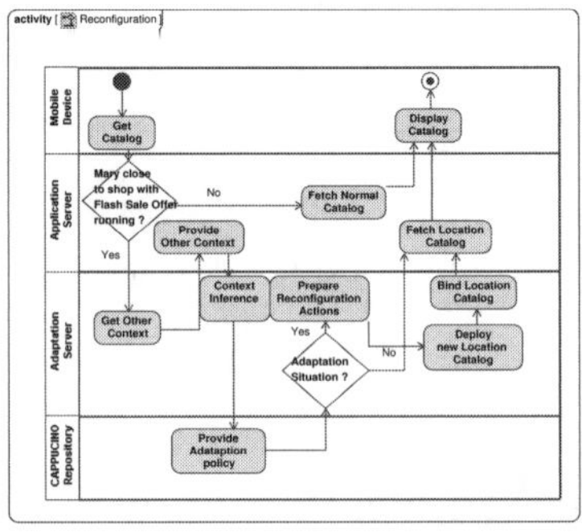

Figure 5.10: Reconfiguration of a mobile device.

5.6 Related Works

This section compares our contribution with regards to the current state-of-the-art related to the definition of platform supporting the self-adaptation of service-oriented architectures.

MUSIC (25) defines a component-based platform supporting the seamless adaptation of mobile application according to *Quality of Service* variations. MUSIC integrates a versatile support for discovering, negotiating, and binding to remote services available in the surroundings of a mobile device. Unless CAPPUCINO, MUSIC adopts a device-centric approach where the device decides the adaptation to perform by applying a utility-based decision making mechanism (17). CAPPUCINO offers a scalable adaptation model where the decision-making is under the control of an adaptation domain, which can decide to reason on the server-side or to deploy local and efficient control loops within the mobile devices.

The QuA project (12) investigates middleware-managed adaptation, which means that services are specified by their behavior, and then planned, instantiated and maintained automatically by middleware services in such a way that the behavioral requirements are satisfied throughout the service life-time. Central in this approach, is a planning framework that searches through the space of possible configurations, and selects one that fits the current context of the application. The planning framework depends on the availability of meta-information about services that has not yet been instantiated, such as meta-information provided by techniques like MDA and ADL. While QuA integrates multiple implementation platforms at the adaptation middleware level, CAPPUCINO offers a versatile programming model based on SCA to leverage the adaptation middleware and thus offer a uniform model for the description and the adaptation of ubiquitous service-oriented architectures.

Adaptive Service Grids (ASG) (18) and VieDAME (20) are initiatives enabling dynamic compositions and bindings of services for provisioning adaptive services. In particular, ASG proposes a sophisticated and adaptive delivery life-cycle composed of three sub cycles: *planning*, *binding*, and *enactment*. The entry point of this delivery life-cycle is a semantic service request, which consists of a description of what will be achieved and not which concrete service has to be executed. VieDAME proposes a monitoring system that observes the efficiency of BPEL processes and performs service replacement automatically upon performance degradation. Compared to our CAPPUCINO platform, ASG and VieDAME focus only on the planning per request of service compositions with regards to the properties defined in the semantic service request. Thus, both approaches do not support a uniform session-based adaptation of both client-side and server-side services as our solution for ubiquitous applications does. Nevertheless, CAPPUCINO provides an extensible infrastructure to integrate ASG and VieDAME adaptive services and thus support the dynamic enactment of service workflows.

CARISMA is a mobile peer-to-peer middleware exploiting the principle of reflection to support the construction of context-aware adaptive applications (4). In particular, services and adaptation policies are installed and uninstalled on the fly. CARISMA can automatically trigger the adaptation of the deployed applications whenever detecting context changes. CARISMA uses utility functions to select application profiles, which are used to select the appropriate action for a particular context event. If there are conflicting application profiles, then CARISMA proceeds to an auction-like procedure to resolve (both local and distributed) conflicts. However, CARISMA fails to integrate the diversity of service providers and legacy services. Nonetheless, the auction-like procedure used by CARISMA could be integrated in the CAPPUCINO control loop as a particular decision-making service.

Finally, R-OSGi extends OSGi with a transparent distribution support (23) and uses jSLP to publish and discover services (24). The communication between a local service proxy and the associated service skeleton is message-based, while different communication protocols (e.g., TCP or HTTP) can be dynamically plugged in. In contrast to R-OSGi, the discovery and binding frameworks of CAPPUCINO are open to support a larger range of discovery and communication protocols.

5.7 Conclusion

Ubiquitous environments strongly constrain the design and development of Context-aware Web Services. An adaptive runtime in such environments should be based on a control loop enclosing i) an adaptive execution kernel, ii) distributed context management facilities, iii) a context and situation adaptation analysis, iv) an adaptation policies and reconfiguration planning, and v) a reconfiguration engine orchestrating deployment and reconfiguration actions. The first contribution reported in this chapter is the CAPPUCINO runtime that provides the building blocks of a distributed control loop and relies on CBSE principles to ensure adaptation to service-oriented ubiquitous systems. Moreover, the use of SCA as a common component model enables CAPPUCINO to perform adaptations of its own structure in order to better satisfy the requirements of the supported Web Service and therefore the end-user.

The second contribution of the chapter is FraSCAti, an execution kernel conforming to the SCA distributed component model for Web Services. SCA promotes independence from programming languages, interface definition languages, communication protocols, and non functional properties. The FraSCAti execution kernel additionally enables the execution of SOA applications with advanced properties, such as the ability to reconfigure assemblies of SCA components at runtime.

The third contribution of this chapter builds upon the context manager COSMOS to build a lightweight middleware solution enabling the scalable and efficient dissemination of context information. SPACES acts as a communication middleware, which is responsible for disseminating the context information between distributed physical entities. This distribution is implemented in CAPPUCINO as an HTTP connector. By considering the context dissemination as a software connector, the realization of the composition of context information into software components is not impacted by the distribution concerns.

As a matter of future work, we plan to improve the scalability of the CAPPUCINO runtime by enabling the deployment of hierarchical control loops. In particular, the adaptation server should deploy a lightweight full-fledged local control loop within the mobile device in order to tolerate local autonomic adaptation like facing disconnection situations. The mobile devices could then react and reconfigure the application autonomously when leaving a given adaptation domain—i.e., when loosing the connectivity with the adaptation server. In addition, adaptation domains could also be grouped as clusters to control a set of applications. Thus, a cluster of adaptation servers will belong to a common adaptation domain, which is dynamically reconfigured by the enclosing adaptation server. Another perspective consists in integrating the concerns of *Quality of Service* (QoS) in the development of context-aware Web Services. We foresee that this integration will be achieved at the application level and at the middleware level by reflecting the *Quality of Context* (QoC). Therefore, we plan to integrate the dynamic negotiation of QoS and QoC agreements during the adaptation process.

References

[1] Tim Berners-Lee, Roy T. Fielding, and Larry Masinter. Uniform Resource Identifier (URI): Generic Syntax. `http://www.ietf.org/rfc/rfc3986.txt`, January 2005.

[2] David Booth, Hugo Haas, Francis McCabe, Eric Newcomer, Michael Champion, Chris Ferris, and David Orchard. Web Services Architecture. Working Group Note, February 2004.

[3] Don Box, David Ehnebuske, Gopal Kakivaya, Andrew Layman, Noah Mendelsohn, Henrik Frystyk Nielsen, Satish Thatte, and Dave Winer. Simple Object Access Protocol (SOAP), May 2007. `http://www.w3.org/TR/SOAP`.

[4] Licia Capra, Wolfgang Emmerich, and Cecilia Mascolo. CARISMA: Contex-Aware Reflective mIddleware System for Mobile Applications. 29(10):929–945, October 2003.

[5] David Chappell. Introducing SCA. white paper, Chappell & Associates, July 2007.

[6] Denis Conan, Romain Rouvoy, and Lionel Seinturier. Scalable Processing of Context Information with COSMOS. In *Proceedings of the 6th IFIP WG 6.1 International Conference on Distributed Applications and Interoperable Systems*, volume 4531 of *Lecture Notes in Computer Science*, pages 210–224, Paphos, Cyprus, June 2007. Springer-Verlag.

[7] Douglas Crockford. RFC 4627 — The Application/json Media Type for JavaScript Object Notation (JSON). IETF RFC, 2006.

[8] Francisco Curbera, Matthew Duftler, Rania Khalaf, William Nagy, Nirmal Mukhi, and Sanjiva Weerawarana. Unraveling the Web Services Web: an introduction to SOAP, WSDL, and UDDI. *Internet Computing, IEEE*, 6(2):86–93, 2002.

[9] Pierre-Charles David, Thomas Ledoux, Marc Léger, and Thierry Coupaye. FPath and FScript: Language Support for Navigation and Reliable Reconfiguration of Fractal Architectures. *Springer Annals of Telecommunications, Special Issue on Software Components — The* FRACTAL *Initiative*, 64(1–2):45–63, January/February 2009.

[10] Roy T. Fielding. *Architectural Styles and the Design of Network-based Software Architectures.* PhD thesis, University of California, Irvine, 2000.

[11] Areski Flissi, Jérémy Dubus, Nicolas Dolet, and Philippe Merle. Deploying on the Grid with DeployWare. In *Proceedings of the 8th International Symposium on Cluster Computing and the Grid (CCGRID'08)*, pages 177–184, Lyon, France, May 2008. IEEE.

[12] Eli Gjørven, Romain Rouvoy, and Frank Eliassen. Cross-Layer Self-Adaptation of Service-Oriented Architectures. In *Proceedings of the 3rd International Middleware Workshop on Middleware for Service Oriented Computing (MW4SOC)*, pages 37–42, Leuven, Belgium, December 2008. ACM Press.

[13] Salim Hariri, Bithika Khargharia, Houping Chen, Jingmei Yang, Yeliang Zhang, Manish Parashar, and Hua Liu. The autonomic computing paradigm. *Cluster Computing*, 9(1):5–17, 2006.

[14] Xiaoming Hu, Yun Ding, Nearchos Paspallis, Pyrros Bratskas, George A. Papadopoulos, Paolo Barone, and Alessandro Mamelli. A Peer-to-Peer Based Infrastructure for Context Distribution in Mobile and Ubiquitous Environments. In *Proceedings of 3rd International Workshop on Context-Aware Mobile Systems (CAMS'07)*, Vilamoura, Algarve, Portugal, November 2007.

[15] IANA. MIME Media Types. `http://www.iana.org/assignments/media-types`, March 2007.

[16] Jeffrey O. Kephart and David M. Chess. The Vision of Autonomic Computing. *IEEE Computer*, 36(1), January 2003.

[17] Jeffrey O. Kephart and Rajarshi Das. Achieving Self-Management via Utility Functions. *IEEE Internet Computing*, 11(1):40–48, 2007.

[18] Dominik Kuropka and Mathias Weske. Implementing a Semantic Service Provision Platform:ĂŤ Concepts and Experiences. *Wirtschaftsinformatik Journal*, 1:16–24, 2008.

[19] Matthieu Leclercq, Vivien Quéma, and Jean-Bernard Stefani. DREAM: a Component Framework for the Construction of Resource-aware, Reconfigurable MOMs. In *Proceedings of the 3rd Workshop on Adaptive and reflective middleware (ARM'04)*, pages 250–255, Toronto, Ontario, Canada, 2004. ACM.

[20] Oliver Moser, Florian Rosenberg, and Schahram Dustdar. Non-intrusive Monitoring and Service Adaptation for WS-BPEL. In *Proceedings of the 17th International Conference on World Wide Web (WWW)*. ACM, 2008.

[21] Open SOA. Service Component Architecture, November 2007. `http://www.osoa.org/display/Main/Service+Component+Architecture+Home`.

[22] Carlos Parra, Xavier Blanc, Laurence Duchien, Nicolas Pessemier, Rafael Leano, Chantal Taconet, and Zakia Kazi-Aoul. *Dynamic Software*

Product Lines for Context-Aware Web Services, chapter 3. Chapman and Hall/CRC, 2010.

[23] Jan S. Rellermeyer, Gustavo Alonso, and Timothy Roscoe. R-OSGi: Distributed Applications through Software Modularization. In *Proceedings of the*, volume 4834, pages 1–20, Newport Beach, CA (USA), November 2007.

[24] Jan S. Rellermeyer and Markus Alexander Kuppe. jSLP, 2009. `http://jslp.sourceforge.net`.

[25] Romain Rouvoy, Paolo Barone, Yun Ding, Frank Eliassen, Svein Hallsteinsen, Jorge Lorenzo, Alessandro Mamelli, and Ulrich Scholz. *MUSIC: Middleware Support for Self-Adaptation in Ubiquitous and Service-Oriented Environments*, volume 5525 of *Lecture Notes on Computer Science Hot Topics*, chapter 2, pages 164–182. Springer-Verlag, 2009.

[26] Romain Rouvoy, Denis Conan, and Lionel Seinturier. Software Architecture Patterns for a Context Processing Middleware Framework. *IEEE Distributed Systems Online (DSO)*, 9(6):12, June 2008.

[27] Lionel Seinturier, Philippe Merle, Damien Fournier, Nicolas Dolet, Valerio Schiavoni, and Jean-Bernard Stefani. Reconfigurable SCA Applications with the FraSCAti Platform. In *Proceedings of the 6th IEEE International Conference on Service Computing (SCC'09)*, Bangalore Inde, 2009. IEEE.

[28] Roy Sterritt, Manish Parashar, Huaglory Tianfield, and Rainer Unland. A Concise Introduction to Autonomic Computing. *Advanced Engineering Informatics*, 19(3):181–187, July 2005.

Chapter 6

Building Context-Aware Telco Operator Services Based on Web Services Technologies

Alejandro Cadenas, Antonio Sanchez-Esguevillas, and Belen Carro

Abstract

Operators are always looking for ways to increase the added value for the services they can offer to their customers. At the time of writing, context-aware services are one of the ways that are being investigated to do so. In such services, the optimum interoperability among service platforms is critical due to the variety of elements at the service layer that are involved in context-aware service delivery. In order to achieve such interoperability with minimum development efforts, Web services technology is considered a must to make such services happen in a minimum time to market. In this chapter, the principles to deploy context-aware services in operator networks are presented. In addition, a successful case is described in detail, implemented with the principles presented in this chapter.

6.1 Introduction

Traditionally, the Telco operators have based the development and deployment of Telco (which means telephony-oriented) operator services on Intelligent Network (IN) technologies. However, Web services are currently one of the implementation options most widely deployed by the main worldwide telecom operators in order to integrate different systems of the service layer.

In a nutshell, Web services are based on atomic transactions between entities, processes, and functions, called services. These can then be deployed and integrated among any supporting applications and run on basically any platform.

Such success is due to several factors; they can be summarized in the following main points that describe the advantages of the Web services in Telco networks:

- Due to the architectural design of Web services, the different entities involved can be a service provider, a service requestor to a service broker. The standard communication protocols and such simple broker-request architectures simplify interoperability.
- Given that in order to implement that only a WebServer framework is required, service platforms usually support HTTP interfaces in a native way. Accordingly, the adaptations to Web services/SOAP are usually cheap and fast.
- Are easy to deploy. Given that Web services are using SOAP/HTTP as transport layer, no significant changes are required on the operators' transport networks to open the ports to HTTP traffic.
- Version updates or adaptations are not too costly usually. The Web services internal structure can be adapted specifically to the service platform data model, or to any specific use. Usually the version updates are composed of new parameters in the WSDL, or new values for them. Such upgrades from the previous Web service version are quick to be deployed.
- During the development phase, there are a lot of support and engineering tools, available libraries, as well as significant existing expertise in the development community.

Web service technologies involve message serialization mechanisms on the sender side, and message parsing mechanisms on the receiver side. Generally speaking, such mechanisms make this technology unsuitable for real-time Telco operator interfaces due to the inherent delays to compose/decompose the messages (specially the XML messages that are included in the Web services messages). Such statements may not apply in certain optimized execution environments, platforms, and transport networks. A key performance indicator to optimize in Telco service networks is the signalling delay, given that the services must be invoked in real time. In fields like third-party service delivery where the delay is not a critical aspect (within certains limits), Web service is the technology to consider.

Having said that, if the delay issue is solved (by using high performance platforms, or by keeping the simultaneous subscribers generating signaling under control) Web services are being considered not only to provide some Telco capabilities or to integrate in Telco networks the services provided by a third-party: Web services are considered also as the main implementation option to deploy contextual services in Telco networks.

This comes from the fact that laboratory implementations of contextual services are deployed in a vertical and monolithic service platform. In such a platform, the contextual service is executed, as well as the contextual processing of the information itself. Such contextual information is usually gathered by a set of sensors, or information providers deployed all over a specific area (out of which there is no contextual service available) or onto a specific platform, like the application server of a third-party service provider or even the user terminal.

The diversity of Telco service layer architecture environments is remarkable, at different levels, including:

- Different platforms involved. The service provider and requestor/consumer can be located or running in platforms with different capabilities. A user terminal can host one of such entities, as well as a Telco application server or a Web server located in a domain external to the operator service layer.

- Different access network technologies. The diversity of the connectivity options exists mainly at the user terminal side. User devices can access service elements through different access options, including cellular, fixed, etc.

- Different service delivery domains. Not only the Telco services or the user devices shall be considered in these upcoming architectures, but also the services provided by entities external to the Telco operator. Accordingly a mechanism that allows interoperating those domains will be a must.

So, based on these previous technological environmental characteristics, the objective of the Telco operators is to provide contextual services globally available to users anywhere, anytime, regardless of the terminal device. Within this objective, Web services are a critical aspect to consider in order to build a feasible system with reasonable time-to-market figures, cross-layer transaction orchestration, or interdomain interoperability.

At the time of writing, some research lines are currently open by the main Telco operator players, focused on making use of the great flexibility of the Web services to design, develop, and make commercial a set of contextual services from the Telco service layer to commercial. Such work is mainly on the research area, although some significant steps have been taken on the deployment side. Such progress is presented in the following sections of this chapter.

6.2 Operator Network Architectures

6.2.1 Legacy Networks

The traditional architecture deployed by the Telco operators is basically composed at a high level by a device layer, where the different user devices are to be located. At a next level on top of the previous one, the access network can be composed of several access technologies, either fixed (copper line, FTTH, xDSL, etc.) or mobile (cellular radio, WiFi, WiMAX, etc.). Such a layer provides access to the service layer through a core network. The service layer is the uppermost one.

Traditionally the services provided by the uppermost layer are tightly coupled to the type of device or access technology used by the user. That is due to different reasons, but two can be remarked:

1. The service is usually tightly coupled to the user device (the services available at a fixed phone are not the same as those available at the mobile).

2. Traditionally there is no connectivity between user devices and service layers not related to that service layer. Accordingly the fixed phones have no connection with the platforms providing services to mobile phones.

6.2.2 Convergence and Horizontal Service Architectures

The architectural solution to the problem of the coupling between the access networks and the user devices can be solved by implementing a domain between the different access networks and the service layer, that becomes common for any of the access networks involved. That way, all the devices will be accessing the same service layer. The user can experience the same service regardless of the user terminal.

Such an intermediate layer is defined by NGN architectures, orchestrating the different transactions from/to the service layer elements to/from different access networks. The most common NGN implementation option is IMS (IP Multimedia Subsystem), specified by 3GPP (1).

The implementation of the NGN ultimately means that the service layer is composed of a set of highly specialized service modules that can interwork among them, in such a way that the vertical monolithic service implementation of the traditional architectures becomes obsolete. That concept is known as horizontality of the service layer.

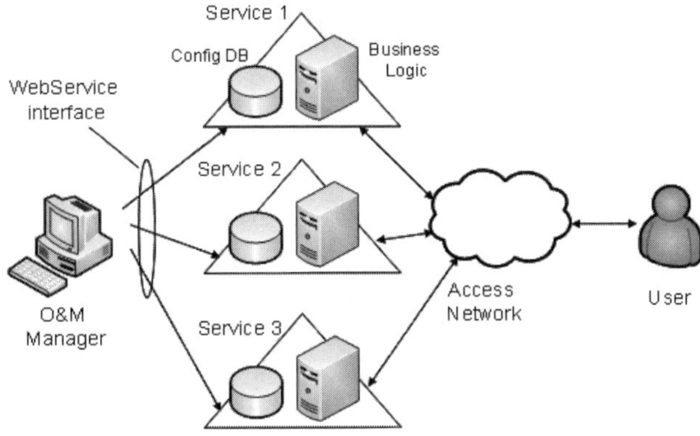

Figure 6.1: Service provisioning over Web service interfaces.

6.3 Web Services in Operator Networks

At the time of writing, the Web services are a common tool in operator networks, due to the technical reasons presented in the previous sections.

Given the inherent delays associated to Web services, most of the existing implementations of Web services in Telco networks are oriented to the provisioning or configuration of service platforms through Machine-to-Machine interfaces implemented over SOAP procedures. In such procedures the small delay introduced is rather acceptable. In addition, the atomicity of the transactions and the flexibility of the XML structure embedded in the SOAP contained make the Web services a perfect protocol to update or synchronize databases, add or delete users in profile databases, etc.

Generally speaking, the Web service technology is oriented to the rapid development of interfaces among elements not traditionally connected, keeping in mind that such interfaces should usually not be part of the Real Time signalling associated with telephony services. Examples are the third-party applications developed by external providers. Those services do not belong to the Telco service layer secure domain and need to be integrated with it in a seamless fashion.

6.3.1 Telco Services

Some examples of Web services in Telco networks are the provisioning services, where a Web service client generates provisioning transactions and sends those to a platform to be provisioned. Potentially any platform can be provisioned through this procedure.

In Figure 6.1 an example is presented. In this example the platform provi-

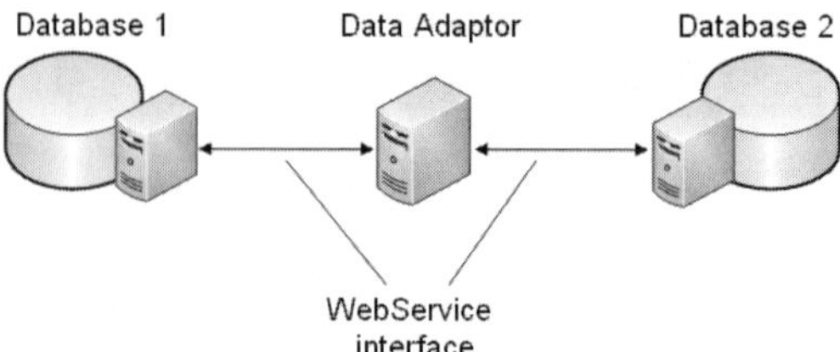

Figure 6.2: Synchronisation between data repositories via a Web service-based Adaptor.

sioned is a service node in charge of handling the subscribers of a Telco operator, including the services that the users have subscribed. Such a database would require being updated every time a new user is generated, every time an existing user needs to be deleted from the system, and finally, every time that an existing user subscribes or unsubscribes to a service.

Web services are also used to synchronize periodically service nodes with configuration or provisioning information. The typical example is two separate systems that have traditionally been operational in a separate mode, but due to update or network evolution reasons, need to synchronize the (for example) subscriber information. In this case, a Web services procedure can be implemented to synchronize specific data fields, or even to perform data type translation. This situation is depicted in Figure 6.2.

Other applications in Telco networks are the data query and retrieval of information like vocal announcements stored at IVRs (Interactive Voice Response systems) or vocal services like ring back tones, etc. Similar procedures would apply in those cases.

6.4 Context in Operator Networks

Context-aware paradigms have been explored for a long time in controlled environments, usually with coverage range limitations (like university buildings or campus areas). In these controlled environments, usually ad hoc sensor networks are deployed. A specific service platform manages the sensor signals, converts them into contextual information and provides the service.

Such functional elements may not be located in the same platform but the dependency among them makes the whole system behave just like if they were running in the same hardware.

Operators are currently facing the challenge of getting added value services. The context-aware evolution path is one of the areas being explored due to different reasons; some of them can be summarized as follows:

- The technological skills of the average user of Telco services are increasing significantly. This means that the subscriber is able to use more advanced communication services. Such skills ultimately means that the subscriber's request for personalized services, specific to the user or the user's situation.

- The diversity of situations in which the average modern subscriber of communication services may experience is also increasing. This means that the subscriber is not just able to use more advanced services, the subscriber will actively request for added-value services if they are specific to the actual user's needs.

- The amount of connected devices that the average user can use during the day is also important. The evolution on both the processing and memory capabilities of the mobile device is especially remarkable. Those aspects are progressively decreasing the gap between the mobile device itself and a laptop or a personal computer, and eventually going beyond these, like sensing capabilities, accelerometer, device-orientation sensor, GPS, etc.

All such aspects, as well as some others of a different nature, make the Telco contextual services a very active research area, where operators worldwide are capitalizing significant research efforts.

6.4.1 Definition of Context in Telco Networks

Context is a concept so broad that some limits to its meaning need to be defined. Accordingly the definition of context will be specific to the purpose inherent of the Telco services. Other scopes might require different definitions of context.

Taking the starting point of Dey's definition of context (2), a particularization of that can be made, in order to refer that to Telco networks, services and subscribers. That definition can be:

Any information about the situation of a user that may affect the potential ability to request, accept, establish, maintain, reconfigure, terminate, or reject any type of communication session with other entity (subscriber or application).

Such information will include the different user profiles or preferences, stored in service or database nodes owned by the Telco operator or by any other third-party provider that may affect the way the Telco services are provided.

6.4.2 Types of Context in Operator Networks: History and Current Situation

As it becomes quite clear in this book, context is a concept that has existed actually for a long time, in one way or another, but whose formalism is relatively new. It has been considered in different aspects of personal communications, but not necesarily following the context-aware formalism or paradigm.

Based on the context definition provided in the previous section, in a Telco environment, context will be extracted/generated from/by elements connected to the Telco network, and that can be used (either in raw format or after a processing to agregate that) by service layer elements to enhance the service trigger or execution.

In mobile communication systems, like GSM cellular system, whose wide deployment boosted during 1990s, some aspects are already considered about the instant situation of the user that are considered before or during the establishment of the personal communication session, or even while the voice call is ongoing. Those aspects can be the following:

- User situation.
  - Connectivity status: The connectivity status of a given mobile was stored by the HLR database, as well as some other information like the VLR or Visited Network where the user is connected to.
  - Radio Link situation: coverage, signal-to-noise ratio, macro- and microdiversity conditions. All those aspects affect the quality of the signal received at the mobile station and accordingly are considered while the call is ongoing to eventually drop the call, or while the call is to be established, in order to abort its establishment.
  - User position information: based on the position of the user, the call can be routed via one specific GSM base station or a different one. Different enhancement algorithms to increase the capacity of the radio networks have been proposed based on such aspects.
- Other users situation. Emergency call requests by users in the same cell or area, that may trigger a preemption of an existing call in order to have enough capacity to route the emergency call.

In 3G/UMTS networks, where the mobile terminal may get access to PS (packet switched) or CS (circuit switched), similar information is stored on the circuit switched side. But on the packet switched side, the capabilities of the end devices are required in order to negotiate the capabilities of the session to be established. Accordingly an establishment of a videoconferencing session is not requested if at least one of the involved mobile devices does not support the appropriate radio bearers for that.

Additional information like transmission power from/to the mobile device is obtained, but processed in a very local segment of the communication path.

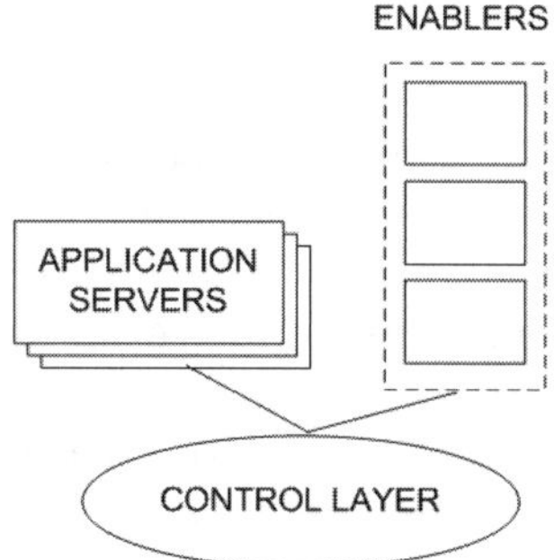

Figure 6.3: Enablers.

Accordingly, such contextual information is obtained and processed by the same entity (in the case of the transmission power of radio cellular networks that would be the transmitter side, i.e., mobile station and base station), in the same way the current deployment architectures of context-aware research works are being proposed.

Such concepts have evolved much further with the deployment of NGN networks, like the architecture proposed in IMS (4). In GSM/UMTS networks the information captured by the access networks is considered at the access network layer. But at the IMS domain itself, some types of information are captured at different levels and are strongly associated with the concept of context.

6.4.3 NGN Enablers

NGN architectures, and specifically IMS, introduce the concept of the enabler (3). Such an entity belongs to the service layer of the architecture, as presented in Figure 6.3.

The function of such elements is based on the horizontality of the service layer, as presented in previous sections. The enablers aim to provide information to the service layer modules to enhance and enrich the functionality that they may provide, but not to provide a service itself that the subscriber may request.

Enablers will centralize the processing of the information that is relevant to potentially any service located in the service layer. This way the associated processing of such type of information can be accessed and reused by any application.

Presence

Some context information is already being handled by NGN Telco networks, as is the case of presence information, through the Presence Server element of the NGN service layer architecture, specified by Open Mobile Alliance (OMA), *http://www.openmobilealliance.org/*. Such information is used by service layer elements to enhance their added value to the subscriber.

This network enabler is a suitable element for early deployments of context-aware Telco services, given that the information that can be stored by the Presence Server for a given subscriber alias (any entity defined for such subscriber, like a mobile device, or a fixed one, etc.) is very flexible.

The important drawback to keep in mind in such deployments is that it is not possible to perform aggregation or processing of different contextual stimuli within the Presence Server. Instead, a mere notification of the new presence status is reported by the subscriber device. Any processing of the inputs coming from different devices or sensors will need to be performed in a separate platform or the native functionality of the presence service would be jeopardized by the additional processing load generated to aggregate or process context information.

The context information that is most commonly captured in current contextual deployments is line presence. It is a key extension of presence information that specifies the situation of each user device and the accessibility of the user himself through each one of the user devices associated to him/her.

Location

Maybe the first type of information actually associated to the user's context itself is the location of the user. This information can be obtained by the cellular network via the LBS (Location Based Services) (5) that the radio technology support, existing different procedures to do so, Cell of Origin (CoO), Time of Arrival (ToA), Angle of Arrival (AoA), and others (6).

During the last years, the implementation of GPS receivers in the mobile devices has become common, in such a way that the location of the user does not depend on the capabilities of the radio access network. As an alternative technique, it becomes possible to implement a virtual CoO procedure bypassing the operator infrastructure, just by progressing the CellID of the cellular base station that the mobile device is attached to a Web server where that CellID is translated into actual latitude and longitude of the geographical position of the user, as the Google API GoogleGEARS implements (7). Such technique relies on the fact that it is not really necessary to handle location information with sensibility of meters, but of hundreds of meters, to provide most of the location-aware services.

Accordingly, OMA specifies an enabler that obtains and stores the location of the users in the same way that the Presence Server does with the availability. Such element implements a standard interface based on SIP (8) specially adapted to manage the query and reports of user locations in a centralized way. That information would then be stored in a location enabler.

Relationship with Other Users

A key aspect of the user's context that traditionally is not widely considered is the other user's context, especially the context of those users that keep a specific relationship with the user under study. Accordingly, the context of user's relatives, work colleagues, friends, etc. will impact at different levels the context of a given user. In order to consider those contributions properly, a way to handle the different permissions is to use the context information by other users, and the relationships among them need to be considered.

That relationship management function may be implemented in a vertical way, in which the same service element that has the information from all users is also applied the different information using sharing policies. That is the case of the social networks like MySpace, Facebook, Xing, and some others.

However, in order to make it possible to share such information, a generic element to centralize such management is required at the service layer. One of the best candidates to do so is the XDMS (XML Document Management Server) element (9). Such an element has the function of storing specific XML documents for each subscriber. In such XML documents, different information about such a user can be included, for instance:

- Identifiers of other users that can get access to information specific to a given user.
- Personal information of any kind about the users. Such information can be preferences, service profiles, or even an extended information repository about application-specific subscriber profiles.
- A list of friends or buddies for applications that may require list management.

This element implements XCAP protocol (10) (carried over HTTP transport protocol), whose main function is to retrieve, delete, modify, or add a document, element, or attribute associated to a given user.

In any case, the XDMS server is not being used to keep the other user's context, as far as the authors are aware of at the time of writing.

6.5 Deployment of Context Aware Services at Telco Layer

6.5.1 Introduction

If the context aware services are developed and deployed by the Telco operator, and located in the Telco service layer, the advantages that can be identified are the following:

1. The information about the contextual situation of the user is much more complete, given that the amount of providers of contextual information that can send their information to the application layer is much higher. That is due to the potential global coverage of the service.

2. The information obtained from a given user is much more robust, as the same type of information (location, ambient information, current activity, etc.) can be captured through different means or sensors. It is then possible to detect spurious signals that can be due to malfunction of sensors or providers.

3. The information about other users whose context is also modelled can be considered to modify the actual situation of a given user. That is based on the ubiquitous nature of the service layer of the Telco operator; it shall be accessible regardless of the location and access network.

4. The traditional context-aware services are tightly coupled to the actual sensing devices (contextual providers). Out of the coverage range of the context capturing devices, no contextual service can be provided, as the actual situation of the user cannot be estimated. Based on the Telco operator service layer, that is globally accessible from a variety of access networks, such dependency of the context-aware services on the location of the user is eliminated. Through an adequate deployment of context capturing devices, with appropriate connectivity for each one of them, the Telco operator service layer can obtain context information, associated to each user, just by querying different sensing devices.

Based on such reasons, the deployment of context-aware services from the Telco operator network is a sensible objective.

Such deployment can be conducted in two basic ways (11):

First, the same philosophy as in vertical service contextual platforms can be followed. The context-aware service architectures deployed in a lab or limited environments can be exported to Telco services. If that option is followed, the limitations of a vertical solution are also imported to the Telco architecture. Such experiences in Telco deployments usually present the limitations of an ad hoc context-capture system and a service highly oriented to specific context information. In any case, such implementations could eventually be appropriate for some particular customers or local services.

On the other hand, horizontal service layer capabilities are to be considered for truly global deployment of contextual services from the Telco operator. This second implementation option is presented in the following sections.

However, in the reasons presented previously to justify the deployment of context-aware services from the Telco operator environment, there is a common technical foundation. That is the connectivity among the different involved domains. Such domains are two: the service layer in which the context-aware services are executed, and the domain in which the context is captured.

6.5.2 Requirements on the Connectivity

This connectivity can be very diverse, and communication protocols properly defined are required.

In the case of a contextual deployment of services from a Telco operator, such connectivity options shall fulfill the following requirements:

1. Flexibility.
2. Simplicity in the implementation.
3. Minimum impact on the transport network in terms of configuration of the different network nodes.
4. Lack of limitations to be carried over different physical/transport protocols.
5. Minimum effort required to upgrade versions.
6. Small round-trip delay is preferred, although real-time performance is not strictly required. That will depend on the specific application, but in most cases this is the case.

As can be seen, such requirements can be fulfilled by a Web services implementation. Accordingly, the Web services option is a clear candidate to be considered as a technical foundation onto which the deployment of contextual services by Telco operators can be based.

6.5.3 Horizontal Architectures

Telco service networks are evolving towards horizontal architectures, as has been described in previous sections of this chapter. The main reason is information leverage among different services and service layer elements and reductions on development costs.

Context is a type of user information that is not yet commonly used in Telco operator networks, although that is widely known for some time already in research and lab environments.

It is critical to understand that the context information becomes much more useful in a horizontal architecture of the service layer, as potentially any service can take advantage of contextual information from the subscribed users. Only difference among services is the actual type of context that each service is interested in.

The fundamental issue to solve is the acquisition of the context information that is specific to each service and the mechanisms to make it possible for the services to request the notifications of the context update and to receive them. Each service will be requesting/receiving the type of context of specific interest for the service.

6.5.4 Service Orchestration

The contextual information has a deep relationship with recommendation, discovery, or personalization of services.

But once the service has been discovered, recommended, or personalized, based on the specific user context and the reference context that the service is associated with, the service needs to be executed.

During execution of the context-aware service, problems may arise from the following facts:

- Simultaneous or pseudo-simultaneous execution of services with similar functionality, if no confirmation is required from the user. That may lead to inconsistent states of some services. In the Telco service case that is especially critical, as most of the Telco services will control the user terminal in high priority aspects (like 3PCC, or third-party call control services).

- Even if the functionality of the proactive contextual services to be triggered is different and the state machines of different services do not collide, the progressive execution of different services during a given small period of time may lead to a high signalling load of contextual notifications. That would eventually mean a problem in terms of consumed bandwidth and processing horsepower at the involved elements. Currently such a problem is being addressed in a formal way by the team that writes this book chapter. Optimum identification is implemented with the following generic procedures (at a high level):

    1. (t=t0). At a given moment a change in the contextual situation of the user occurs. Such modification of the user's context needs to be notified.

    2. Instead of notifying the update instantaneously, a delay is introduced before the notification is triggered. The actual delay (t1) is a parameter to be optimized, but it is considered to be around the order of few seconds. During that period of time some other

modifications of the context associated with the user might take place.

3. (t=t0+t1). Once the security period expires, a contextual notification is generated, reporting the final contextual state. That way the transient modifications are discarded, and only the potentially permanent situations generate a contextual notification.

- The service execution may imply service concatenation, as several separate services may be subscribed to notifications about changes on the same type of context information, and some of those services could potentially change as well as the context of the user. Such concatenation may be based on a subsequent contextual discovery of services to be executed by the initial service, or not. The main implications would be the contextual information management by the services. Additional details fall out of the scope of this chapter. This is currently one of the hot research topics in the context area.

6.6 A Commercial Implementation Case

6.6.1 The Corporate Market

The contextual implementation based on Web services presented is oriented to the corporate Telco market. Such a market has key characteristics that make it perfectly suitable for such an evolution:

- Implementations with a cost required are acceptable if those mean enhancements are for the enterprise with respect to its competitors in terms of productivity, production optimization, etc.
- End-user's acceptance barriers to new technologies are less important than in a residential market.
- Specific developments that may be eventually required are a viable option.

The global objective of the implementation described in the following sections is to be able to capture the Telco context of a corporate user that potentially can hold a variety of devices (fixed, mobile, PC-based VoIP softphone, etc.). That information can be used in many operator telephony services.

Specifically, the context situation described in the next subsections is the capacity to receive an incoming call successfully. If such information is properly captured, it can be reused by a service whose objective is to handle on-net/off-net terminating calls according to the contextual situation of the employee of the customer enterprise. This will include routing of calls incoming to the fixed device to the mobile device, as well as from one employee to another.

Such routing shall be performed based on contextual information of the destination employee as well as user-defined policies. Such business intelligence is part of the Telco service itself and falls out of the scope of this book.

This contextual service needs to be provided from the operator service layer given that it has an inherent convergent nature (i.e., involves terminals from different access neworks). The call shall be routed seamlessly among different device types, both fixed and mobile.

In order to design, develop, and optimize such system, Web service technologies are chosen in the implementation presented. Web services are considered to carry the context information of the user as well as to orchestrate the context-aware service delivery.

6.6.2 Context Information Capture

The context information is captured by the phone devices, namely the fixed phones and mobile phones, in a first phase, although the integration of PC based Softphones is considered for a second phase. Such information is the line presence that indicates whether the employee is actually available at the fixed or mobile phone.

Such availability status is easy to obtain by the mobile network, through the registration status of the mobile phone. That information is kept at the HLR jointly with the visited network where the user is connected to. The information is updated every time the user changes the registration status of the mobile device.

However, that availability status information is not easy to obtain from a fixed telephony network, where the fixed phones of the customer enterprise are connected to a PBX (Private Branch Exchange), located at the customer's premises.

Both types of user's context, fixed and mobile, shall be obtained and processed in order to compose useful and robust context information about the user.

Once captured, the context information is sent to a centralized element at the horizontal service layer. Any service can then make use of such information, available at the centralized element.

The global architecture is depicted in the Figure 6.4.

In this figure the different networks are included, naming the PLMN (Public Land Mobile Network) and PSTN (Public Switched Telephony Network).

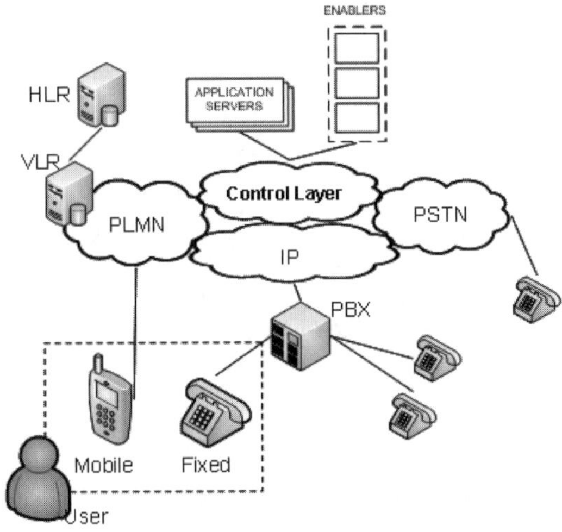

Figure 6.4: Network architecture.

6.6.3 Exporting Context to the Operator Network: System Proposed

In order to capture the user's context from the fixed domain (including VoIP PC-based softphones), the deployed architecture is depicted at a high level as shown in Figure 6.5. The different components are introduced in this section. In this figure the horizontal service layer at the operator's network will have access to the call control capabilities of the PBX over the fixed extensions through the network elements depicted in the diagram of Figure 6.5.

In order to simplify the global architecture and decrease the impact due to the different PBX manufacturers that may exist in a given operator's network, a hierarchical structure is proposed for the Business Gateway, in such a way that two separate gateway layers will exist: EBG (Enterprise Business Gateway) and Context Enabler. That structure is depicted in Figure 6.5, where:

- EBG is the Enterprise Business Gateway, whose main function is to perform protocol adjustment from a fully standard Web service-based protocol to a PBX vendor-specific and reverse. Accordingly, EBG will act as a Context Provider entity. Multiple EBGs will exist, potentially a separate EBG per PBX. The function of this element is to capture the context of the subscriber and progress that information over a proper Web service event-based interface. In addition, the commands received from the services at the service layer, over Web service interface will also be adapted to the specific protocol implemented by the PBX element, in order to execute the commands properly received from the service

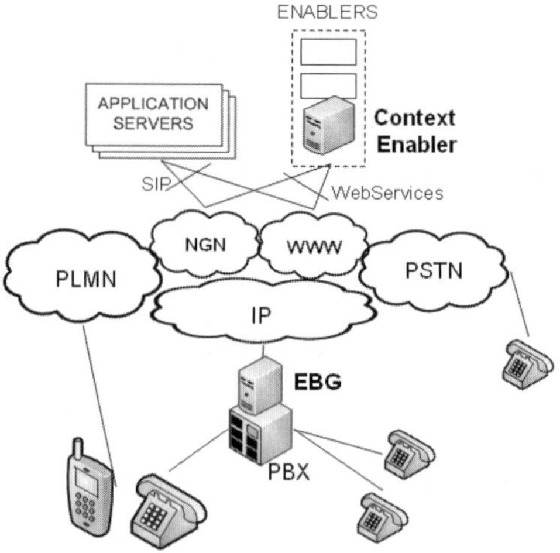

Figure 6.5: Business Gateway based architecture.

layer elements.

- The interface between the Context Enabler and EBG shall be standard protocol. Due to the connectivity and flexibility aspects detailed in previous sections, this connection is Web-services based.
- The main function of the Context Enabler is to aggregate the context of the subscribers. Accordingly, this element will be queried by the different entities of the service layer (the application servers) and will provide the proper information back. Such interaction between the application server and the Context Enabler may be implemented through different means, as shown in Figure 6.5. The interaction may be done over the SIP interface, if the service is a Telco IMS service or has an IMS interface. On the other hand, the interaction between the Context Enabler and the service may happen also over a Web-service interface, if the service is located in a different domain, is developed/operated by a third-party, etc. In the specific implementation presented, given that it is an interface completely included in the operator domain, it is based on SIP protocol.

Protocols Used

The protocol between EBG and the Context Enabler needs to be standard, easy to implement, and able to carry all the information and primitives required.

The protocol proposed for that purpose is the CSTAv3 (Computer Supported Telecommunication Application version 3), specified by the European Computer Manufacturers Association (ECMA, *http://www.ecma-international.org*). CSTAv3 is a SOAP-based protocol that can implement Web services (WS-CSTA, (12)) to carry the CSTA primitives over a HTTP transport layer. The WS-CSTA protocol's objective is twofold:

1. Report to a CTI application (the application layer hosted in the operator network) the selected events that take place at the PBX. Such events are the contextual reports that will be aggregated at the Context Enabler into consistent Telco context information.

2. Carry the commands from the context-aware Application Server to the PBX entity in a consistent way, including all the relevant parameters. The purpose of such commands is to execute the different media-related procedures (call forward, hold, retrieve, diversion, etc.) according to the service business logic.

Some main manufacturers of PBXs already include the support of CSTAv3 in their roadmap for their future products, but such implementations of the protocol is usually not complete and require adaptations like the one implemented at the EBG platform.

The Web service interface is basically composed of two different mechanisms, presented in the following subsections.

A. Events

The events are reports of updates of the status captured by the EBG that are reported to the Context Enabler through the HTTP network. These events can be triggered every time a specific event takes place, or periodically if the information needs to be refreshed.

An example of a CSTAv3 event sent from EBG to the Context Enabler is the notification that the user is no longer on the fixed phone, that is, the call has been released. Such a report is the call-cleared-event and the XSD description of the XML document to be embedded in the SOAP container is shown in the following figure (not all optional tags are included).

```
<xsd:element name="CallClearedEvent">
<xsd:complexType>
<xsd:sequence>
<xsd:element ref="csta:monitorCrossRefID"/>
<xsd:element name="clearedCal" type"csta:ConnectionID"/>
<xsd:element ref="csta:correlatorData" minOccurs="0"/>
<xsd:element ref="csta:userData" minOccurs="0"/>
<xsd:element ref="csta:cause"/>
<xsd:element ref="csta:mediaCallCharacteristics" minOccurs="0"/>
<xsd:element ref="csta:callCharacteristics" minOccurs="0"/>
<xsd:element ref="csta:callLinkageData" minOccurs="0"/>
</xsd:sequence>
</xsd:complexType>
```

```
</xsd:element>
```

B. Services

The services are command messages sent from the application layer to the EBG through the HTTP network in order to trigger the execution of some media-related procedure. Those commands can be triggered from the application layer in a asynchronous way, depending on the business logic executed on the specific application.

An example is the CSTAv3 command sent from the application to the EBG is the clear-call command, whose purpose is to trigger the clearance of a specific call, identified in the message itself. The XSD description of the XML document to be embedded in the SOAP container is shown in the following figure.

```
<xsd:element name="ClearCall">
<xsd:complexType>
<xsd:sequence>
<xsd:element name="callToBeCleared" type="csta:ConnectionID"/>
<xsd:element ref="csta:userData" minOccurs="0"/>
<xsd:element ref="csta:extensions" minOccurs="0"/>
</xsd:sequence>
</xsd:complexType>
</xsd:element>
```

Advantages of the System

The Web service-based architecture implements a seamless evolution of the enterprise services into a complete horizontal structure. The services can be made contextual and fully oriented to the corporate user.

Contextual services can be provided to corporate customers by combining the information about the mobile (available at the PLMN, also accessible from the service layer in NGN architectures) and the fixed corporate extensions. Such services can be, among others: presence-based intelligent routing with PBX extensions as possible destinations, hosted messaging service for fixed corporate extensions, and many others. It is important to note that this system does not mean anchoring of the session in any device or access network, as the service layer is common for both worlds.

Implementation Aspects of the System

Based on the previous details, any Telco service can take advantage of the current situation of a user just by accessing the Context Enabler to get such information. If such a system is to be implemented in IMS-compliant architecture, and the granularity of the contextual information captured is not too detailed, it is a sensible approach to integrate such contextual information into the IMS Presence Server, in order to have a single point where such type of information is stored. That aggregation in the OMA Presence Server also has the key advantage that there is a whole SUBSCRIBE/NOTIFY procedure to query/retrieve the information from the Telco services or any application that implements SIP.

As has been presented previously, the context-aware Telco corporate applications will need to control the calls, at several levels, like call routing, third-party call control, etc. Such procedures are already in place on the mobile side, in which the serviced layer has full control of the mobile activity and calls. On the corporate fixed side, the applications will interact with the EBG element. That needs to be done via the Web services interface implemented at the EBG. If the application implements such an interface, a direct interaction can be implemented.

However there are Telco applications that are heavy, and adding new interfaces to those is difficult, regardless of its nature. In those cases, the Context Enabler itself may behave as a protocol adaptor between the network protocols that are implemented by the applications, and the EBG, acting as a Web services-SIP gateway.

6.6.4 Procedures

The signalling procedures proposed for the system are listed and described in the following sections.

Heartbeat

The Heartbeat functionality will be implemented between each CSTA peer, namely EBG and Context Enabler or the application. Its purpose is to provide the adequate robustness to the system in case of failure of a given CSTA network element.

The context information is inherently subject to expire, depending on the nature of such information. If the connection is lost between the context providers (EBG in this case) and the context processing element (Context Enabler in this case), the last contextual information for the users will not be reliable after a short period of time (as an update of the user's context might happen and would not be reflected in the Context Enabler due to the connection failure). If critical decisions are to be made by the applications based on the fact that the user is available at a given device, is talking over the mobile or over the fixed phone, etc., such information needs to be obtained in a reliable way. That is why the heartbeat mechanism needs to be put in place.

This functionality is implemented via SystemStatus CSTA message, exchanged between the peers. Given that the Web service interface does not imply any state machine running at the platform, it shall be designed at the different platforms that are supposed to follow this heartbeat procedure via application-level state machines.

The proposed scenario in the system is depicted in the following Figure 6.6.

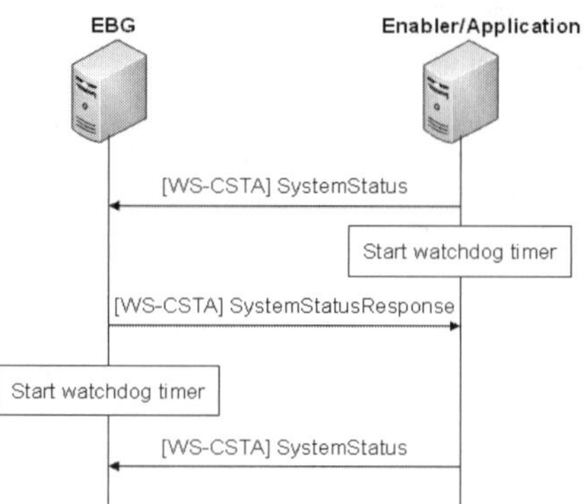

Figure 6.6: Heartbeat mechanism.

The watchdog timers are configurable parameters at the platforms involved, and need to be optimized depending on both the transport network used to interconnect them and the nature of the application itself.

The XML embedded in the SOAP container for the SystemStatus and SystemStatusResponse message will basically contain the ID and the status of the Application or the element sending the message, and eventually the value of the watchdog timer to be applied for the next round before the next system status message is sent by the receiving platform. That can be used to optimize on the fly the watchdog timer value.

Telco Context Monitoring Procedures

The monitoring functionality at the context provider will progress to the interface with the Context Enabler all the signalling related to the specific events that are taking place at a given user device so that the application layer can take the corresponding action in return. In the case of the mobile extension, as it is an interface within the Telco operator service layer, such an interface can be SIP-based or other operator network protocols, but on the fixed domain, the context provider of the PBX extensions is the EBG and the interface to be implemented is Web services.

Two separate types of monitoring can be differentiated. The difference relies basically on the type of procedure that is reported to the upper service layer over the Web service interface:

1. Event monitoring. This type of monitoring will be used by Monitoring Function (MF) applications that will take actions over, for example, incoming calls towards the user's extension. The generated Web service reports are triggered every time the EBG detects a change in the contextual situation of the extension. Such events are on the following list; this is not exhaustive:

- User hangs up the phone after a call.
- User picks up the phone and is about to dial a number.
- User puts a call on hold.
- User reads a message on the text display that some fixed extensions have, depending on the specific capabilities of the telephone system.
- Other scenarios may apply.

2. Routing monitoring. This type of monitoring is considered to be especially critical and applies when an incoming call arrives to the user device. In that case such an event is reported to the Context Enabler in order to progress such information to whichever service may control the incoming calls to the user, to take specific decisions about the actual destination of such call. Due to the particularity of the service itself, the reporting of the incoming call is considered separately from the previous type of events.

These two types of events will be required by different services platforms in the network. That is why in the following scenarios these two separate types of procedures are presented separately, although they will be happening in parallel depending on whether the event or routing monitoring functions are started for a given extension.

In order to start the event notification procedures at the context provider, in the case of the Web service interface (EBG case), the Context Enabler will send a Monitor request to track call control and other activities from the PBX itself. This request will request the EBG to send all the notification events (new incoming call, established call, etc.) to the control function.

The signalling flow interchanged is displayed in Figure 6.7.

The sequence diagram is described as follows:

1. The Context Enabler or Contextual Telco Application sends a Web service CSTA Monitor Start message to the EBG (Context Provider) in order to start receiving contextual events for a specific user device. The XML definition is shown as follows:
<xsd:element name="MonitorStart">
<xsd:complexType>
<xsd:sequence>
<xsd:element name="monitorObject" type="csta:MonitorObject"/>
<xsd:element name="requestedMonitorFilter" type="csta:MonitorFilter" minOccurs="0"/>
<xsd:element name="monitorType" type="csta:MonitorType" minOccurs="0"/>
<xsd:element name="requestedMonitorMediaClass" type="csta:MonitorMediaClass" minOccurs="0"/>
</xsd:sequence>

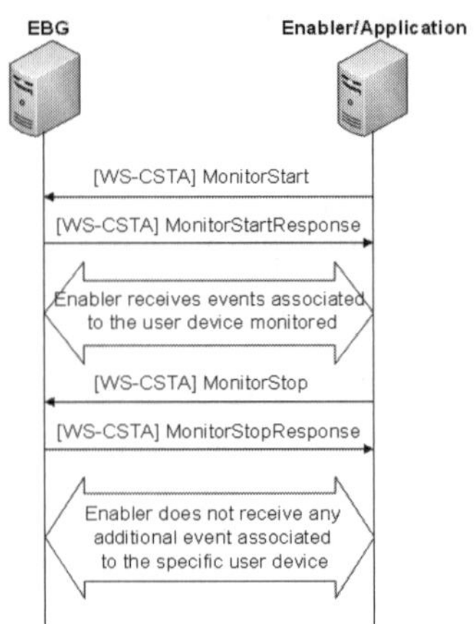

Figure 6.7: Monitoring mechanism.

```
</xsd:complexType>
</xsd:element>
```

It can be noticed that the identifier of the specific user device that is to be monitored is included in the Web service message in the tag monitorObject.

2. Upon reception, the EBG will interpret the MonitorStart message and will send the proper message to the PBX. Such activities are proprietary and depend on the specific PBX implementation and manufacturer. The communication PBX-EBG is outside of the scope of this document.

3. The EBG will reply with a MonitorStartResponse primitive. The XML definition is shown as follows:

```
<xsd:element name="MonitorStartResponse">
<xsd:complexType>
<xsd:sequence>
<xsd:element ref="csta:monitorCrossRefID"/>
<xsd:element name="actualMonitorFilter" type="csta:MonitorFilter" minOccurs="0"/>
<xsd:element name="actualMonitorMediaClass" type="csta:MonitorMediaClass" minOccurs="0"/>
<xsd:element name="monitorExistingCalls" type="xsd:boolean" minOccurs="0"/>
</xsd:sequence>
</xsd:complexType>
</xsd:element>
```

The monitoring session established via these procedures is usually stopped only in case any maintenance action is to be carried out on any of the equipment involved (PBX, EBG, Context Enabler, etc.). Therefore, the Monitor Start, Monitor Stop procedures must be a part of the controlled start/stop protocols of the elements of the platform or the involved application server itself. The same signalling flow will be applied for the MonitorStop. Such procedure will then close the Web service procedures that report the contextual changes of the fixed terminal from EBG to Contextual Enabler.

Routing

Such procedures are focused on selecting the right destination for an incoming call to the user device. As in previous versions, we will focus on the fixed side, that is the EBG will be acting as Context Provider.

This type of contextual monitoring can be understood as a specific use case or service for the contextual events described in the previous section. However, it is considered as a separate type of monitoring because of the following two reasons:

1. The criticality of the service is very high, as it controls the routing of a user communication session.
2. In order to avoid inconsistencies, the network system described only has one service in the network with such functionality.

The scenario execution is simple. When a call arrives to the PBX (coming from a public network or internal from another extension), a specific Web service notification is sent to the Context Enabler, that reports of such an event to the Service platform in charge of selecting the destination of the call. Such service can consider, based on specific business logic, a different destination if the user is not at the fixed phone. The actual destination can be selected based on similar criteria (the alternate user is not at the phone neither), or some other (according to an agenda service he/she is on holidays, etc.). The flow is detailed as follows.

1. The PBX gets a session establishment request, that may arrive from external public networks or from internal extension. The session establishment progress is stopped and the Context Provider (EBG) is informed about that. A notification is generated from EBG to the Context Enabler platform. Such notification is a RouteRequest event embedded in the Web service message. The RouteRequest XML to embed in the Web service message is shown as follows.
   <xsd:element name="RouteRequest">
   <xsd:complexType>
   <xsd:sequence>
   <xsd:element ref="csta:crossRefIdentifier"/>

```
<xsd:element name="currentRoute" type="csta:CalledDeviceID"/>
<xsd:element ref="csta:callingDevice" minOccurs="0"/>
<xsd:element name="routedCall" type="csta:ConnectionID" minOccurs="0"/>
<xsd:element name="priority" type="xsd:boolean" minOccurs="0"/>
<xsd:element name="replyTimeout" type="xsd:long" minOccurs="0"/>
<xsd:element ref="csta:mediaCallCharacteristics" minOccurs="0"/>
<xsd:element ref="csta:callCharacteristics" minOccurs="0"/>
<xsd:element name="routeCallInfo" type="csta:ConnectionInformation" minOccurs="0"/>
<xsd:element ref="csta:subjectOfCall" minOccurs="0"/>
</xsd:sequence>
</xsd:complexType>
</xsd:element>
```

2. The Context Enabler gets such information and reports that to the service that has requested reports about incoming calls. Different working procedures are available at this point, given that the interface is an operator network one, but for simplicity, in the system described, a Web service RouteRequest is sent from the Context Enabler to the service platform.

3. The Service performs internal processing based on the specific business logic (configured by the user, in this case), and selects a destination for such session. The service notifies that new destination to the EBG. That notification can be performed via a direct notification if the Service implements a Web service interface or via Context Enabler if that is not the case. In this case the service triggers a Web service command (RouteSelect) to the EBG, with the information about the destination to send the session to (the destination number for the call).

4. The PBX routes the session to the new destination received from the service. When that is performed the contextual monitoring events are generated as per the regular procedure shown in previous sections. For instance, a contextual notification will be generated from EBG to the Context Enabler, to notify that the user has no pending call.

5. Additional re-routing procedures may also be triggered at the new destination of the session.

6. When the session is finally established, the Context Enabler will get the appropriate notification of that via the Web service interface. At that moment, the Service can be notified by the contextual enabler and clean the dialogue register.

This flow is depicted in Figure 6.8.

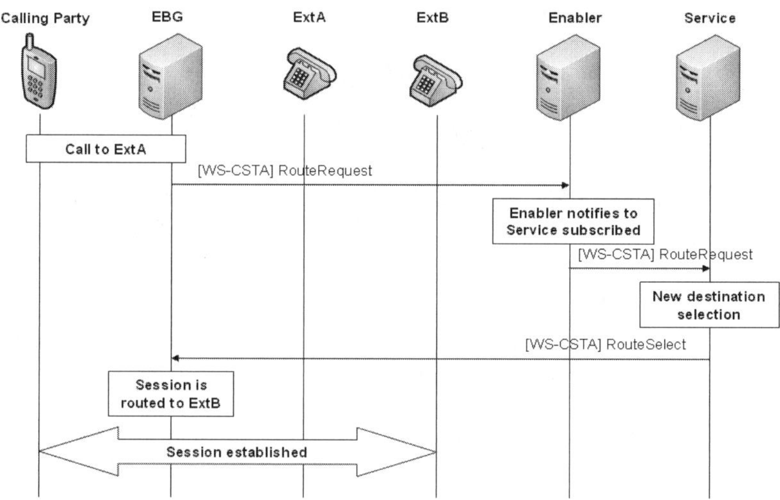

Figure 6.8: Monitoring mechanism.

6.6.5 Integration with External Applications

The system described in this implementation study can easily be extended by performing integration not only with the traditional Telco telephony services like the ones presented in the previous sections. Integration can easily be performed with other added value services located either at the operator network service layer or at external domains, developed by third-party providers.

Such integration can be performed through the different ways described so far. The two basic options described are the following:

1. Interface between the Service Platform and the Contextual Enabler based on standard protocol interfaces, i.e., SIP, or similar protocols, in the case of mobile networks that can be Intelligent Network protocols if real-time call control functionality is required.

2. Interface between the Service Platform and the Contextual Enabler based on Web service interface, over SOAP/HTTP. This option is perfectly suitable to integrate the services located in a domain different than the operator service layer, like standard Web Servers, including Web2.0 applications.

Some examples are provided in the next subsections.

Personal Agenda service

The personal agenda of the user can be located in a Web Application Server accessible through the Internet, in a Web-based calendar service. That application server will implement a Web service interface that can be used by the Context Enabler to request the availability information of a given user. Such information is very useful contextual information in order to know the user's availability to be used by other services.

Messaging Service

A messaging service, either Telco (SMS, MMS) or not provided by the operator, like new services or microblogging services very common in Web2.0 environments (Twitter, etc.) can be enriched with context information of the users that are generating or receiving the message (geopositioned news specific to the position of the user, messages posted by a buddy attached to a specific location that are only delivered to the destination under specific contextual circumstances, etc.). The integration of those services with the Contextual Enabler can be performed with standard operator network protocols, but the flexibility of the Web service interface is perfectly suitable for Store-and-Forward services like messaging, that do not have tight time requirements.

Social Networks Applications

Social networks are excellent consumers of contextual information, as news, messaging, and generic social content can be associated to the specific context situation of the receiver, being that context obtained somewhere else (like at the Telco network). However, the social networks are extremely relevant as Context Providers, as these are forums where the users post specific information about the current activity, appointments, different degrees of relationship between users, etc. Such information is fundamental contextual information to be progressed to a Contextual Enabler through a Web service interface, by using the specific API provided by each social network.

6.7 Conclusions

Web service technologies are a sensible option to implement interfaces among Telco service layer network elements that need to meet certain requirements of rapid development, flexibility for future protocol updates, and adaptability to the transport network.

Such characteristics make the Web service the proper technology to manage the flow of contextual user information within the Telco operator service layer domain and between the Telco service layer and external entities like third-party services or elements that capture raw context information.

Such a direction is being actively explored by main telecom operators. In this chapter an implementation case study is provided. A particular case is applied to the enterprise convergent telecom segment and relies on the deployed elements to capture the context of both fixed and mobile devices associated with a given corporate user. A Web service (CSTAv3/SOAP/HTTP) implementation is deployed at the fixed domain to export the situation of the fixed user device to a context server that manages the information and reports of specific user situations to Telco services that will provide specific functionality. Some Telco services are described in low-level detail like the user session routing service. Additionally, some added value services that can be enriched with contextual Telco user information are also provided.

The system presented has been deployed and installed in a limited amount of big corporations at the time of writing this chapter, according to the details presented in previous sections. In this live trial, the performance of the Web service interface has proven to be more than enough to handle the amount of subscribers and signalling traffic generated and the reliability of the solution is fulfilling the highest availability figures required for Telco services. Such results, jointly with the minimum development times (due to the flexibility and modularity of the Web service interface), is a clear justification of the right choice of the Web service technology.

Web services will continue to be explored in Telco deployments to explore the usage of contextual information by Telco services as well as some other service platforms located in other domains with whom the Telco service layer will be integrated.

References

[1] 3GPP TS 23.228. *IP Multimedia Subsystem, (IMS): Stage 2.*

[2] Dey, A. K. *Understanding and using context*, Personal and Ubiquitous Computing Journal, 5(1), 5-7. 2001.

[3] Konstantinides, C., and Charalambous, C.D. *Supervisory and Notification Aggregator Service Enabler in a Fixed Mobile Convergent Architecture.* ISWCS 2007. 4th International Symposium on Wireless Communication Systems, 2007, pp. 737-741.

[4] Agrawal, P., Yeh J.-H., Chen J.-C., and Zhang T., *IP multimedia subsystems in 3GPP and 3GPP2: overview and scalability issues*, Communications Magazine, IEEE, vol. 46, no. 1, pp. 138-145, January 2008.

[5] Li, Q., and Cao, J. *LBS in digital city context.* Proc. SPIE 6045 (2005).

[6] Kealy A., Winter S., and Retscher G., *Intelligent location models for next generation location-based services*, Journal of Location Based Services, vol. 1, no. 4. (2007), pp. 237-255.

[7] *Google Gears API*, http://code.google.com/intl/en/apis/gears/.

[8] Rosenberg J., Schulzrinne H., Camarillo G., Johnston A., Peterson J., Spark R., Handley M., and Scholer E., *Session Initiation Protocol (SIP).* IETF RFC 3261, June 2002.

[9] OMA recommendation: *XML Document Management Architecture. OMA-AD-XDMV1_0-20060612-A.*

[10] M. Isomaki, and E. Leppanen, *An Extensible Markup Language (XML) Configuration Access Protocol (XCAP) Usage for Manipulating Presence Document Contents.* IETF RFC 4827, May 2007.

[11] Cadenas, A., Hermida, A., Arias, A., and Serna, J., *Distributed PBX gateways to enable the hosted enterprise services architecture in a NGN scenario*, Innovations in NGN: Future Network and Services, 2008. K-INGN 2008. First ITU-T Kaleidoscope Academic Conference, vol., no., pp. 203-210, 12-13 May 2008.

[12] Standard ECMA-348 *Web Services Description Language (WSDL) for CSTA Phase III.*

Chapter 7

Using SOC in Development of Context-Aware Systems: Domain Model Approach

Katarzyna Wac, Pravin Pawar, Tom Broens, Bert-Jan van Beijnum, and Aart van Halteren

Abstract The drive for user-centric systems in combination with an emergence of wireless networking and increasing computing capabilities of mobile devices propel the development of Context-Aware (CA) systems. Moreover, nowadays, for the development of computer systems, the Service-Oriented Computing (SOC) paradigm became popular as it advocates use of services to support the development of rapid, low-cost, interoperable, evolvable, and distributed applications. Being a system development paradigm, the SOC can provide fundamental building blocks for developing ubiquitous CA systems. However, context-awareness itself is a relatively new and scattered research area. Due to the complex challenges encountered in a development of CA systems (such as context acquisition or the multifaceted nature of context), current research initiatives address only some specific problems, abstracting from others. Nevertheless, a holistic view of context-awareness encompasses numerous relevant cross-cutting research and development aspects. This holistic view is presented by means of a domain model of context-awareness proposed in this chapter. We reveal state-of-the-art of the individual research (sub-) domains and we analyze how they fit in the proposed domain model and how they map onto the widely used layered model for CA systems development. SOC could be effectively used to develop services catering to each (sub-) domain and SOC fundamentals such as interoperability and service composition could be further exploited to develop a complete CA system. This is shown in a case study of the Amigo project. Therefore, the contribution of this chapter is multifold. It provides a domain view of context-awareness to the newcomers in this field and highlights opportunities for SOC practitioners in the context-awareness domain. Moreover, it also provides a holistic view of context-awareness to the existing researchers, and helps practitioners by means of the Amigo project case study to apply the SOC concepts and the context-awareness domain model in the development of CA systems.

7.1 Introduction

With the emergence of ubiquitous computing systems, the *Service-Oriented Computing* (SOC) system development paradigm gains interest due to its inherent use of services to support the development of rapid, low-cost, interoperable, evolvable, and massively distributed applications (11). On the other hand, we observe advances in mobile devices, like enriched computation resources, increased storage and communication capabilities, and device miniaturization. These factors enable the development and deployment of sensor-rich environments, which, propelled by the need to provide user-centric applications, contribute to a development of so-called *Context-Aware* (CA) systems. The CA systems use user context (e.g., location, time) to provide relevant information and/or services to this user, where relevancy depends on the user's task (40). In general, SOC can provide fundamental building blocks for developing ubiquitous CA systems. Several technologies have emerged to realize the SOC: Web Services (WS), Jini or uPnP. Instead of focusing on a particular technology to realize SOC, in this chapter, we focus on a SOC as a paradigm to develop CA systems. Research on the context-awareness paradigm is ongoing in multiple research domains. However, due to the inherent complexity of developing generic context-awareness systems, the majority of the research addresses specific problems of interest, abstracting from others. To facilitate the structured development of context-aware systems, a widely adopted layered model of context-aware systems is often used (39, 2, 55, 60, 4); it divides a context-aware system into a physical, context data, semantic, inference, and an application layer, thereby facilitating separation of concerns between context acquisition and its usage. There exist a number of context-aware computing systems (e.g., PACE, AWARENESS, SOCAM) developed along this layered model. Some of these use the concepts of SOC for interconnecting services between different layers. However, a layer may involve concepts from various research domains. For example, the context acquisition layer may include research inputs from sensor networks, actuator networks, discovery of context sources, authentication and authorization of users, and context sources. To make the connection between research domains and layers of a CA system explicit, in this chapter we propose a context-awareness domain model, which provides a comprehensive view of the research domains involved in the area of context-awareness and their research objectives. We argue that the researchers specialized in these domains should develop specialized services using their (sub-)domain expertise and make these services available to the other researchers in other (sub-)domain. This would enable, using the fundamentals of SOC, a rapid development of modular, global, and scalable context-aware systems. To illustrate this, we provide a case study of the EU *Amigo* project. By developing and deploying specialized domain services and their interlinking using the SOC principles, the Amigo project has delivered successful context-aware system targeted for the networked home environment. The contribution of this chapter is multifold. It provides a domain view of context-awareness to the newcomers in this field, highlights opportunities for SOC practitioners in the CA domain. Moreover, it also provides a larger view of CA to the existing researchers, and helps practitioners by means of the Amigo project case study to apply the SOC concepts and the CA domain model in the development of CA systems. This chapter is structured as follows: Section 2 presents existing definitions and concepts related to context and context-awareness. The definition of SOC and its implementation technologies such WS, Jini, or uPnP are provided in Section 3. Section 4 presents layered model for CA systems. It analyzes the SOC applicability to the layered model and emphasizes the need to consider a domain view of context-awareness. Following this, Section 5 introduces the domain model of CA systems, indicates its research domains and presents our view on why such a model should be combined with the SOC paradigm to develop modular, global, and scalable CA systems. This section also specifies interconnections between the common layered model and proposed domain model of context awareness. Section 6 presents a case study of the Amigo EU research project. Based on this case study we derive research guidelines for using SOC in the development of CA systems. Section 7 presents conclusive remarks.

7.2 Context and Context-Awareness

7.2.1 Introduction

This section discusses basic concepts and fundamentals of the Context Awareness (CA) and Service Oriented Computing (SOC) paradigm by citing a number of research papers in these areas. Section 2.2 is on the relationship of context-awareness with other research domains. Section 2.3, Section 2.4 and Section 2.5 discuss definition of context, categorization of context, and properties of context; respectively. Section 2.6 illustrates on the definition of context-awareness. Section 2.7 introduces SOC and Section 2.8 briefly describes some of the SOC implementation technologies.

7.2.2 Relationship of Context-Awareness with Other Research Domains

Research on context-awareness originated in the early 90s. Context-awareness is still a relatively scattered research area and consequently domain definitions and concepts are not yet finalized. There is often a confusion regarding the relation between context-awareness and other computing paradigms, such as ubiquitous computing, pervasive computing, ambient intelligence, location-based computing, calm computing, and intelligent or smart computing, or emotional computing. All these computing paradigms have their main objective of delivering personalized user-centric applications and services to users, employing mobile and embedded devices integrated seamlessly in the daily life of users. One may say that context-awareness has a more restricted scope than other research paradigms. However, as this paper also demonstrates, context awareness has far reaching applicability. According to us, context awareness plays an important role in paving the way for other paradigms in realizing personalized applications (see Figure 7.1).

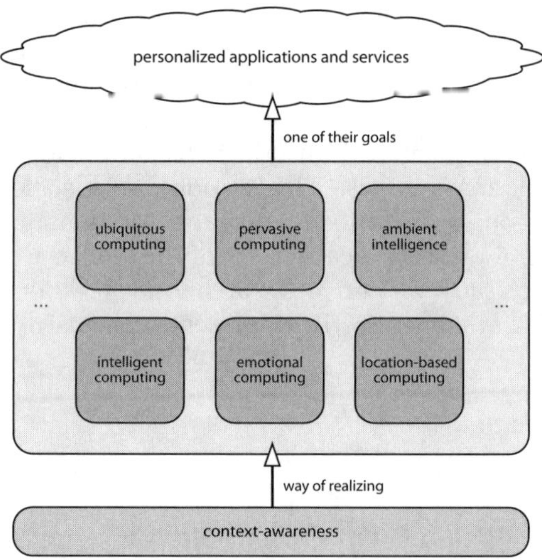

Figure 7.1: Context-awareness and its relation with other research domains.

7.2.3 Definitions of Context

Based on the scope of developed context-aware applications, many research groups propose definition of context based on their own perception. First, context has been defined by Schilit and Theimer (97) as "location, identities of nearby people and objects, and changes to those object" for the purpose of their location-based office applications. Lamming and Flynn (65) describe context as any information stored in a personal office-oriented "diary" serving as a basis for their "forget-me-not" office application.These first definitions defined context by its use. The research on context-awareness conducted by Schmidt et al. (99) resulted in a more abstract definition of context as "the situation the user is in" which has a unique name and set of features (i.e., attributes) with specific values. The sources of context as indicated in (99) are vision, audio, motion, location sensor, and bio-sensors that are used in a light-sensitive PDA display application. However, according to Dey, definition of context based on a particular example has a limited scope and it is difficult to apply such a definition to a broader range of applications (39). Therefore, he proposes a more generic definition of context as "any information that can be used to characterize the situation of an entity, where the entity is a person, place, or object that is considered relevant to the interaction between a user and an application, including the user and application themselves" (40). After Dey, there have been multiple attempts to define context in the context-awareness research community. For example, Ebling et al. (44), for the purpose of their online presence application, define context as aspects of the physical world (e.g., temperature) and conditions in a virtual world (e.g., presence). Gray and Salber (49) indicate that context is spatio-temporal information concerning a user. Prekop and Burnett (89) take a broader perspective than Gray and define context as any information concerning not only a user but also his activity. Bradley and Dunlop (14), similarly to Dey, focus on context definition including the user and his applications, while Goslar et al. (48) narrow down this definition towards a user and his enterprise applications (e.g., e-mail). At the same time, the research on context definition's shifted focus from a user and his activities towards entities participating in these activities. For instance, Wei et al. (121) define context as any information concerning user's mobile device and its capabilities, as well as the networks used and their characteristics. Chalmers (30) define context as information relevant to the user along the timescale; the context which is "now" and the "past" context, Gloss (47) defines context as the network availability in space and time, while van Bunningen et al. (112) define context as a situation in which user's accesses a database. van Kranenburg et al. (115) and Khedr and Karmouch (60) further broaden the definition of context including in it information concerning a user, his psychical locations change in time, as well as change in user's physical and social networks. da Costa et al. (37) define context as a storage, network, power, and memory parameter of user's devices. Dey's definition is at the moment the most referenced one; it is a basis for our further research.

7.2.4 Categorization of Context

Gray and Salber (49) categorize context as sensed (by means of context sensors) or explicit given as a user input, and they demonstrate both context types in their museum-tour application. Goslar et al. (48) categorize context as a hard context, i.e., retrieved using hardware (context) sensors, and soft context, i.e., retrieved using software sensors from the application's database. They demonstrate the use soft and hard context for enterprise applications. Bradley and Dunlop (14) categorize context as meaningful and incidental. Meaningful context is implicitly related to the user's primary high-level goal (e.g., to catch a train), whereas incidental context is concerned with incidental occurrences that are normally unrelated to the user's primary high-level goal (e.g., being caught in a sudden downpour). They demonstrate this categorization in their application for visually impaired persons. A view on context proposed by Prekop and Burnett (89) categorizes context in an external context such as location, proximity to other objects, temperature, time and lighting levels, while internal context includes personal events, communication, emotional and physical state. They demonstrate their categorization in the frame of a workshop-organization application. The term internal context was specifically proposed by them to extend context-awareness into cognitive domains, such as decision making. Wei et al. (121) categorizes context into static and dynamic context, depending on how fast it changes over time, and demonstrate it in an active network platform performing context-aware user handovers. Thomson et al. (108) and Khedr and Karmouch (60) categorize context as simple (e.g., location information), composite (as composed from simple information) or a collection of similar context information. Coppola et al. (35) categorize context as a concrete (primitive) and abstract (inferred from primitive), which can be private (i.e., available to a user) or public (available to everyone). Furthermore, Razzaque et al. (91) categorize context based on "viewpoint" of the application at hand, and this viewpoint can be conceptual, measurement-based, or temporal. A conceptual viewpoint categorizes context in a primary or secondary context (as related to the application's primary high-level goal or secondary goals). For a measurements-based viewpoint, they categorize context as a continuous variable, enumeration, state value, or description. For a temporal viewpoint as static or dynamic context, depending on how fast it changes over time. According to Shishkov (101) context is categorized as physical, measurable by hardware sensors (e.g., location, light, sound, or movement), whereas the logical context such as user's goals, work context, or business processes needs to be inferred from the physical context or specified by the user. Dockhorn Costa et al. (42) categorizes context as either intrinsic or relational with respect to the entity being characterized. Intrinsic context belongs to the essential nature of an entity and does not depend on the relationship with other entities (e.g., location of a building), while relational context depends on the relation between distinct entities (e.g., containment relation between rooms entities in this building).

7.2.5 Properties of Context

There exist multiple approaches defining properties of context, tailored to particular research needs. For example, according to Gray and Salber (49), as context is time-and-space-oriented for a given entity, it is distributed over an area or throughout a (geographical) space or a unit of time. Hence, context has different space coverage and different resolution in that space, and it has its own time frame, frequency of appearance (can be transient or persistent) and repeatability in time (can be static or dynamic). As context provides a measure of certain environmental property in space and time, could also be characterized by the range of errors in the measurement, i.e., spatial accuracy and time span. Similarly, Ebling et al. (44) have indicated possible "inaccuracies and uncertainty" related to context-aware systems, and in their research they apply Quality of Information (QoI) metrics like freshness and confidence to the context data. Broens (17) emphasize the dynamic nature of availability and quality of context (QoC), which makes the development of context-aware applications complex. In general, context can be provided as an explicit user input or can be sensed (i.e., be an implicit input from the user perspective), and in a generic context-aware systems the emphasis is put on the latter functionality (49). In such a case, context characteristics depend on the nature of context sensors and how these sensors get information about this environment. Banavar and Bernstein (5) characterize context sensors as "inaccurate" and van Bunningen et al. (112) present them as "small constrained devices." Both of them indicate sensors distribution in the user environment and possible sensors mobility. Moreover, in most of the cases, sensors are subject to noise or even failure, and as they sample environmental phenomena, their output is unreliable and represents only an approximation of those phenomena. Finally, the process of interpretation of sensed context can be also a subject to ambiguity and approximation. Hence, in most cases, the sensed context is imperfect and variable quality information. They indicate that any context information needs to be associated with QoC metrics. The final explicit definition of properties of context and an exhaustive list of its attributes do not exist in the context-awareness research community. In many cases, researchers, provide QoC metrics definitions, adapted to their research objectives. For example, Buchholz et al. (25) model context properties like precision, probability of correctness, trustworthiness, resolution, and up-to-datedness. Salden et al. (94) model context's availability, fidelity, integrity, regulatory, and price. Razzaque et al. (91) indicate that QoC attributes can be related to the user's subjective measures, like context credibility or timeliness, or to objective measures, like context source information or context collection method. Huebscher and McCann (56) do not define QoC metrics, but indicate its use by service's utility function defined over the QoC attributes like precision or refresh rate. Broens (17) uses multiple notions of QoC while proposing dynamic bindings of the CA system to context sources. Widya et al. (122) define QoC metrics like freshness, availability, service cost, and they propose a min-max-plus algebraic QoC model to model these metrics. Similarly, Sheikh et al. (100) describes QoC in terms of precision, freshness, temporal and spatial resolution, and probability of correctness.

7.2.6 Context-Awareness

There are many definitions of context-awareness proposed by researchers, however the final definition has not been agreed upon. For example, Schilit et al. (98) define context-awareness as location-awareness; Lamming and Flynn (65) define it as an awareness of the user's activities in an office environment; Schmidt et al. (99) propose a more abstract definition of context-awareness, as an awareness of a "situation the user is in." However, according to Dey (40) these definitions, defining context-awareness based on specific objectives of the application at hand, are difficult to abstract from and apply to a broader range of applications. Therefore, he proposes a more generic definition of a context-aware system as "the system (that) uses context to provide relevant information and/or services to the user, where relevancy depends on the user's task" (40). Nevertheless, there are definitions of context-awareness adapted towards the specific research objectives. For example, Henricksen et al. define context-aware applications as the applications that "adapt to changes in the environment and user requirements" (55) for the purpose of their PACE context-aware middleware, while van Bunningen et al. (112) defines context-awareness as an awareness of a "situation under which user's database access happens." Similarly, da Costa et al. (37) define context-awareness as system self-reflection on storage, network, power, and memory parameters. Similar to his definition of context, Dey's definition of context-awareness is the most referenced one at the moment.

7.3 Service-Oriented Computing

In the past few years, the IT industry has experienced a paradigm shift from the component-oriented and object-oriented computing to Service Oriented Computing (SOC). In this chapter, we adopt the principles of SOC presented in (80, 81, 57). SOC is a computing paradigm that utilizes services as fundamental building blocks for the development of distributed systems. Services could be described, published, discovered, and dynamically assembled for developing massively distributed, interoperable, and evolvable systems. Because of these characteristics, any piece of code and any software component could possibly be reused and transformed into a service that could be made available in the network. SOC relies on the *Service-Oriented Architecture* (SOA) that provides a support infrastructure to interconnect a set of services, whereas each service accessible through standard interfaces and messaging protocols. The SOA distinguished three roles: *service provider*, *service registry*, and *service user* (client). The interactions between them involve publish, find and bind operations, as shown in the Figure 7.2.

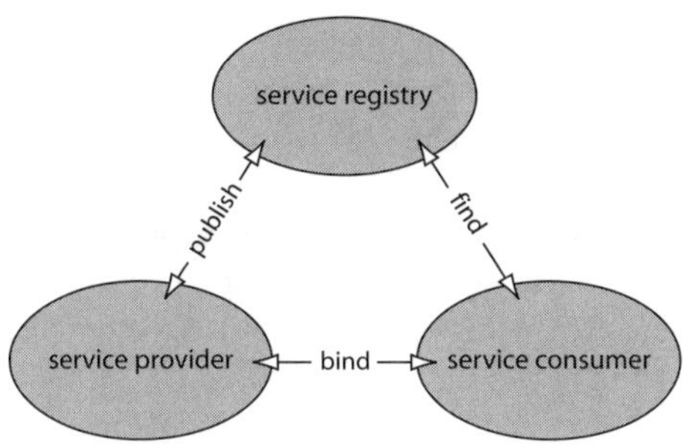

Figure 7.2: Components of the Service-Oriented Architecture.

These roles and operations act upon the service artifacts: service description and service implementation. A service provider is responsible for publishing a description of its services in the service registry. A service user finds a suitable service in the service registry and binds to this service in order to utilize the service. SOC can be implemented with use of technologies like Web Services, Jini, uPnP, or SLP, as we present in the following subsections.

Web Services A basic architecture for the WS model consists of the following five layers (113): the transport layer exchanges request-response messages using protocols like HTTP(S). The packaging layer uses XML-based SOAP protocol to package the messages to be exchanged. The information layer carries the SOAP messages and provides functionality to encode and decode these messages. The services layer uses Web Services Description Language (WSDL) to describe a SOAP/XML Web service. The discovery layer publishes the information about Web services and provides a mechanism to discover the available Web services through the Universal Description, Discovery, and Integration (UDDI) specification. The Web service model permits different distributed services to run on a variety of software platforms and architectures, and moreover allows them to be written in different programming languages. For example, OSGi programming framework supports creation of Web Services using Java. The Microsoft .NET framework provides support for the creation of the Web services using C++ and C# languages. This feature of the Web services provides interoperability in a heterogeneous environment.

Jini The Jini system architecture consists of three categories: infrastructure, programming model, and services (119). The Jini infrastructure is built on top of the Java Remote Method Invocation (RMI) system and provides a set of components that enables building a federated Jini system, while the services are the entities within the federation. The basic infrastructure consists of the discovery protocol and the lookup service. The discovery protocol allows an entity wishing to join the Jini federation to find a lookup service. The lookup service acts as a service registry where the service provider publishes the services and the service user could find the service. The service provider discovers the Jini lookup service and registers a service proxy with it. The service proxy implements all the interfaces of a service and contains the logic to communicate with the service. It is also downloadable from the network. The Jini client discovers a lookup service and requests a desired service either by the service interface or description. If this type of service is registered with the lookup service, the corresponding service proxy will be returned to the client. To utilize the service, the client instantiates a service proxy and invokes the implemented interface methods. The programming model includes models for leasing, event notification, and transactions. The Jini federation refers to the informal group of services and their clients that use the Jini-defined interaction patterns.

Universal Plug and Play UPnP is specification for service presentation and discovery published by UPnP forum. The UPnP architecture allows peer-to-peer networking of PCs, networked appliances, and wireless devices. The UPnP architecture supports zero configuration networking. An UPnP compatible device from any vendor can dynamically join a network, obtain an IP address, announce its name, convey its capabilities upon request, and learn about the presence and capabilities of other devices. The UPnP architecture (72) specifies six phases of interaction. In the addressing phase a device obtains an IP address. Discovery phase let's the devices become aware of each others existence. The description refers to learning details about devices and the offered services. The control phase lets invoking the offered services. The eventing phase is used to notify the changes in the device states. The presentation phase could be used to know the status of the devices and control them using Web pages. The UPnP architecture heavily leverages the Internet components including IP, TCP, UDP, HTTP, SOAP, and XML. Similar to the Web services, any operating system and programming language could be used to build UPnP products.

Service Location Protocol SLP is a service discovery protocol that allows computers and other devices to find services in a local area network without prior configuration. SLP is used by devices to announce services on a local network. SLP has three different roles for devices. A device can also have two or all three roles at the same time. User Agents (UA) are devices that perform service discovery on behalf of client software. Directory Agents (DA) are devices that aggregate services information. They are used in larger networks to reduce the amount of traffic and allow SLP to scale. The existence of DAs in a network is optional, but if a DA is present, UAs and SAs are required to use it instead of communicating directly. Services are advertised using a Service URL, which contains the service's location: the IP address, port number and, depending on the service type, path. Client applications that obtain this URL have all the information they need to connect to the advertised service. The actual protocol the client uses to communicate with the service is independent of SLP (53). SLP also uses Internet technologies such as TCP, UDP, and DHCP to support various operations.

7.4 Layered Model of Context-Aware Systems

To facilitate a structured development of context-aware systems, models for architectures of the context-aware systems have been proposed in the literature. For example Dey uses a layered approach in designing his context toolkit (38), Henricksen et al. in PACE (55), similarly Wegdam in AWARENESS (120), Coppola et al. in MoBe (35), Gu et al. in SOCAM (50), and Khedr and Karmouch in the ACAI middleware (60). A major similarity in these diverse proposals is their approach following a layered model, as shown in the Figure 7.3 based on after Ailisto at al. (2) and Baldauf et al. (4). This model divides a context-aware system into five layers: physical, context data, semantic, inference, and an application layer. In some approaches the middle layers are presented together, resulting in three, instead of five layers architecture. This is the case for a MoBe, SOCAM, ACAI, and AWARENESS middleware. Nevertheless, all the approaches represent at large the same overall functionality of the system and its layers, which we summarize in Figure 7.3.

At the physical layer, there are context sensors (or other context objects), which may provide usable context information by producing output in a raw format, like, user's vital signs readings in the analogue format derived out of sensors placed on his body, or WLAN access points RF signal strength. These context sensors can be physical, virtual, or logical (67). Physical sensors capture environmental data, e.g., light, visual context, audio, motion/acceleration, location, touch, temperature, and physical attributes. Virtual sensors get context data from software applications or services, e.g., electronic calendar or e-mails, or by tracing user mouse movements. Finally, logical sensors combine physical, logical, and other information obtained for example from the databases to solve higher tasks. Concerning the functionality of this layer, among the others, Razzaque et al. (91) indicate that the context-aware system architecture needs to express sensed context in terms of context quality attributes and properties of the sensors that supply context data. At the context data layer, the raw context data is retrieved and pre-processed for storage. This layer contains drivers for physical sensors and APIs for virtual and logical sensors (67). Continuing our example, user's ECG signs would be derived out of physical layer data or user's location coordinates are computed

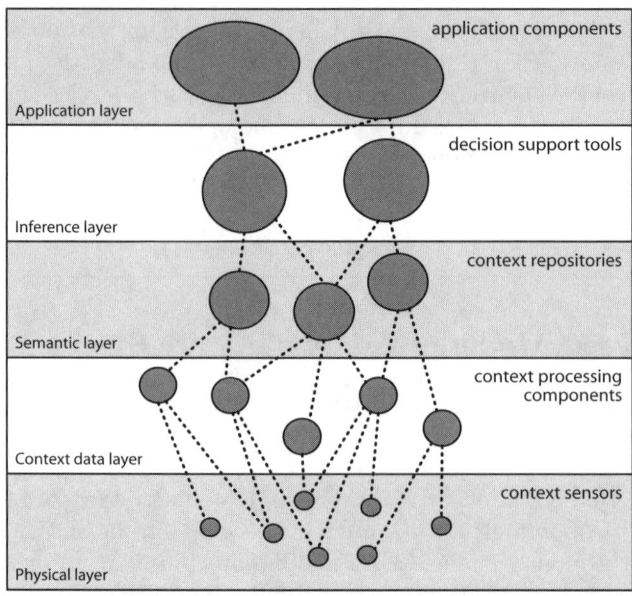

Figure 7.3: Layered model of context-aware systems.

based using the triangulation technique. The semantic layer contains objects that semantically reason upon the context data and transform it into the form meaningful for further inferring context, and store it for later retrieval, for example by deriving user's heart rate from the ECG signs. The semantic layer objects could also indicate that the coordinates computed from RF signals hint that the mobile terminal is in the user's home environment. The inference layer uses correlated real-time information transformed at the semantic level together with the inference rules (possibly dynamically learned) to make an educated guess of what the user (either man or machine) is doing and what kind of services he/she might want. At this layer, the inference object (agent) infers that the user, a young epileptic patient is at this moment at home. Moreover, as the user is being treated under a continuum healthcare program (e.g., Epilepsy Safety System tele-monitors her health condition (116, 16)), this layer, based on the user's heart rate pattern (categorized it as normal or abnormal) will predict or infer if the user has/will likely to have a seizure. This layer also needs to monitor and continuously infer user's voluntary care givers' location at any given time. Finally, the inference layer offers (in a synchronous or asynchronous way) inferred context data to the application layer. At this layer, the inferred context data is used to react on changes in the context in order to provide specific service or other usable result for the service end-user. In our example, in case of the predicted seizure the system will notify the user to stop current activities and to lie down safely. In case the seizure is inferred, the system will notify a healthcare practitioner in healthcare center and user's nearest available voluntary care giver to bring help. The layered architecture of context-aware systems facilitates separation of concerns between context acquisition and pre-processing (as background services) and its usage at the application layer to provide personalized services to end-user. At any higher layer, this architecture facilitates the combination of input, i.e., context information from one or more objects on any of the lower layers. This architecture supports an iterative context-aware systems design; emphasizes modularity and composability of context-handling components and apt allocation of responsibilities. Hence, it is adapted by many researchers to design their context-aware systems.

7.5 Domain Model for Context-Awareness

7.5.1 Introduction

In this section we propose a domain model applicable to the context-awareness research area. The context and its complex nature, as presented in section 2, are in the center of attention of all the researchers. As Figure 7.4 shows, while analyzing research on context-awareness, we identify four research domains and thirteen corresponding subdomains. The domains relate to the research on context and its characteristics, context management, interactions with context-aware service user and cross-cutting aspects of a context-aware systems. Along this section we present the current state of the art in these domains and subdomains is highlighted with numerous examples. The SOC (and particularly WS) paradigm is emphasizes, whenever applicable.

7.5.2 Context Modelling

Within the research on context and its characteristics, we distinguish research on theoretical foundations for context-awareness and context modelling, which may serve further as a basis in the development of context-aware systems.

Theoretical Foundations The foundations for context-aware systems contain research on, for example, context-aware action systems proposed by Yan and Sere (124). These systems extend the traditional action systems with the context information being incorporated into the action's guard conditions. The major concepts of context-aware action systems modelled by Yan and Sere are: a parallel composition concept, a prioritizing composition concept and a concept of nesting of activities performed by a context-aware system. Padovitz et al. (79) present multi-attribute utility theory for a sensor fusion in context-aware environments. The proposed Context Spaces Theoret-

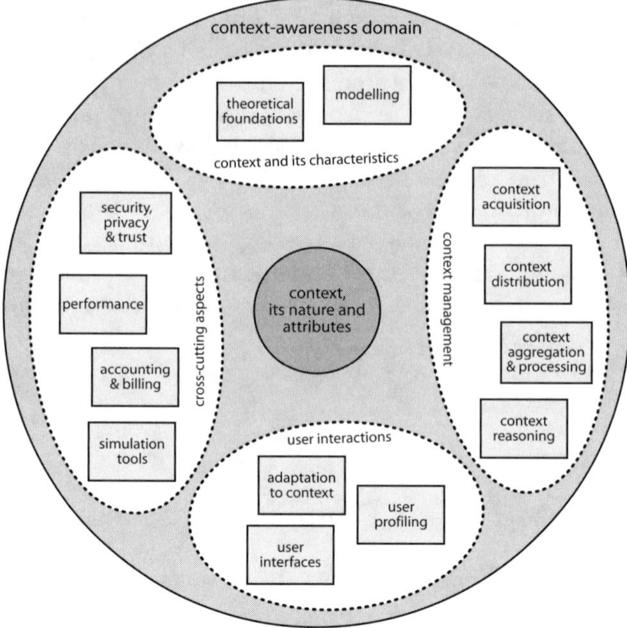

Figure 7.4: Context-awareness domain model.

ical Model (CSTM) aims to model the nature of context and context-aware situation, hence it defines a context attribute, context state (as a vector of attributes), situation space (real-time situation described by context attributes) and relevance and contribution functions, operating on context attributes and expressing the association of context with a given situation. The data fusion technique proposed by Padovitz et al. (79), takes the information represented by the CSTM, and in order to facilitate a decision regarding the occurrence of a given situation, it computes a degree of support for the occurrence of a given situation and compares it with support for other alternative situations.

Context Modelling The research on context modelling aims in an abstract representation of the context and context-awareness related concepts. For example, context and its attributes are modelled in a time and space dimension by Gray and Salber (49). Chan et al. (31) model (context-based) events and event's notifications in their context-aware middleware. Harper (54) proposes a context model for a web browsing; the model includes a Web-specific hypertext containing user-specific context information presented within the page. More generically, Khedr and Karmouch (60) and van Bunningen et al. (112) model context using simple or composite context-information data structures associated with information attributes. Dockhorn Costa et al. (41) model intrinsic and relational context of an entity. Semantics plays an important role in the interpretation of context and its changes as driving forces behind context-aware system's actions. Therefore, modelling of semantics is another challenging aspect of context modelling. Capturing the semantics of context is primarily done by using ontologies, i.e., particular models representing the nature and relationships of context. Bouquet et al. (13) propose a model for ontologies in context-aware systems. They express the model in the Web Ontology Language (OWL, (126)). Maamar and Narendra (68) takes the use of OWL further by proposing ontology-based context resolving techniques for composing of Web services. Belhanafi et al. (8) and Gu et al. (50) use the OWL to demonstrate their context model, serving as a basis to share, understand, and reason upon the context information. Khedr and Karmouch (60) and van Kranenburg et al. (115) propose representation of context with the use of context foundational, core, and application ontologies. Moreover, Guiling et al. (52) propose a generic ontology-based model for context query, matching, and context-based policies. Concerning semantics and context-awareness, Banavar and Bernstein (5) indicate challenges in the semantic modelling of context, context-aware services, devices, user tasks, and knowledge representation. Strang and Linnhoff-Popien (103) indicate the set of challenges, like distributed composition, partial validation, richness of quality of context information, incompleteness and ambiguity, to be tackled when modelling context. They categorize context models as a key-value, markup scheme-based, graphical (e.g., Unified Modelling Language, (78)), object-oriented, logic- and ontology-based. They indicate the latter models as the most promising for the future. Similarly, Razzaque et al. (91) indicate set-theory-based, directed-graph based and first-order-logic based context models, emphasizing the necessity of modelling user's preferences and profiles in comprehensive data structures exposing the dependency relations between the user's preferences and profiles. As we already indicated in Section 2, the Quality of Context (QoC) plays an important role in context-aware systems. Therefore, the QoC modelling is also tackled by multiple researchers. Buchholz et al. (25) tackle the QoC with conjunction to Quality of Device (QoD) and Quality of Service (QoS), while Salden et al. (94), Razzaque et al. (91), Huebscher and McCann (56), Widya et al. (122) and (100) model the attributes associated with the QoC models (extensive details in Section 2). Besides the research upon the context models, there exists also research upon the models of context-awareness-related concepts. For example, Winograd (123) identifies three different context-management models: widgets for a process-centric view, networked services for a service-centric view, and blackboard-models for a data-centric view. Moreover, Broens et al. and Pawar and Tokmakoff propose a model for context-aware service discovery (18) (82), while Ravedy and Issarny (90) propose a model for the context-aware service description, and Funk et al. (46) propose a model for context-aware service composition.

7.5.3 Context Management

Context management and its frameworks are researched when developing infrastructural support for context-aware systems. There exist multiple context management infrastructures as proposed, for example by Schilit and Theimer (97), Dey et al. (40), Chen and Kotz (32), Capra et al. (27, 28), Moran and Dourish (73), Ailisto et al. (2), Banavar and Bernstein (5), Pokraev et al. (86), Huebscher and McCann (56), Khedr and Karmouch (60), Coppola et al. (35), Gu et al. (50), Kiani et al. (61), Belhanafi et al. (8), Henricksen et al. (55), van Kranenburg et al. (114), or Baldauf et al. (4). These context management infrastructures are organized along the layered context-aware system architecture, as presented in Section 2. Therefore, these frameworks provide all (or a combination of) context management functionalities like context acquisition, distribution, processing, and reasoning.

Context Acquisition The context acquisition functionality serves as a basis for any context-aware system. In the research on context acquisition, context sensors and networks of these, play an important role. Generally, all research concerning the development of infrastructures for context-aware systems results in proposing some application-specific context-sensors. For example, Schmidt et al. (99) indicate context sensors as vision, audio, motion, location, and specialized bio-sensors. Lieberman and Selker (67) propose more generic categories for sensors being physical, virtual, or logical. Yang et al. (125), for the purpose of their context-aware e-learning environment, propose a context acquisition mechanisms serving as a basis of (e-learning) content discovery in the user's specific e-learning environment.

Context Distribution As acquired context can be distributed, Cohen and Raz (34) study the requirements for a context distribution system encompassing context creation, maintenance, and management. They conclude that a modular approach, in which context information is provided by (distributed) context brokers, is the most effective way of dealing with a context distribution. Sygkouna et al. (105) propose, and theoretically study decentralized mechanisms for searching of distributed context based on context usage patterns. Pawar et al. (84) argue that the context distribution in pervasive environments should address concerns such as the distributed and intermittent nature of context sources, mapping offered and desired QoC, representation of context and context source capabilities to support reasoning, and users' privacy enforcement. The Context Distribution Framework (CDF) proposed by them provides a service-oriented infrastructure for the context-aware applications hosted on a mobile device. Broens et al. (19) proposes a context binding transparency to simplify the development of context-aware applications. This transparency and corresponding infrastructure mechanisms hide the complexity of creating and maintaining context bindings between a context-aware application and dynamically available context sources with dynamic quality. Concerning the context distribution, on one hand, there exist research on mechanisms for context and context-aware services description and registration by context providers and, on the other hand, there exist research on (distributed) context and context-aware services discovery, selection, negotiation, and remote acquisition by the users. Strang (102) propose a centralized Advanced Mobile Service Registry for services registration and discovery. Similarly, Doulkeridis et al. (43) propose Context-Aware Service Directory in Mobile Environments for services registration and discovery. Furthermore, Carter and Vukovic (29) focus on location-based discovery of Web services, while Tian et al. (109) propose a QoS-aware specification, registration, discovery, selection, and invocation of mobile Web services. Concerning more generic approaches, Broens et al. (18) propose context-aware services discovery protocol while using ontologies to capture semantic of the user's query. (Pawar and Tokmakoff (82) propose the concept of context-aware persistent service discovery, whereas in the event of a context change or the appearance of new services, if a more suitable service is found, then the client is notified, and given a possibility to switch to it. Furthermore, Raverdy and Issarny (90) propose context-aware service discovery platform that encompasses diverse distributed repository-based (e.g., Ariadne, OSGi) and broadcast-based (e.g., SSDP, SLP, UPnP, Jini, Web services) service discovery protocols.

Context Aggregation and Processing Context aggregation and processing are referred to as an aggregation and processing of context information pieces relevant for the provided context-aware service. As we emphasize in Section 2, context aggregation and its further processing plays a critical role to success of a context-aware service delivery. Guanling and Kotz (51) propose a theoretical, graph-based abstraction for collecting, aggregating, and disseminating context information. Information itself is modelled as events produced by context sources, flowing through a directed acyclic graph of event-processing operators, and delivered to subscribing applications. Similarly, Buchholz et al. (24) propose an autonomous system (called CoCo) for acquisition and processing context information at the system runtime; the CoCo is based on graph-oriented language and incorporates the semantics of the acquired and processed information. In many cases, the research on context processing leads to research on QoC metrics and models, conducted for example by Buchholz et al. (25), Salden et al. (94), Razzaque et al. (91), Huebscher and McCann (56) or Widya et al. (122) (research on QoC has been extensively presented in section 2).

Context Reasoning Context synthesis and reasoning is tackled by researchers working on decision support systems for context-aware systems. This research encompasses context inference (in a reactive manner) and context prediction (in a proactive manner) methods. For its purpose, the use of data mining techniques is encouraged by, for example, Chen and Kotz (32) who uses Bayesian networks, Peddemors et al. (85) who uses neural-networks and Mähönen et al. (70) who uses decision trees. Analogously, Sun et al. (104) applies the rule-based reasoning mechanisms upon the context information, while Agostini et al. (1) extends it with the ontological reasoning. Furthermore, Dockhorn Costa et al. (42) proposes use of the Event-Condition-Action rules for context reasoning, while Kocaballi and Kocyigit (62) propose a granular best-match algorithm to reason upon the context. Widya et al. (122) propose min-max-plus algebraic model for context (and its QoC) reasoning purposes. Chalmers (30) emphasizes that context reasoning, and especially the appropriateness of actions performed by a context-aware system on behalf of a user, needs to be learned by the system along its usage. To facilitate the system's learning path, he emphasizes the need for occasional feedback from the user upon the performed actions. Furthermore, Feng et al. (45) indicate that reasoning should always be done while taking not only the current user's and system's context, but also the historical context. Van Kranenburg et al. (115) research the reliability of "grounded" context reasoning; reasoning that is properly justified and strongly reflects the user needs and preferences. Also there exist more application-oriented context reasoning techniques. For example, Bardram (6) research upon context reasoning techniques used in a healthcare environment, and provides a generic mechanism for reasoning on context information to determine user's, i.e., a healthcare professional activities. Similarly, Choi and Shin (33) indicate use of neural-networks-based methods for context reasoning module in home services environments.

Context Management For the context management category as a whole, Feng et al. (45) address the impact of the user-centricity requirement on context-information management strategies. It implies that the context of data queries, queries' constraints, queries' result measurements, and result delivery should be user-centered. Furthermore, van Bunningen at al. (112) indicate several implications of context-awareness on context-related data management. These encompass user-oriented implications and system-oriented implications. The user-oriented implications concern challenges upon the system's personalization, security, proactiveness, and traceability of context data. The system-oriented implications encompass challenges such as dynamic unreliable storage, computation and communication capabilities, the need for context metadata, for context learning and reasoning mechanisms, as well as dealing with different representations of context. Although the ontology-based context management components proposed by Sbodio and Thronicke (96) and a generic Context Management Framework proposed by Sanchez et al. (95) or van Kranenburg et al. (114) overcome a major part of the above-mentioned system-oriented challenges, there is still need for an extensive research upon overcoming the user-oriented implications.

7.5.4 User Interactions

In user-interaction research, we distinguish research on user profiling, user interfaces, and services' personalization based on adaptation to context.

User Profiling Many researchers indicate the need for advances in user-profiling research, particularly in establishing a research community agreement if the user's profile is part of the context or not. For example, Bellavista et al. (9) for the purpose of their research indicate user preferences as part of the context information. However, Niemegeers and Heemstra de Groot (76) indicate the need for generic user profiles in the user's personal area networks, before advanced personalized context-aware applications can be developed. Similarly, Takahashi et al. (106) propose a generic agent-based methodology upon deriving user's requirements for context-aware applications in home environments. Moreover, Bottazzi et al. (12) propose user's requirements analysis, where the user is a member of an ad hoc group using context-aware services.

User Interfaces The research on user interfaces focuses mainly on unobtrusive personalized intuitive interfaces, following the user and his needs, as proposed for example by Wagner et al. (118). Generally, researchers indicate that usability of user interfaces is a critical factor for the system's acceptance. Chalmers (30) researches sociological and philosophical aspects of disappearing user interfaces in case of advances context-aware tools and services. He emphasizes the importance of the tool's learning phase upon its disappearance, and an occasional appearance for a feedback from the user. Similarly, Toninelli et al. (111) implements a service dedicated for user's interface, enabling the user to simply and intuitively specify service queries at a high level of abstraction.

Adaptation to Context It encompasses research on enforcement of context-aware behavior by a context-aware system to fulfil the system's user demands. Lieberman and Selker (67) indicate a challenge of system's adaptation to context. Moreover, they indicate the challenge of getting context-sensitivity levels right, before the user gets discouraged to use the system exhibiting unpredictable and inappropriate behavior. Similarly, Banavar and Bernstein (5) indicate the necessity of the validation of user experience while using a context-aware application The validation results could be used further for service reconfiguration along the user's needs. Furthermore, Kwan et al. (63) implement the application's code adaptation for context-aware services and define the Adaptation Quality Index in determining the service adaptation's satisfaction level as given by the service user.

7.5.5 Cross-Cutting Aspects

Cross-cutting aspects of infrastructures supporting delivery of context-aware services, are user research. Particularly, there is much research on an infrastructures' genericness and on the requirements analysis for such infrastructures. Ebling et al. (44) indicate the privacy, scalability, extensibility, synchrony, and quality of context as a major indispensable design issues for a system delivering context-aware services. Concerning the system requirements analysis, Capra et al. (27) argues that the existing infrastructures for mobile devices are not suitable for delivering context-aware services due to their processing, storage, and communication shortcomings. As a solution, they propose use of a self-reflection mechanism, implying that the applications running on a mobile device needs to be aware of their execution environment.

The information about the device and the application's execution environment they model is an application's metadata. Getting the concept of an application's awareness even broader, Banavar and Bernstein (5) indicate the need for semantic modelling of services, devices, user tasks, knowledge representation, infrastructure and its re-configuration, in order to get generic, but still efficient context-aware services infrastructures. In the same line of thinking, Roman et al. (92) give an example of runtime specification of requirements for such an infrastructure with use of software agents monitoring the context of services and service users. Moreover, Broens et al. (21) focuses more precisely on requirements for a generic application framework supporting the development of context-aware applications in mobile environments. Regardless of the research tackling context-aware systems and context management as a whole (Section 3.2), we distinguish research upon at least four emerging system's cross-cutting aspects, like system's security, privacy and trust, performance management, accounting and billing and diverse system's simulation tools.

Security, Privacy and Trust The research on these aspects in context-aware systems encompasses, among the other aspects, research on authentication and authorization, privacy, trust, and policies enforcement in these systems. Many researchers, like Ebling et al. (44), Banavar and Bernstein (5) and then van Bunningen et al. (112) indicate privacy and security among the most critical challenges in development of context-aware services. As the possible solutions for these, Covington et al. (36) propose context-based access control and policy enforcement mechanisms, while Nishiki and Tanaka (77) propose a distributed agents-based system for policy enforcements, network federation, and dynamic access control to context-aware services. Hulsebosch et al. (58) developed context-aware user authentication and authorization methods controlling access to context-aware services. The critical philosophical opinion of Brey (15) on freedom and privacy in context-awareness, or comprehensive study of Tatly et al. (107) upon the security challenges in mobile business (i.e., robustness, easiness-to-use of security architecture ensuring users privacy, anonymity, confidential communication, and secure payment schemes), further indicate the importance of the security-related objectives to be researched in the frame of the research on context-awareness. Promising solutions may be provided by Neisse et al. (75) researching upon the context-aware distributed trust management architecture linking trust with security and privacy issues, or by the Trust-based Security Infrastructure developed by Le Xuan et al. (66), and targeting similar results as Neisse.

Performance The research on performance of context-aware system focuses mainly on two aspects considerably influencing this performance. Namely, it focuses on the performance of underlying heterogeneous networking infrastructures supporting the execution and delivery of context-aware services, and on the service user's mobility support. With respect to these two aspects, the context-awareness domain benefits a lot from earlier, noncontext-awareness related work. For example, already Imielinski and Badrinath (59) indicate disconnection management as a very important requirement for any mobile service. Similarly La Porta et al. (64) indicate mobility management and wireless communications as major challenges for mobile service performance, and they give general guidelines addressing these challenges. These guidelines include, among the others, an asymmetric design of application protocols and use of network-based proxies, which perform complex functions of pre-fetching or caching of a critical data on behalf of mobile users. Capra et al. (28) propose a self-reflection mechanism as a support for context-aware services, i.e., service's awareness of device characteristics and service's execution environment, based on which the service adaptation can be performed. For the efficiency of a context distribution to mobile devices, Pawar et al. (84) propose transparent off-loading of resource intensive context manipulation from the mobile device to the infrastructure in the fixed network. There exists research on performance evaluation, and particularly on performance improvements for services that are mobile, but not necessarily context-aware. For example, Tian et al. (110) considers mobile Web services and indicates messages compression methods as a way to improve the service performance, while Pratistha et al. (87) proposed an architecture that scales-down overhead associated with mobile Web services improving their performance. Tian et al. (109) propose QoS-aware mobile Web services specification, discovery, selection and invocation, as well as adaptive compression of messages in order to improve performance of services. For better performance of mobile context-aware services, Strang (102) proposes that all services need to be registered in a service registry, which is to be situated in a fixed network infrastructure, and therefore accessible in spite of mobile service disconnections and limited connectivity. Moreover, for better overall performance of services, Wei et al. (121) proposes a platform for (context-aware) active networks that can facilitate mobile service delivery by runtime adaptation of network parameters. Moreover, Chan et al. (31) proposes reconfiguration and migration of context-aware mobile services along the service chain in order to achieve better service performance. Similarly, Pratistha and Zaslavsky (88) proposes a dynamic migration of (context-aware) Web services to different hosts in order to improve the performance of the delivered services. Furthermore, Doulkeridis et al. (43) exhibits performance increase of mobile services that use context-aware service discovery mechanisms, while showing low overhead-costs of that mechanism. Mei et al. (69) address the problem of minimizing the context processing delay in a context-aware application by optimally assigning the heterogeneous computation resources. The solution proposed by them transforms the assignment problem into a path-searching problem and is inspired by the previous work done on the similar problems in the area of parallel computing. Capra et al. (28) indicate that the state of resources available for context-aware mobile service, like power, needs to be taken into overall service utility function to maximize this resource utility while maximizing performance of a context-aware service. More generically, Ryan and Rossi (93) propose a set of metrics to capture software and resource utilization and performance attributes (e.g., power or memory usage) impacting the performance of context-aware mobile services. Moreover, van Bunningen et al. (112) focuses on the context-information management system's performance for a delivery of reliable context-aware services. Similarly, Bellavista et al. (9) and then Wac et al. (117) indicate that the performance of the service and underlying wireless networks is important context information that should be monitored along the service delivery. To benefit from the availability of the multiple IP networks available to a mobile device (multi-homing), Pawar et al. (83) propose a context-aware mobile middleware that exploits multi-homing by selecting the most performing wireless network among the available ones. Furthermore, Broens et al. (22) emphasize influence of performance of context sources binded at service runtime on the overall performance of a context-aware service. Scalability is also an important aspect of context-aware system's performance. Buchholz and Linnhoff-Popien (23) classify the context-aware systems according to their scalability needs, and identify context and service performance challenges with regard to their numerical, geographical, and administrative scalability.

Accounting and Billing The research on accounting and billing is not covered well in the context-awareness domain. For example, Mihovska and Pereira (71) and Bhushan et al. (10) provide requirements for accounting, billing, and charging of next-generation context-aware services, while Wei et al. (121) just indicate the need for a context-aware user charging model, without further providing solutions for these.

Simulation Tools There is ongoing research on this aspect facilitating development of operational context-aware systems. Examples of such tools are tools for simulation of context, simulation of sources of context, or tools for verification and validation of context-aware applications. Bylund and Espinoza (26) distinguish two types of the context simulation tools: a simulation suite that simulates context values and semi-realistic simulation environments. They propose a Quake game-based 3D simulator. Similarly, Barton and Vijayaraghavan (7) propose a 3D simulator for simulating the user's environment, from which context is generated and acquired at the system runtime. Morla and Davies (74) propose network-parameters-related context-changes simulator. Broens and van Halteren (20) research a Java-based framework for simulating heterogeneous sources of context.

7.5.6 Interconnection Between the Models

The layered context-awareness system model has 5 layers (Section 2), while in the context awareness domain model we have identified thirteen research subdomains grouped in four domains (Section 3). As presented in Figure 7.5), we identify that the research concerning context modelling (and particularly theoretical foundations and modelling) is spanned across the top four or the five-layers, while the system's cross-cutting aspects (like security or performance), are spanned across all the five-layers of the system model. Furthermore, research on context management is distributed over the different layers inside the system model. Namely, research on context acquisition spans over the physical layer and context data layer (bottom two) layers. Research on context distribution spans over the context data layer and the semantic layer. Research on the context aggregation and processing spans over the middle three layers: the context data layer, the semantic layer, and the inference layer. Research on context reasoning spans over the inference layer, and the application layer (upper two layers). Research related to interactions with a user of a context-aware system spans over the upper layers of the model. Namely, research on user profiling spans over the top three layers: the semantic layer, the inference layer and the application layer. Research on services' personalization based on adaptation to context spans over the upper two layers: the inference layer and the application layer. Finally, research on user interfaces spans over the top, application layer.

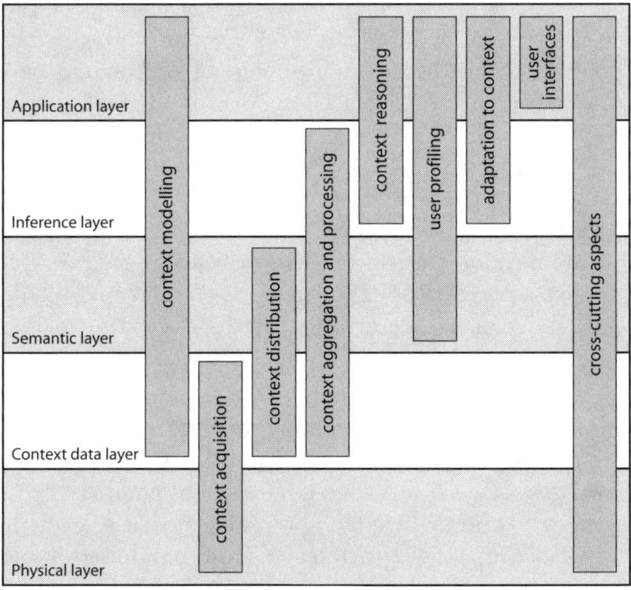

Figure 7.5: Interconnections between layered system model and context-awareness domain model.

The layered context-aware system model is suitable as a basis for a development of context-aware systems, while the context-awareness domain model is revealing research domains, their research objectives as well as possible applications of the conducted research. The presented interconnections between these models may serve as a starting point for newcomers to get a systematic overview of the context-awareness research domain, for researchers working in one domain to get acquainted with the developments in other subdomains, and for practitioners to identify shortcomings of their context-aware system and to identify possible ways of overcoming them using various interdomain associations depicted herewith.

7.6 Application of Domain Model in the Amigo System

In this section we argue that to take advantage of the power of SOC, the researchers specialized in different domains should develop specialized services using their domain expertise and make these services available to the other researchers to help building modular, scalable, and interoperable CA SOC systems. This is shown in an example by the case study of the EU Amigo project, which has developed SOC-based open source software for the context-aware intelligent networked home environment. The Amigo programming and deployment framework has a support for both .NET/OSGi programming as a basis for the application or a component development using Web Services. The SOC adds significant value to the Amigo project software, mainly because of the loose coupling of the software components, discovery and composition mechanisms, services interoperability, and a possible choice of familiar SOC technology by the developers. An overview of SOC-based services organized as per the context-awareness domain model in the Amigo project is shown in the Figure 7.6. A concise description of these services follows. For more details, we refer to a number of documents and deliverables of the Amigo project as listed on the project Web-site (3).

Context and its characteristics A number of context related concepts in the Amigo project are based on the Amigo-S ontology. A networked home theory in Amigo suggests that the Amigo environment comprises rooms, devices, services, persons, and the knowledge about these entities and the associated context is modelled using the principles of Ontology based knowledge representation.

SOC based middleware The Amigo SOC based middleware consists of two principle components. The interoperable service discovery and interaction middleware provides application transparent mechanisms for the discovery and interaction middleware protocols that execute on the network. It basically translates the incoming/outgoing messages of one protocol into messages of another, target protocol. The supported protocols are UPnP, SLP, and WS-Discovery that use SOAP and RMI as the underlying mechanisms for the exchange of messages. The Semantic Service Description, Discovery, Composition, Adaptation, and Execution (SDCAE) middleware uses the Amigo-S vocabulary to enable integration of heterogeneous services into the higher level services based on their SOC-based abstract specification.

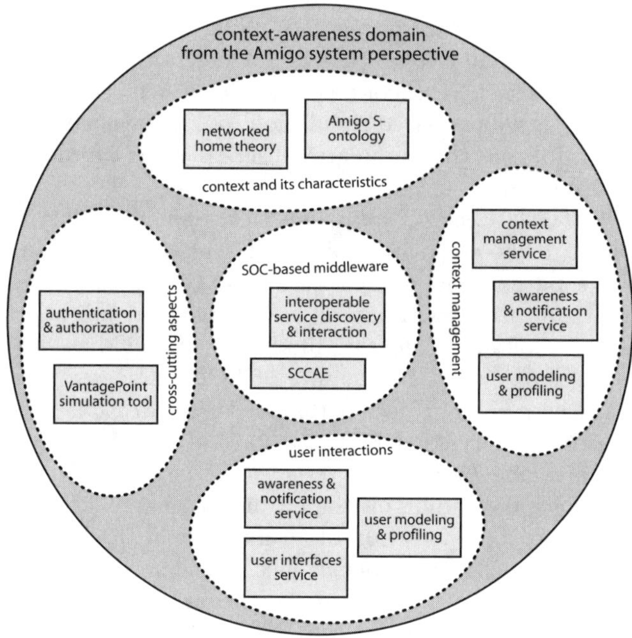

Figure 7.6: Enabling SOC in the Amigo System via the CA domain model

Context Management In this domain, the Amigo context management service acquires information coming from various sources, such as physical sensors, user activities, and applications to subsequently combine or abstract these pieces of information into the context information to be provided to context-aware services and applications. The Awareness and Notification Service (ANS) keeps track of changes in various types of context, for example activities and presence of people and further makes available this context to the applications by notifying them about the context changes. The User Modelling and Profiling Service (UMPS) tailors information presentation to the context and provides reasoning support to predict user's future behavior.

User interactions The user interface service provides several interface related services supporting interaction via specific modalities (e.g., speech, GUI, gesture). From the user interactions perspective, the ANS provides notifications with appropriate intensity, based on the user's preferences and current context. The UMPS constructs, maintains, and exploits the user models and profiles representing users preferences for the adaptation of the user interface features and information presentation features in a multi-user environment.

Cross-cutting aspects The authentication and authorization service encapsulates the communication and cryptographic primitives that are used for the device/user registration, authentication, and authorization with the centralized security service. The accounting and billing service offers a basic service for managing An IP Detail Record (IPDR) documents. The IPDR provides information about Internet Protocol-based service usage and other documents. The VantagePoint is a Java-based simulation tool that visualizes, queries, and edits Amigo-S ontologies that simulate a user-specified physical environment using easy-to-use GUI.

An Example of SOC-Based Application Development in the Amigo Project

As an analogy to the claim we make in this paper, various services in the EU Amigo project are developed by a number of researchers from a number of participating organizations with expertise in a particular field. Hence, this is a good example to follow the domain model-based SOC-based context-aware applications development. An example of SOC-based services/application development in the Amigo project is that of *Crisis Response Service* (CRS). The Crisis Response Service (CRS) is a part of the Amigo project and is responsible for invoking *Emergency Response Services* (ERS) such as a police patrol or ambulance depending on the nature of emergency situations. Such a situation could be, e.g., detection of an intruder in the home or an elderly patient gets unconscious. On the occurrence of such a situation, CRS searches for the closest available ERS based on the nature of the emergency, location of the emergency, and location of the ERS. The application makes use of the following Amigo *Intelligent User Services* (IUS): *Context Aware Service Discovery* (CASD), *Awareness and Notification Service* (ANS), and *Context Management Service* (CMS). An application developer interested to invoke CRS could do so by means of one of the following two means:

Discover and Use Emergency Response Services Directly from CRS: This approach allows the application developer interested to invoke ERS by directly invoking *Discover and Use Emergency Response Service* function of the CRS. In this case, CRS uses the CASD service to obtain a reference to the requested ERS such as an ambulance, police patrol car, or fire brigade by taking into account the context information of the home (e.g., time, location) and also the context of the ERS (e.g., location, speed).

Subscribe to CRS with ANS: In this case, CRS is made aware of any crisis situation in the home by subscribing to ANS with specific rules to be aware of the emergency condition. The ANS notifies the CRS when the rule condition is satisfied. For example, CRS specifies a rule in ANS to be notified when an intruder is detected. On detecting the intrusion, the anti-Intrusion context source provides certain context data to ANS so that the rule is satisfied. Once the CRA obtains the notification from ANS for intrusion detection, it uses the CASD service to obtain a reference to the appropriate ERS such as a patrol car, taking into account the context information of the Intrusion (e.g., time, location) and also the context of the ERS (e.g., location, speed).

Furthermore, CRS could be used as an application or as a service. In the application mode, the CRS reads information about the ANS rules and required ERS directly from the framework properties, while in the service mode, it provides the interfaces to invoke it as a service.

7.7 Conclusive Remarks

Service-Oriented Computing (SOC) is one of the most popular computer systems development paradigm because it makes use of the services to support the development of rapid, low-cost, interoperable, evolvable, and massively distributed applications. In this paper, we present how the SOC paradigm could be effectively used to develop domain model-based *Context-Aware* (CA) systems. To this aim, we have presented an overview of both, CA computing and SOC areas. Based on an extensive study of CA computing literature, we have proposed a domain model of context-awareness. As shown in this chapter, this model consists of four main domains and various subdomains. These domains (and corresponding subdomains) are: context fundamentals (theoretical foundations, context modelling), context management (context acquisition, context distribution, context aggregation and processing, context reasoning), user interactions (user profiling, user interactions, adaptation to context), and cross-cutting aspects (simulation tools, accounting and billing, performance, security, privacy, and trust). The current state of the art in these domains and subdomains is highlighted with numerous examples. We have also mapped this domain model onto a pre-existing layered model of CA systems development. We argue that to take advantage of the power of SOC, the researchers specialized in these domains should develop specialized services using their domain expertise and make these services available to the other researchers to help building a modular, scalable, and interoperable CA systems. This is shown in example by the case study of the EU Amigo project, which has developed SOC-based open source software for the context-aware intelligent networked home environment. This chapter has various contributions in this area. It provides a domain view of context-awareness to the newcomers in this field and highlights opportunities for SOC practitioners in the CA domain. Moreover, it also provides a larger view of CA to the existing researchers, and helps practitioners by means of the Amigo project case study to apply the SOC concepts and the CA domain model in the development of CA systems.

Acknowledgments

This research has been carried out in frame of the Dutch Freeband AWARENESS project (BSIK grant number-5902390).

References

[1] A. Agostini, C. Bettini, and D. Riboni. Loosely coupling ontological reasoning with an efficient middleware for context-awareness. In *2nd International Conference on Mobile and Ubiquitous Systems: Networking and Services (MobiQuitous05)*. IEEE Press, 2005.

[2] H. Ailisto, P. Alahuhta, V. Haataja, V. Kyllönen, and M. Lindholm. Structuring context aware applications: Five-layer model and example case. In *The Ubicomp Workshop on Concepts and Models for Ubiquitous Computing*, Sweden, 2002.

[3] Amigo. Amigo: Ambient intelligence for the networked home environment, 2008. http://www.amigo-project.org/.

[4] M. Baldauf, S. Dustdar, and F. Rosenberg. A survey on context-aware systems. *International Journal of Ad Hoc and Ubiquitous Computing*, 2006.

[5] G. Banavar and A. Bernstein. Software infrastructure and design challenges for ubiquitous computing applications. *Communications of the ACM, Special Issue: Issues and Challenges in Ubiquitous Computing*, 45(12):92–96, 2002.

[6] J.E. Bardram. Applications of context-aware computing in hospital work: examples and design principles. In *ACM Symposium on Applied Computing*, Nicosia, Cyprus, 2004.

[7] J. Barton and V. Vijayaraghavan. Ubiwise, a ubiquitous wireless infrastructure simulation environment. Technical report, Dept. of Computer Science, 2002.

[8] N. Belhanafi, C. Taconet, and B. Guy. Camido, a context-aware middleware based on ontology meta-model. In *Workshop on Context Awareness for Proactive Systems*, pages 93–103, 2005.

[9] P. Bellavista, A. Corradi, R. Montanari, and R. Stefanelli. Context-aware middleware for resource management in the wireless internet. *IEEE Transactions on Software Engineering*, 29(12), 2003.

[10] B. Bhushan, S. Steglich, B. Mrohs, and C. Rack. Life cycle management of personalized and context-aware mobile services goals, requirements and interfaces. In *IEEE International Conference on Services Computing (SCC06)*, page 319, 2006.

[11] M. Bichler and K. J. Lin. Service-oriented computing. *IEEE Computer*, 39(3):99–101, 2006.

[12] D. Bottazzi, A. Corradi, and R. Montanari. Enabling context-aware group collaboration in manets. In *Autonomous Decentralized Systems (ISADS05)*, page 310, 2005.

[13] P. Bouquet, F. Giunchiglia, F. van Harmelen, F. Serafini, and H. Stuckenschmidt. C-owl: Contextualizing ontologies. In *Second International Semantic Web Conference (ISWC03)*, volume LNCS 2870, pages 164–179, Las Vegas, Nevada, USA, 2003. Springer Verlag.

[14] N. Bradley and M. Dunlop. Towards a multidisciplinary user-centric design framework for context-aware applications. In *1st UK-UbiNet Workshop*, London, UK, 2003.

[15] P. Brey. Freedom and privacy in ambient intelligence. *Ethics and Information Technology*, 7(3):157–166, 2005.

[16] T. Broens. Supporting the developers of context-aware mobile telemedicine applications. In R. Meersman, Z. Tari, and P. Herrero, editors, *On the Move to Meaningful Internet Systems 2005: OTM Workshops, Ph.D. Student Symposium*, pages 761–770, Agia Napa, Cyprus, 2005. Springer Berlin / Heidelberg, LNCS 3762.

[17] T. Broens. *Dynamic context bindings: Infrastructural support for context-aware applications*. Phd, University of Twente, the Netherlands, 2008.

[18] T. Broens, S. Pokraev, M. van Sinderen, J. Koolwaaij, and P. Dockhorn Costa. Context-aware, ontology-based service discovery. In *Ambient Intelligence*, pages 72–83. Springer Lecture Notes 3295, 2004.

[19] T. Broens, D. Quartel, and M. van Sinderen. Towards a context binding transparency. In *13th Open European Summer School Dependable and Adaptable Networks and Services (EUNICE07)*, Enschede, the Netherlands, 2007.

[20] T. Broens and A. van Halteren. Simucontext: Simulating context sources for context-aware applications. In *International. Conference on Networking and Services (ICNS06)*, Silicon Valley, USA, 2006.

[21] T. Broens, A. van Halteren, M. van Sinderen, and K. Wac. Towards an application framework for context-aware m-health applications. *International Journal of Internet Protocol Technology*, 2(2):109–116, 2007.

[22] T. Broens, M. van Sinderen, A. van Halteren, and D. Quartel. Dynamic context bindings in pervasive middleware. In *IEEE Middleware for Pervasive Computing Workshop (PerWare07)*, 2007.

[23] S. Buchholz and C. Linnhoff-Popien. Towards realizing global scalability in context-aware systems. In *1st International Workshop on Location- and Context-Awareness (LoCA05)*, Oberpfaffenhofen, Germany, 2005.

[24] T. Buchholz, M. Krause, C. Linnhoff-Popien, and M. Schiffers. Coco: dynamic composition of context information. In *1st International Conference on Mobile and Ubiquitous Systems: Networking and Services (MOBIQUITOUS04)*, page 335. IEEE Press, 2004.

[25] T. Buchholz, A. Kupper, and M. Schiffers. Quality of context: What it is and why we need it. In *10th Workshop of the HP OpenView University Association (HPOVUA03)*, Geneva, Switzerland, 2003.

[26] M. Bylund and F. Espinoza. Testing and demonstrating context-aware services with quake iii arena. *Communications of the ACM*, 45(1):46–48, 2002.

[27] L. Capra, W. Emmerich, and C. Mascolo. Reflective middleware solutions for context-aware applications. In *REFLECTION '01: Proceedings of the Third International Conference on Metalevel Architectures and Separation of Crosscutting Concerns*, pages 126–133, London, UK, 2001. Springer-Verlag.

[28] L. Capra, S. Zachariadis, and C. Mascolo. Q-cad: Qos and context aware discovery framework for adaptive mobile systems. In *IEEE International Conference on Pervasive Services (ICPS05)*, 2005.

[29] A. Carter and M. Vukovic. A framework for ubiquitous web service discovery. In *6th International Conference on Ubiquitous Computing*, Nottingham, England, 2004.

[30] M. Chalmers. A historical view of context. *Computer Supported Cooperative Work*, 13(3-4):223–247, 2004.

[31] A. Chan, S.N. Chuang, J. Cao, and H.V. Leong. An event-driven middleware for mobile context awareness. *The Computer Journal*, 47(3):278–288, 2004.

[32] G. Chen and D. Kotz. A survey of context-aware mobile computing research. Technical report, Dept. of Computer Science, Darthmouth College, 2000.

[33] J. Choi and D. Shin. Research and implementation of the context-aware middleware for controlling home appliances. *IEEE Transactions on Consumer Electronics*, 51(1):301–306, 2005.

[34] R. Cohen and D. Raz. An open and modular approach for a context distribution system. In *IEEE/IFIP Network Operations and Management Symposium (NOMS04)*, volume 1, page 365. IEEE Press, 2004.

[35] P. Coppola, V. Della Mea, L. Di Gaspero, S. Mizzaro, I. Scagnetto, A. Selva, L. Vassena, and P. Zandegiacomo Riziò. Mobe: A framework for context-aware mobile applications. In *Workshop on Context Awareness for Proactive Systems (CAPS 2005)*, Helsinki, Finland, 2005. Helsinki University Press.

[36] M. J. Covington, P. Fogla, Zhan Zhiyuan, and M. Ahamad. A context-aware security architecture for emerging applications. In *18th Annual Computer Security Applications Conference*, page 249, 2002.

[37] C.M. da Costa, M. da Silva Strzykalski, and G. Bemard. A reflective middleware architecture to support adaptive mobile applications. In *ACM Symposium on Applied Computing*, pages 1151–1154, Santa Fe, New Mexico, 2005. ACM Press.

[38] A. Dey. The context toolkit: Aiding the development of context-aware applications. In *Workshop on Software Engineering for Wearable and Pervasive Computing*, Limerick, Ireland, 2000.

[39] A. Dey. *Providing Architectural Support for Context-Aware Applications.* Phd, Georgia Institute of Technology, 2000.

[40] A. Dey, D. Salber, and G. Abowd. A conceptual framework and a toolkit for supporting the rapid prototyping of context-aware applications. *Human Computer Interaction Journal*, 16(2-4):97–166, 2001.

[41] P. Dockhorn Costa, J.P. Andrade Almeida, L. Ferreira Pires, G. Guizzardi, and M. van Sinderen. Towards conceptual foundations for context-aware applications. In *3rd International Workshop on Modeling and Retrieval of Context (MRC06)*, Boston, US, 2006.

[42] P. Dockhorn Costa, L. Ferreira Pires, M. van Sinderen, and T. Broens. Controlling services in a mobile context-aware infrastructure. In *2nd Workshop on Context Awareness for Proactive Systems (CAPS06)*, Kassel, Germany, 2006.

[43] C. Doulkeridis, N. Loutas, and M. Vazirgiannis. A system architecture for context-aware service discovery. In *International Workshop on Context for Web Services (CWS05) at the 5th International and Interdisciplinary Conference on Modeling and Using Context*, Paris, France, 2005.

[44] M. Ebling, G. Hunt, and H. Lei. Issues for context services for pervasive computing. In *Advanced Topic Workshop Middleware for Mobile Computing at the IFIP/ACM Middleware Conference*, Heidelberg, Germany, 2001.

[45] L. Feng, P. Apers, and W. Jonker. Towards context-aware data management for ambient intelligence. In *15th International Conference on Database and Expert Systems Applications*, Zaragoza, Spain, 2004.

[46] C Funk, C Kuhmunch, and C. Niedermeier. A model of pervasive services for service composition. In *On the Move to Meaningful Internet Systems 2005: OTM Workshops*, pages 215–224, 2005.

[47] B. Gloss. System architecture of a mobile message transport system. In *11th Open European Summer School - Networked Applications (EUNICE05)*, pages 38–45, Madrid, Spain, 2005.

[48] K. Goslar, S. Buchholz, A. Schill, and H. Vogler. A multidimensional approach to context-awareness. In *7th World Multiconference on Systemics, Cybernetics and Informatics (SCI03)*, 2003.

[49] P. Gray and D. Salber. Modelling and using sensed context information in the design of interactive applications. In *8th IFIP International Conference on Engineering for Human-Computer Interaction (EHCI01)*, volume LNCS 2254, pages 317–335. Springer-Verlag, 2001.

[50] T. Gu, H.K Pung, and D.Q. Zhang. A service-oriented middleware for building context-aware services. *Journal of Network and Computer Applications*, 28(1):1–18, 2005.

[51] C. Guanling and D. Kotz. Context aggregation and dissemination in ubiquitous computing systems. In *4th IEEE Workshop on Mobile Computing Systems and Applications*, page 105. IEEE Press, 2002.

[52] W. Guiling, J. Jinlei, and S. Meilin. A context model for collaborative environment. In *10th International. Conference on Computer Supported Cooperative Work in Design*, page 1. IEEE Press, 2006.

[53] E. Guttman. Service location protocol: automatic discovery of ip network services. *IEEE Internet Computing*, 3(4):71–80, 1999.

[54] S. Harper. Middleware to expand context and preview in hypertext. In *ACM SIGACCESS Conference on Computers and Accessibility*, pages 63–70, Atlanta, GA, USA, 2004. ACM Press.

[55] K. Henricksen, J. Indulska, T. McFadden, and S. Balasubramaniam. Middleware for distributed context-aware systems. In R. Meersman, Z. Tari, and P. Herrero, editors, *On the Move to Meaningful Internet Systems 2005*, volume LNCS 3760, pages 846–863, Agia Napa, Cyprus, 2005. Springer Berlin / Heidelberg.

[56] M. Huebscher and J. McCann. An adaptive middleware framework for context-aware applications. *Personal and Ubiquitous Computing Archive*, 10(1):12–20, 2005.

[57] M. Huhns and M. P. Singh. Service-oriented computing: Key concepts and principles. *IEEE Internet Computing*, 9(1):75–81, 2005.

[58] R. Hulsebosch, A. Salden, M. Bargh, P. Ebben, and J. Reitsma. Context sensitive access control. In *SACMAT '05: Proceedings of the tenth ACM symposium on Access control models and technologies*, pages 111–119. ACM, 2005.

[59] T. Imielinski and B.R. Badrinath. Data management for mobile computing. *SIGMOD Record*, 22(1):34–39, 1993.

[60] M. Khedr and A. Karmouch. Acai: Agent-based context-aware infrastructure for spontaneous applications. *Journal of Network and Computer Applications*, 28(1):19–44, 2005.

[61] S.L. Kiani, M. Riaz, S. Lee, and Y.K. Lee. Context awareness in large scale ubiquitous environments with a service oriented distributed middleware approach. In *4th International Conference on Computer and Information Science (ICIS05)*, Las Vegas, US, 2005. IEEE Computer Society.

[62] A. B. Kocaballi and A. Kocyigit. Granular best match algorithm for context-aware computing systems. In *International Conference on Pervasive Services (ICPS06)*, Lyon, France, 2006.

[63] V. Kwan, F. Lau, and C. Wang. Functionality adaptation: A contextaware service code adaptation for pervasive computing environments. In *IEEE/WIC International Conference on Web Intelligence*, Halifax, Canada, 2003.

[64] T.F. La Porta, K. Sabnani, and R. Gitlin. Challenges for nomadic computing: Mobility management and wireless communications. *Mobile Networked Applications*, 1(1):3–16, 1996.

[65] M. Lamming and M. Flynn. "forget-me-not" intimate computing in support of human memory. In *International Symposium on Next Generation Human Interface*, pages 150–158, 1994.

[66] H. Le Xuan, P. Tran Van, G. Pho Duc, Z. Yonil, L. Sungyoung, and L. Young-Koo. Security for ubiquitous computing: Problems and proposed solution. In *12th IEEE International Conference on Embedded and Real-Time Computing Systems and Applications*, page 110, 2006.

[67] H. Lieberman and T. Selker. Out of context: Computer systems that adapt to, and learn from, context. *IBM Systems Journal*, 39(34):617–632, 2000.

[68] Z. Maamar and N.C. Narendra. Ontology-based context reconciliation in a web services environment: From owl-s to owl-c. In *Workshop on Web Services and Agent-Based Engineering (WSABE)*, New York City, USA, 2004.

[69] H. Mei, P. Pawar, and I. Widya. Optimal assignment of a tree-structured context reasoning procedure onto a host-satellites system. In *16th International Heterogeneity in Computing Workshop (HCW07)*, Long Beach, CA, USA, 2007.

[70] P. Mähönen, M. Petrova, J. Riihijärvi, and M. Wellens. Cognitive wireless networks: Your network just became a teenager. In *INFOCOM06*, Barcelona, Spain, 2006. IEEE.

[71] A. Mihovska and J. M. Pereira. Location-based vas: Killer applications for the next-generation mobile internet. In *12th IEEE International Symposium on Personal, Indoor and Mobile Radio Communications*, volume 1, pages B–50, 2001.

[72] B.A. Miller, T. Nixon, C. Tai, and M.D. Wood. Home networking with universal plug and play. *IEEE Communications Magazine*, 39(12):104–109, 2001.

[73] T. Moran and P. Dourish. Introduction to this special issue on context-aware computing. *Human-Computer Interaction*, 16(2,4), 2001.

[74] R. Morla and N. Davies. Modeling and simulation of context-aware mobile systems. *IEEE Pervasive Computing*, pages 48–56, 2004.

[75] R. Neisse, M. Wegdam, and M. van Sinderen. A distributed context-aware trust management architecture. In *4th International Conference on Pervasive Computing (PERVASIVE06)*, 2006.

[76] I. Niemegeers and S. Heemstra de Groot. Fednets: Context-aware ad-hoc network federations. *International Journal on Wireless Personal Communications*, 33:319–325, 2005.

[77] K. Nishiki and E. Tanaka. Authentication and access control agent framework for context-aware services. In *Symposium on Applications and the Internet Workshops*, page 200, 2005.

[78] OMG. Unified modeling language, 2007. http://www.uml.org/.

[79] A. Padovitz, S. Loke, A. Zaslavsky, B. Burg, and C. Bartolini. An approach to data fusion for context awareness. In *5th International Conference on Modelling and Using Context (CONTEXT05)*, Paris, France, 2005. Springer-Verlag.

[80] M.P. Papazoglou and D. Georgakopoulos. Service-oriented computing (introduction). *Communications of ACM*, 46(10):24–28, 2003. 944233.

[81] M.P. Papazoglou, P. Traverso, S. Dustdar, and F. Leymann. Service-oriented computing research roadmap. Technical report, European Union Information Society Technologies, 2006.

[82] P. Pawar and A. Tokmakoff. Ontology-based context-aware service discovery for pervasive environments. In *IEEE International Workshop on Services Integration in Pervasive Environments (SIPE06)*, Lyon, France, 2006. IEEE Press.

[83] P. Pawar, B.J. van Beijnum, A. Peddemors, and A. van Halteren. Context-aware middleware support for the nomadic mobile services on multi-homed handheld mobile devices. In *12th IEEE Symposium on Computers and Communications (ISCC07)*, Aveiro, Portugal, 2007. IEEE Press.

[84] P. Pawar, A. van Halteren, and K. Sheikh. Enabling context-aware computing for the nomadic mobile user: A service oriented and quality driven approach. In *IEEE Wireless Communications and Networking Conference (WCNC07)*, Hong Kong, 2007. IEEE Press.

[85] A. Peddemors, H. Eertink, and I. Niemegeers. Communication context for adaptive mobile applications. In *3rd IEEE International Conference on Pervasive Computing and Communications Workshops (PERCOMW05)*, pages 173–177, 2005.

[86] S. Pokraev, J. Koolwaaij, M. van Setten, T. Broens, P. Dockhorn Costa, M. Wibbel, P. Ebben, and P. Strating. Service platform for rapid development and deployment of context-aware, mobile applications. In *International Conference on Webservices (ICWS05)*, Orlando, CA, USA, 2005.

[87] D. Pratistha, N. Nicoloudis, and S. Cuce. A micro-services framework on mobile devices. In *International Conference on Web Services*, Nevada, USA, 2003.

[88] D. Pratistha and A. Zaslavsky. Fluid: supporting a transportable and adaptive web service. In *ACM Symposium on Applied Computing*, Nicosia, Cyprus, 2004.

[89] P. Prekop and M. Burnett. Activities, context and ubiquitous computing. *Computer Communications, Special Issue on Ubiquitous Computing*, 26:1168–1176, 2003.

[90] P.G. Raverdy and V. Issarny. Context-aware service discovery in heterogeneous networks. In *IEEE International Symposium on a World of Wireless, Mobile and Multimedia Networks (WoWMoM05)*, 2005.

[91] M.A. Razzaque, S. Dobson, and P. Nixon. Categorisation and modelling of quality in context information. In R. Sterrit, S. Dobson, and M. Smirnov, editors, *Workshop on AI and Autonomic Communications (IJCAI05)*, 2005.

[92] G. Roman, C. Julien, and A. Murphy. A declarative approach to agent-centered context-aware computing in ad hoc wireless environments. In A. Garcia et al, editor, *Software Engineering for Large-Scale Multi-Agent Systems, Research Issues and Practical Applications*, volume 2603 of *Lecture Notes in Computer Science*, pages 94–109, 2003.

[93] C. Ryan and P. Rossi. Software, performance and resource utilisation metrics for context-aware mobile applications. In *11th IEEE International Symposium on Software Metrics*, page 10 pp., 2005.

[94] A. Salden, B.J. van Beijnum, and I. Widya. Quality of context modelling a prerequisite for viable ad-hoc context-aware service provisioning. Deliverable, whitepaper, Telematica Instituut and CTIT, 2004.

[95] L. Sanchez, J. Lanza, R. Olsen, M. Bauer, and M. Girod-Genet. A generic context management framework for personal networking environments. In *3rd International Conference on Mobile and Ubiquitous Systems: Networking and Services*, page 1. IEEE Press, 2006.

[96] M. L. Sbodio and W. Thronicke. Ontology-based context management components for service oriented architectures on wearable devices. In *3rd IEEE International Conference on Industrial Informatics (INDIN05)*, page 129, 2005.

[97] B. Schilit and N. Theimer. Disseminating active map information to mobile hosts. *IEEE Network*, 8(5):22–32, 1994.

[98] B.N. Schilit, N. Adams, and R. Want. Context-aware computing applications. In *Workshop on Mobile Computing Systems and Applications*, pages 85–90, Santa Cruz, CA, USA, 1994. IEEE Computer Society.

[99] A. Schmidt, M. Beigl, and H. Gellersen. There is more to context than location. *Computers and Graphics*, 23(6):893–901, 1999.

[100] K. Sheikh, M. Wegdam, and M. van Sinderen. Middleware support for quality of context in pervasive context-aware systems. In *IEEE Percom Workshop at 5th Annual IEEE International Conference on Pervasive Computing and Communications*, 2007.

[101] B. Shishkov. Awareness service infrastructure architectural specification of the service infrastructure. Deliverable, whitepaper, University of Twente, 2005.

[102] T. Strang. Towards autonomous context-aware services for smart mobile devices. In *4th International Conference on Mobile Data Management (MDM03)*, volume LNCS 2574, pages 279–293, Melbourne, Australia, 2003. Springer.

[103] T. Strang and C. Linnhoff-Popien. A context modeling survey. In *UbiComp 1st International Workshop on Advanced Context Modelling, Reasoning and Management*, pages 34–41, 2004.

[104] J. Sun, J. Sauvola, and J. Riekki. Application of connectivity information for context interpretation and derivation. In *8th International Conference on Telecommunications (ConTEL05)*, volume 1, pages 303–310, Zagreb, Croatia, 2005.

[105] I. Sygkouna, M. Anagnostou, and E. Sykas. Efficient search mechanisms in a context distribution system. In *20th International. Conference on Advanced Information Networking and Applications (AINA06)*, volume 1, page 6 pp. IEEE Press, 2006.

[106] H. Takahashi, T. Suganuma, and N. Shiratori. Amuse: An agent-based middleware for context-aware ubiquitous services. In *11th International Conference on Parallel and Distributed Systems (ICPADS05)*, Fukuoka Institute of Technology, Japan, 2005.

[107] E. I. Tatly, D. Stegemann, and S. Lucks. Security challenges of location-aware mobile business. In *2nd International Workshop on Mobile Commerce and Services (WMCS05)*, page 84, 2005.

[108] G. Thomson, S. Terzis, and P. Nixon. Towards dynamic context discovery and composition. In *1st UK-UbiNet Workshop*, London, UK, 2003.

[109] M. Tian, A. Gramm, H. Ritter, J. Schiller, and J. Voigt. Qos-aware cross-layer communication for mobile web services with the ws-qos framework. In *34th Annual Conference of the German Informatics Society*, Ulm, 2004.

[110] M. Tian, T. Voigt, T. Naumowicz, H. Ritter, and J. Schiller. Performance considerations for mobile web services. In *Workshop on Applications and Services in Wireless Networks*, Bern, Switzerland, 2003.

[111] A. Toninelli, A. Corradi, and R. Montanari. Semantic discovery for context-aware service provisioning in mobile environments. In *1st International Workshop on Managing Context Information in Mobile and Pervasive Environments*, Agia Napa, Cyprus, 2005.

[112] A. van Bunningen, L. Feng, and P. Apers. Context for ubiquitous data management. In *International Workshop on Ubiquitous Data Management (UDM05)*, Tokyo, 2005.

[113] R. van Engelen. Code generation techniques for developing light-weight xml web services for embedded devices. In *ACM Symposium on Applied Computing*, pages 854–861, New York, NY, USA, 2004. ACM.

[114] H. van Kranenburg, M. S. Bargh, S. Iacob, and A. Peddemors. A context management framework for supporting context-aware distributed applications. *IEEE Communications Magazine*, 44(8):67–78, 2006.

[115] H. van Kranenburg, A. Salden, T. Broens, and J. Koolwaaij. Grounded contextual reasoning enabling innovative mobile services. In *5th Workshop on Applications and Services in Wireless Networks (ASWN05)*, Grenoble, France, 2005.

[116] K. Wac. Towards qos-awareness of context-aware mobile applications and services. In R. Meersman, Z. Tari, and P. Herrero, editors, *On the Move to Meaningful Internet Systems 2005: OTM Workshops, Ph.D. Student Symposium*, volume LNCS 3762, pages 751–760, Agia Napa, Cyprus, 2005. Springer Berlin / Heidelberg.

[117] K. Wac, A. van Halteren, R. Bults, and T. Broens. Context-aware qos provisioning in an m-health service platform. *International Journal of Internet Protocol Technology*, 2(2):102–108, 2007.

[118] M. Wagner, A. Tarlano, R. Hirschfeld, and W. Kellerer. From personal mobility to mobile personality. In *Eurescom Summit 2003*, pages 155–164, Heidelberg, Germany, 2003.

[119] J. Waldo. The jini architecture for network-centric computing. *Communications of ACM*, 42(7):76–82, 1999. 306582.

[120] M. Wegdam. Awareness: A project on context aware mobile networks and services. In *14th Mobile and Wireless Communication Summit*, Dresden, Germany, 2005.

[121] Q. Wei, K. Farkas, P. Mendes, C. Prehofer, B. Plattner, and N. Nafisi. Context-aware handover based on active network technology. In N. Wakamiya, M. Solarski, and J. Sterbenz, editors, *IFIP TC6 5th International Workshop on Active Networks (IWAN03)*, volume LNCS 2982, Kyoto, Japan, 2003. Springer-Verlag.

[122] I. Widya, B.J. van Beijnum, and A. Salden. Qoc-based optimization of end-to-end m-health data delivery services. In *14th IEEE International Workshop on Quality of Service (IWQoS06)*, New Haven CT, USA, 2006. IEEE Press.

[123] T. Winograd. Architectures for context. *Human Computer Interaction Journal*, 16(2,3), 2001.

[124] L. Yan and K. Sere. A formalism for context-aware mobile computing. In *3rd International Symposium on Parallel and Distributed Computing, 3rd International Workshop on Algorithms, Models and Tools for Parallel Computing on Heterogeneous Networks (ISPD-C/HeteroPar04)*, pages 14–21, 2004.

[125] S.J.H. Yang, A.F.M. Huang, R. Chen, T. Shian-Shyong, and S. Yen-Shih. Context model and context acquisition for ubiquitous content access in ulearning environments. In *IEEE International Conference on Sensor Networks, Ubiquitous, and Trustworthy Computing*, pages 78–83. IEEE Press, 2006.

[126] Z. Zuo and M. Zhou. Web ontology language owl and its description logic foundation. In *4th International Conference on Parallel and Distributed Computing, Applications and Technologies (PDCAT03)*, pages 157– 160. IEEE Press, 2003.

Chapter 8

A Pragmatic Approach to Context-Aware Service Organization and Discovery

Jian Zhu and Hung Keng Pung

Abstract It is a challenging task to efficiently discover a relevant business service for end-users over the vast Internet. The "relevance" of the service in a service discovery can be dependent on its location, the preference of a user, or in general, the contexts of the services and users. However, most of the existing service discovery methods have not taken context information and diverse user needs into consideration. We present herewith a practical approach to context-aware service organization and discovery, where semantic attributes of services and user's contexts are taken into consideration during service lookup. Services are organized logically in two users' perspectives: "local" and "global," and issues such as local service administration and privacy are considered in the system design. To further enhance the service matching capability in Business-to-Business (B2B) settings, i.e., to satisfy different perceptions of requestors, an asymmetric way of computing Web service process similarity is proposed. The major contributions of this work are the proposed framework for supporting context-aware service organization and discovery in real time, and the process matching scheme for B2B applications.

8.1 Introduction

The rapid growth of the Internet and network systems has led to an ever-increasing number of businesses participating in e-commerce worldwide. Today, the Web has evolved solely from a repository of data to a collection of complex and heterogeneous services. Furthermore, the permeation of mobile and wireless computing further eases the process of service sharing and accessing. Mobile users may share their application services (e.g., video streaming) at anytime, anywhere through their personal devices, such as handphones and PDAs. Service discovery in this mobile and dynamic environment is thus a challenging task, as the mobility and dynamism of services may greatly affect their relevance to the user.

In existing frameworks, service discovery is mostly achieved by a centralized server, such as UBR (1) and JLS (2) in industrial standards. The registry-style indexing and matching mechanism is efficient for service organization and discovery. However, their ability to scale and its potential single point of failure are of most concern in practice. As such, recently researchers are trying to leverage on emergence techniques such as the Peer-to-Peer (P2P) systems for distributed service discovery. P2P systems can be broadly divided into two categories: unstructured and structured, but both of them have certain limitations: in the former category, the fully distributed approach does not meet the efficiency requirement and queries may not always be resolved; while in the latter category, the Distributed Hash Table (DHT) concept destroys the locality properties of the data (e.g., content and physical) and hence cannot support partial or range search for resources. Indeed, in any practical service framework, the content locality of services (i.e., the depository of service information) is quite important from the perspective of the service provider, as it has implications on service administration and privacy. On the other hand, information of the physical locality (i.e., location of the service) and other contexts of services may play a considerably role in determining the relevancy of these services to the user, as location-based search is often required in service discovery (3).

In this paper, we propose a pragmatic approach to context-aware service organization and discovery, where services are organized logically in two views of the user: "local" and "global": In the local view, services with geographic coordinates are first mapped to the area node in the *geographic tree* (Section 8.3.1), and then mapped to the points on the Hilbert Space Filling Curve (SFC). For the global view, services are indexed at the middleware according to the classifications of their applications; the contexts of these services are maintained by the respective *class server* (Section 8.3.2). To provide the capability of service matching for Business-to-Business (B2B) applications, a process matching scheme for Web services is proposed whose details are described in the second part of this paper. Compared to other work, our framework has the following advantages: (i) Information of the content and physical localities for each service is preserved in the local view. Service discovery within a local scope is location-aware. Location-based range search can be supported effectively at real time; (ii) Service organization in the global view tends to be static and rarely changed in comparison to that of the local view. Service discovery and recommendation in a wider geographic scope can be done efficiently by leveraging on the context information at the class server; (iii) The two-view design enables a greater flexibility in controlling the geographic scope of services by the service providers. They may choose to register with the more trustful local service manager alone so that the presence of their services is only available to a specific local scope for reasons such as to better manage privacy/security threats or as part of their business expansion plan. (iv) The defined process model for process matching captures the structural information of a business process; thus, the behavioral properties of the process can be modeled with weights assigned to various structures of the process for user requirements.

The rest of the paper is organized as follows: Section 16.2 presents an overview of related work done in service discovery and service matching. Section 8.3 introduces the framework for our context-aware service organization and discovery. It discusses the concepts of the motivations behind the two-view design. Several mechanisms for service discovery are presented. A proof of concept for the framework is provided via a prototype implementation. Section 8.4 presents the Web service process matching scheme that performs service matching at the process level. Preliminary experiments have been designed and carried out as part of our verification of the proposed scheme. Finally, Section 15.7 concludes the chapter and outlines our future research directions.

8.2 Related Work

The idea of service discovery has been around for many years. However its research and development efforts have been intensified since the proliferation of the commercial use of the Internet. This is due to its enormous commercial potentials and the technological challenges arisen in the context of automatic service composition. Existing frameworks for service discovery include industry standards and research proposals. However, most industry standards consider either local-scale service discovery ((4) (5)) or global-scale discovery ((1) (2) (6)). The local-scale discovery is efficient and meaningful but not scalable for a cross-area search; while the global-scale discovery seldom consider local-specific issues such as service administration and privacy protection. As for implementation simplicity, most of these frameworks deploy a centralized approach. They rely on the service registry that stores all publicly available service information. The representatives are UDDI from OASIS (1) and Jini from Sun Microsystems (2). The registry-style organization has attributed to the ease of administration control and resource management, which is also superior in local and static service discovery. The disadvantage, however, is the lack of scalability in the design. Moreover, the reliability of such frameworks is often limited by the availability of services in practice if contexts of services are not considered (7). In research proposals, P2P overlay networks, such as

Gnutella (8), Chord (9), CAN (10), have been deployed as the platforms for service discovery. WSPDS (11) utilizes Gnutella as the underlying protocol. The initial version employes probability flooding for service discovery; while the improved version uses semantic-based routing scheme for efficiency purpose. Meteor-S (12) builds a semantic service discovery framework over the JXTA platform (13). It organizes registries of Web services in an unstructured P2P manner. However, for unstructured P2P overlays, two major issues are that (i) they may result in queries not always be resolved; (ii) they do not guarantee any performance bound while scaling. On the other hand, structured P2P overlays solve these issues. Most of these frameworks rely on the DHT concept and assign key ownerships in a predetermined manner. Keys are then looked up efficiently by using the overlay structure. The representative is INS/TWINE from MIT (14), which is based on Chord structure and supports partial matching by hashing different segments of the description (i.e., attribute-value tree) of the resource. However, the structured DHT approach has its own issues: First, most hashing techniques adopted do not consider data semantics; thus, partial or range search are not supported. To solve this issue, semantic P2P approaches are proposed. In (15), the authors map the sequence of service keywords in the d-dimensional CAN space to 1-dimensional linear space by using the Hilbert SFC technique. The purpose is to cluster semantic closed services to support semantic search and reduce routing efforts. Similar ideas have also been proposed in (16), (17) and (18). Second, structured P2P overlays require a high maintenance cost for large-scale deployment. As such, hierarchical P2P systems are further proposed, GloServ (19) and VIRGO (20) are the two representatives. Both of them differentiate services based on their application classes, and the hierarchy structure is constructed accordingly, e.g., *entertainment-music-pop.* Services within each class on each layer form a virtual group that is managed separately. Last, data localities are usually destroyed during the indexing process, especially for the content locality that is extremely important in the business environment. To handle this problem, some approaches such as SkipNet (21) provides controlled data placement and guaranteed routing locality by organizing data primarily by domain names, e.g., *john.microsoft.com/data-name.*

In terms of service matching, the conventional approaches are relying on service registries and based on categories or keywords to search for a relevant service. With the effort of Semantic Web framework, richer and precise descriptions are used to define a service, such as those based on Ontology languages ((22) (23) (24)). By using semantic attributes of a service including its inputs/outputs, the matching can be done at the conceptual level and thus to achieve better accuracy. However, to allow business developers to better describe the service they want and further increase the retrieval precision, the recent work ((22) (23) (25)) underline the needs for the matching of service behaviors by using its process model. A survey study of service process matching can be found in (26). In (25), (27), and (28), the process dissimilarity is computed by deriving the edit distance from deletion/insertion of a node/edge. However, with the recent study on graph distance measure, a high degree of matching precision is underlined. In (29), the authors argue that existing graph distance measures have a low degree of precision because only node and edge information of the graphs are considered. The richness of substructure information should also contribute to the evaluation of graph distance. Indeed, in the area of process matching, less work have been done on the matching of process structure information. A pioneer work is found in (30). The authors extend their previous work in (27) by first converting the process graph into a block tree, where structure information is captured by each block in the tree, and then using a Binary Tree Vector (BTV) to represent the tree. The distance measure is derived from BTVs using the algorithm proposed in (31). However, such a distance metric is not considered as the true similarity measure, and zero distance cannot imply that the two trees are identical as discussed in (31). Besides, this approach requires the tree to be ordered, which may not be true for business processes.

Our framework deploys a hierarchical structured P2P approach for service organization and discovery in the local view. Distinguished from previous work, the hierarchical structure is based on the geographic classification, which we call a geographic tree. Services in the same area are organized through the Hilbert SFC to preserve their localities as well as support location-based range search. In the global view, we deploy class-dependent rule reasoning approach over the context data of services, which is flexible and extensible. For service process matching, our similarity measure considers both the function and structure level information of the process, so that structure weights can be specified to satisfy different perceptions of the requestors.

8.3 System Framework Design

Figure 8.1 shows the overall system framework. Within the framework, the "local" and "global" roles of services are realized separately but integrated. The left side of the figure defines the view for local service organization and discovery, where "local" refers to a specific area or organization domain either by proximity or by geographic classification, such as the shopping mall where the user visiting is perceived as "local" for him. On the right side of the figure, we have the view for global service organization and discovery, where "global" refers to wide-area service collection. The service organization is based on its application class. For each class, information of services is indexed and rules of context reasoning can be designed and modified. The motivation behind the two-view design is that the segregation of the "local" role and "global" role of services has enabled local-specific system design issues such as service content locality and administration control to be addressed in the appropriate

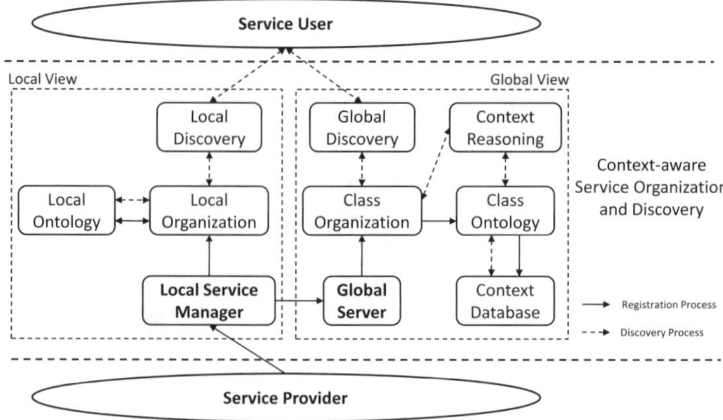

Figure 8.1: System framework overview.

local context; while the global role of services allows the service available in a wider geographic scope to be discovered more efficiently by using its static class information. Those highly dynamic services such as mobile services may exist in the local view only, which would make the candidates of local service discovery be more reliable and up-to-date, while reducing updating overheads in the global view. In the rest of this section, we will elaborate the detailed component design and processes for service registration and discovery in the framework.

8.3.1 Local Service Organization View

Hierarchical P2P Overlay

In reality, many Internet wide-deployed efficient systems use a hierarchical approach, as it not only reduces the number of resources to the manageable size, but also allows autonomy for different parts of the system. In service discovery, the service administrative domain is quite important for privacy and security concerns, i.e., content locality. Thus, in this paper, services in the same proximity or belonging to the same administrative domain (e.g., organization) are considered as local area services, where "local area" may refer to a geographic region or a domain. Furthermore, different areas of services can be organized in a hierarchical manner. We call this hierarchical structure the *geographic tree* and formally define it by using ontology. Nodes in the upper level of the tree cover services of larger areas, e.g., a country; while nodes in the lower level are for smaller areas, e.g., a shopping mall. Figure 8.2 illustrates the local ontology schema and a campus geographic tree.

The geographic tree is predefined at build time but flexible to change at runtime. A new ontology can be added in by different domain administrators

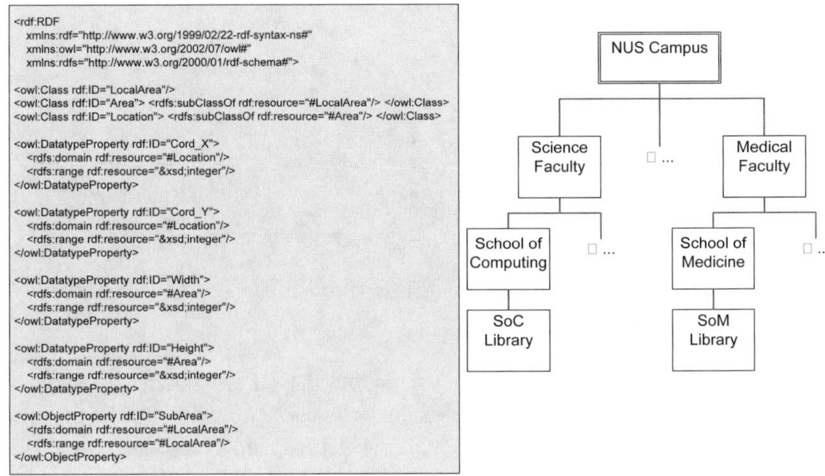

```
<rdf:RDF
   xmlns:rdf="http://www.w3.org/1999/02/22-rdf-syntax-ns#"
   xmlns:owl="http://www.w3.org/2002/07/owl#"
   xmlns:rdfs="http://www.w3.org/2000/01/rdf-schema#">

<owl:Class rdf:ID="LocalArea"/>
<owl:Class rdf:ID="Area"> <rdfs:subClassOf rdf:resource="#LocalArea"/> </owl:Class>
<owl:Class rdf:ID="Location"> <rdfs:subClassOf rdf:resource="#Area"/> </owl:Class>

<owl:DatatypeProperty rdf:ID="Cord_X">
   <rdfs:domain rdf:resource="#Location"/>
   <rdfs:range rdf:resource="&xsd;integer"/>
</owl:DatatypeProperty>

<owl:DatatypeProperty rdf:ID="Cord_Y">
   <rdfs:domain rdf:resource="#Location"/>
   <rdfs:range rdf:resource="&xsd;integer"/>
</owl:DatatypeProperty>

<owl:DatatypeProperty rdf:ID="Width">
   <rdfs:domain rdf:resource="#Area"/>
   <rdfs:range rdf:resource="&xsd;integer"/>
</owl:DatatypeProperty>

<owl:DatatypeProperty rdf:ID="Height">
   <rdfs:domain rdf:resource="#Area"/>
   <rdfs:range rdf:resource="&xsd;integer"/>
</owl:DatatypeProperty>

<owl:ObjectProperty rdf:ID="SubArea">
   <rdfs:domain rdf:resource="#LocalArea"/>
   <rdfs:range rdf:resource="#LocalArea"/>
</owl:ObjectProperty>
```


Figure 8.2: Example of local ontology schema and an instance for a campus.

at each node of the tree. For instance, when a new faculty is built in a specific area of the campus, the campus administrator may update the local ontology by attaching the new branch ontology to indicate such an extension. Meanwhile, local area services are re-organized according to the organization scheme shown in the next few sections.

Local Area Organization: Hilbert SFC & Superpeer

For a specific node in the geographic tree, or a local area, the SFC technique is applied to organize services inside. A SFC is a continuous fractal curve mapped from k-dimensional space to 1-dimensional space. The locality of points in the original space is preserved, meaning that points that are close on the curve are mapped from close points in the k-dimensional space. Among all kinds of SFCs, the Hilbert SFC performs the best as it has better locality-preserving behavior. When describing the Hilbert SFC, an important factor to consider is the degree order, which indicates the granularity in which to address a larger range or a smaller range in the original space, and due to its fractal property, the Hilbert SFC can be constructed recursively from lower granularity to higher granularity. For instance, second order curves can be drawn by scaling down each grid-cell in the first order curve, and different orientations are used to ensure curve continuity (Figure 8.3).

In this paper, service providers with 2-dimensional geographic coordinates are mapped to points on the 1-dimensional Hilbert curve. Although geodesic height can be supported as the third dimension, it is not considered at this moment. Depending on the proximity of local area services, the local geographic map can be further divided into $2^r \cdot 2^r$ grid-cells, where r represents the degree order for the Hilbert SFC. Paper (32) suggests $r = 19$ is sufficient

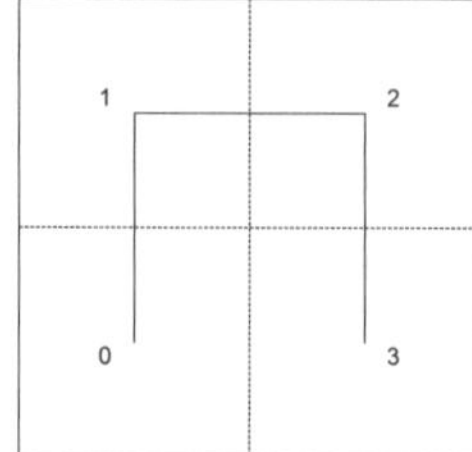

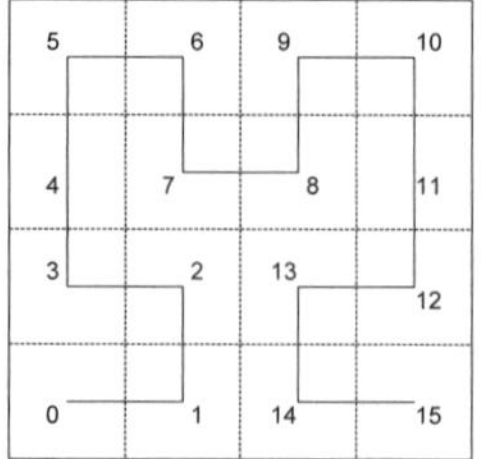

Figure 8.3: The Hilbert SFC: 1st order (left) and 2nd order (right).

to make sure each cell covers less than $1m^2$ on the earth's surface. Since we have the geographic tree as the first tier classification, r value would be much smaller in this case, and in the prototype we use a predefined cell size ($1m^2$) to derive the actual r value. For each grid-cell, at most one service provider can be contained, which is to ensure no duplicated identifier is assigned for service registration. Service provider in the local view is represented by peer form, and each service peer is connected with the two nearest ones on the Hilbert curve in clockwise and anti-clockwise directions, which we call them as the *left neighbor* and *right neighbor*. The service peer with the smallest identifier on the curve is connected with the one with the largest identifier. Therefore, service providers in a local area are organized in a ring topology.

In order to maintain the local topology on the Hilbert curve as well as the geographic tree structure, each area has a designated superpeer. In the context of service framework, the superpeer can be the local service manager (Figure 8.4). Superpeer distinguishes from ordinary peers in two aspects: (i) It maintains the hierarchical structure defined in the geographic tree by setting tree links among themselves, i.e., parent-child link; (ii) It supports the role of service administrator, i.e., defining service access polices and assigning links in the local topology. By organizing superpeers in a hierarchical manner and using tree-based routing scheme for cross-area service discovery, it can achieve high efficiency.

Identifier for Service Provider

Each service peer in the framework has a binary identifier assigned during its registration. The identifier has two parts: *areaID* and *peerID*. The areaID is derived from the geographic tree, which is much like the zip code in mail postings. In the design of the geographic tree, we use parameter d to describe the tree depth and value $2^2 - 1$ for the maximum branch factor, which means areaID has $2d$ bits and every 2 bits represents a level in the geographic tree. For area nodes with depth less than the tree depth, we use 1 as the pending bit in their areaIDs. Therefore, combined with the local Hilbert SFC order r, the identifier for each service peer has a range of $[0, 2^{2d+2r} - 1]$ and is of $2d + 2r$ bits length. Note for peerID, if two service peers have the same values

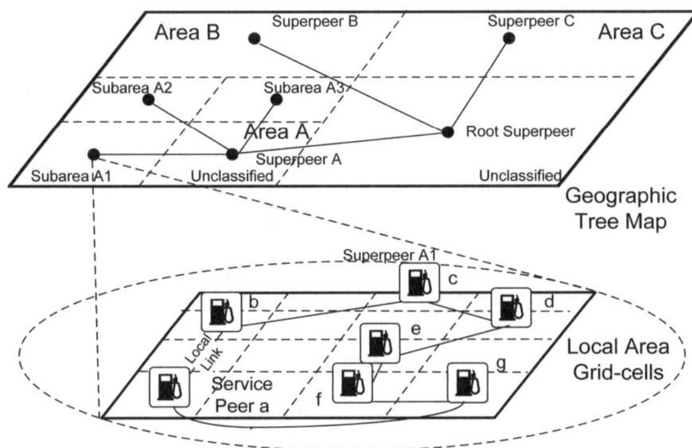

Figure 8.4: Illustration of the two-tier service organization in the local view.

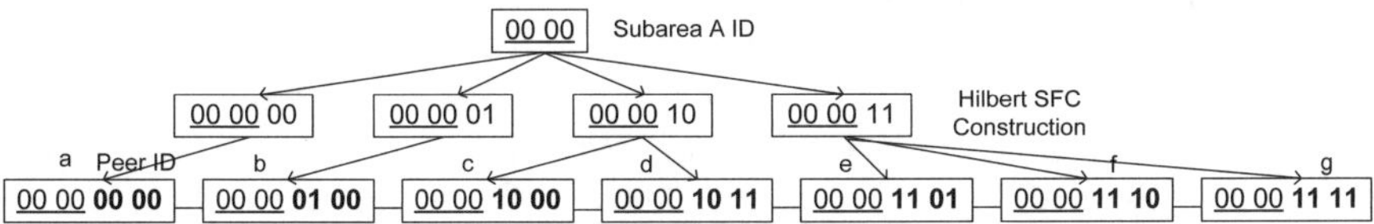

Figure 8.5: Service IDs assigned for each service peer in Figure 8.4.

in the first $2s$ bits, they are said to be contained in the same granularity cell of the sth order of the Hilbert SFC. Figure 8.5 shows the identifiers assigned for each service peer illustrated in Figure 8.4. The areaID is represented in 2^2 bits, since the geographic tree has the depth of 2; while the peerID has 4 bits, since the local area has $2^2 \cdot 2^2$ grid-cells. The figure also shows the recursive construction manner of the Hilbert SFC with 1st order omitted.

Hilbert SFC + Chord & Service Keyword Indexing

On the Hilbert curve, the Chord overlay is deployed to support efficient peer discovery, as Chord is effective in adaptation to peers join and leave, which is required in practice, e.g., for mobile service providers. Every service peer in the system maintains a *finger table* for routing. The finger table is constructed when the peer joins the framework and is updated whenever there is new service peer joining or leaving. For each service peer, there are $2r$ entries for its finger table, where the ith entry contains the first peer that succeeds by at least 2^{i-1} in the peerID space. Note that the first entry refers to the left neighbor of the current peer. Figure 8.6 shows the two finger tables for the illustrated topology. Combining the Hilbert SFC mapping and Chord routing algorithm, by indicating the location of a service peer, we can look it up in $O(log(m))$ hops, where m is the number of peers in the local area.

However, the above strategy is only for searching of a service peer with a given location. In practice, keyword-based search is rather more widely used.

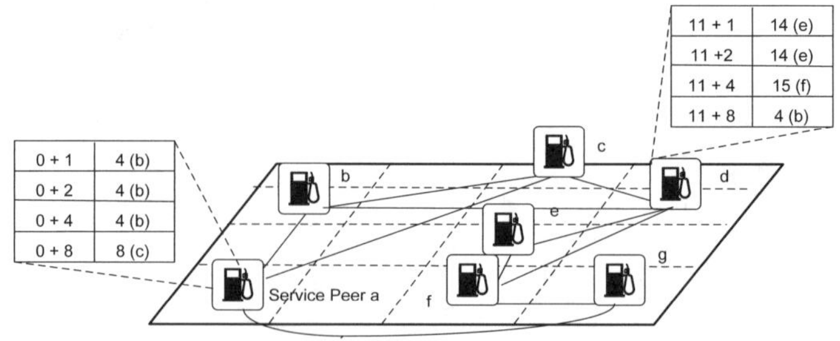

Figure 8.6: Example of the Chord overlay for the local topology.

Here, we further map the keyword space of services to the Hilbert SFC by using consistent hash functions, e.g. SHA-1. Each keyword of the service is hashed to r-bit key and distributed to peers on the Hilbert curve. A key is assigned to the first peer whose identifier is equal to or greater than the key value. However, data replication is necessary in the mobile and dynamic environment, as service peers may join, leave or fail at any time. In this case, we may set the length of the key less than r bits. For instance, with key length equals to $r - 3$, the last 3 bits of the peerID is not considered when distributing keys, and the same set of key data can be shared by 2^3 peers. For the actual searching, the same hash function is applied to retrieve the key for each keyword of the query. The problem is then transformed to the searching of a particular peer on the Hilbert curve, Chord routing can thus be applied.

8.3.2 Global Service Organization View

Hierarchical Class Overlay

In the global view, services are organized according to their application classes (e.g., *Business* class), which are more stable and unlikely to change. Service class hierarchy (e.g., *Business-Shopping*) is built according to the existing standards, e.g., ODP (33). Similar to the geographic tree, we use ontology to specify service classes as well as the class hierarchy, i.e., global ontology, and for each service class, there is a *class server* to do service management and class-specific context reasoning at the middleware side. Besides, there is a *global server* on top of the class servers to do ontology maintenance and handle service registration that will be discussed in Section 8.3.3. Figure 8.7 shows an example for the class hierarchy in the campus scenario.

Context Representation & Management

One major feature in the global view is about the context reasoning for selection, ranking, and recommendation of candidate services based on service and user's real-time contexts. In the local view, we already embed the location

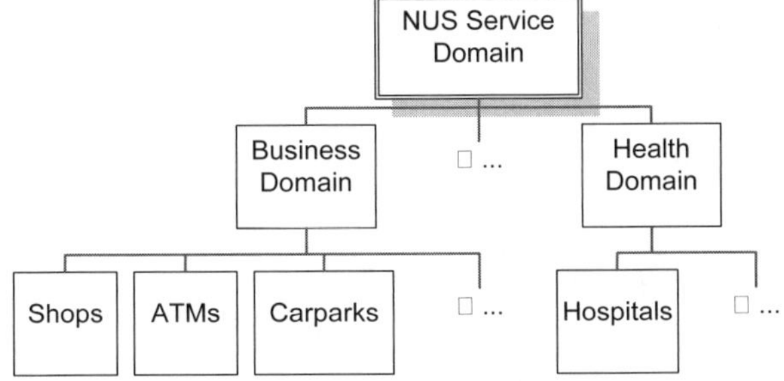

Figure 8.7: Example of a global service class hierarchy.

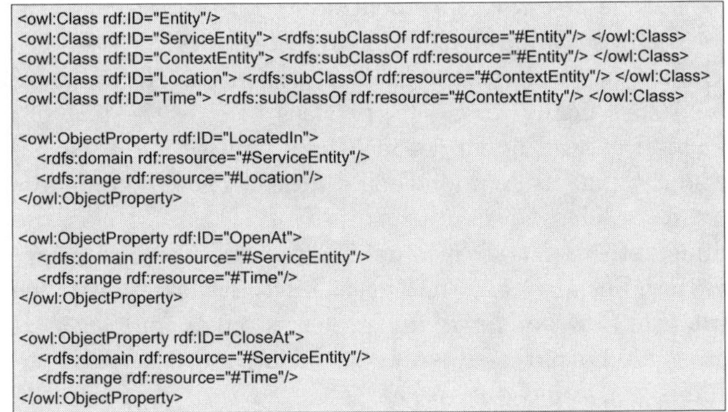

```
<owl:Class rdf:ID="Entity"/>
<owl:Class rdf:ID="ServiceEntity"> <rdfs:subClassOf rdf:resource="#Entity"/> </owl:Class>
<owl:Class rdf:ID="ContextEntity"> <rdfs:subClassOf rdf:resource="#Entity"/> </owl:Class>
<owl:Class rdf:ID="Location"> <rdfs:subClassOf rdf:resource="#ContextEntity"/> </owl:Class>
<owl:Class rdf:ID="Time"> <rdfs:subClassOf rdf:resource="#ContextEntity"/> </owl:Class>

<owl:ObjectProperty rdf:ID="LocatedIn">
   <rdfs:domain rdf:resource="#ServiceEntity"/>
   <rdfs:range rdf:resource="#Location"/>
</owl:ObjectProperty>

<owl:ObjectProperty rdf:ID="OpenAt">
   <rdfs:domain rdf:resource="#ServiceEntity"/>
   <rdfs:range rdf:resource="#Time"/>
</owl:ObjectProperty>

<owl:ObjectProperty rdf:ID="CloseAt">
   <rdfs:domain rdf:resource="#ServiceEntity"/>
   <rdfs:range rdf:resource="#Time"/>
</owl:ObjectProperty>
```

Figure 8.8: Example of context ontology for the *Carpark* service class.

information of services into the organization scheme. Other simple context reasoning (e.g., service workload and user preference) can also be supported, but they are deployed in a lightweight manner to fit the device capability (e.g., memory and power) of the service provider and user. In the global view, all this information is stored and processed at the class server. Context information such as service location, workload, opening time, closing time, and user preference, is formally described using the ontology language, and we differentiate the schema model and the instance model in the implementation: The schema model is used to define the fundamental class property such as the context element type (e.g., *Location*, *Time*) and cost type (e.g., *Cost per Entity*, *Cost per Hour*); while the instance model is used to store the detailed context data associating with each service instance. Figure 8.8 shows part of the context ontology schema.

Once a service is registered in the global view, its related context data is stored in the context database at the corresponding class server. Each context element is stored in a triple form: <*contextName*, *property*, *value*>. However, in practice, values of most context elements are dynamic, e.g., locations for mobile service providers. A simple way to deal with this case is to update the context database periodically or whenever the contexts change. A more practical way is by relying on some context search engine (e.g., SOCAM (34)) to retrieve service contexts at real time. Currently, we adopt the first simple approach in the prototype.

Class-Specific Context Reasoning

The class-specific service organization design allows us to differentiate reasoning rules in each service class. Service classes in the lower level of the class hierarchy may inherit or override rules from those in the upper level. For instance, for the *Carpark* service class, Besides defining rules for determination of candidates based on service specification and user preference, it can inherit general rules such as availability check by using service opening and closing time from the upper level class, e.g., Business class. Recommendation rules can be defined specifically for a service class in case the test of the general rules fails. This is useful and practical, as it will provide reasonable results rather than none to the users upon their queries. The recommendation system is also meant to be class-specific to simplify the task for class rule designer. Figure 8.9 illustrates a set of rules defined for the Carpark class. In the example, *rule1* handles the case that users have preferred cost; *rule2* handles the case that users do not have any preferred cost; and *rule3* recommends the user a possible carpark with range $\leq 1km$ and cost variance ≤ 2 if the requested one is not found due to *rule1*.

8.3.3 Service Registration & Leave

When a new service peer wants to join the framework, it needs to retrieve its areaID and peerID first. We assume the new peer is aware of the local superpeer through either peers in the framework (e.g., WiFi broadcast) or our static global server. However, if currently there is no superpeer in the area, the new peer will take the role. Therefore, depending on the role of

```
[CarParkRule1: (?c rdf:type nussd:ClientEntity) (?c nussd:PreferCost ?pc) (?pc nussd:CostValue ?pcv)
notEqual(?pcv "0"^^xsd:double) (?s rdf:type nussd:CarParkServices) (?s nussd:CarParkServicesCost ?sc)
(?sc nussd:CostValue ?scv) le(?scv ?pcv) -> (?s nussd:CostSatisfy ?c)]

[CarParkRule2: (?c rdf:type nussd:ClientEntity) (?c nussd:PreferCost ?pc) (?pc nussd:CostValue ?pcv)
equal(?pcv "0"^^xsd:double) (?s rdf:type nussd:CarParkServices) -> (?s nussd:CostSatisfy ?c)]

[CarParkRule3: (?c rdf:type nussd:ClientEntity) (?c nussd:PreferCost ?pc) (?pc nussd:CostValue ?pcv)
notEqual(?pcv "0"^^xsd:double) (?s rdf:type nussd:CarParkServices) (?s nussd:CarParkServicesCost ?sc)
(?sc nussd:CostValue ?scv) difference(?scv ?pcv ?dcv) le(?dcv 2) (?s nussd:LocatedIn ?sl)
(?c nussd:LocatedIn ?cl) (?sl nussd:Cord_X ?slx) (?sl nussd:Cord_Y ?sly) (?cl nussd:Cord_X ?clx)
(?cl nussd:Cord_Y ?cly) difference(?slx ?clx ?dx) difference(?sly ?cly ?dy) product(?dx ?dx ?dx2)
product(?dy ?dy ?dy2) sum(?dx2 ?dy2 ?d2) le(?d2 "1000000"^^xsd:double) ->
(?s nussd:RecommendFor ?c)]
```

Figure 8.9: Example rules in the *Carpark* service class.

the peer, two scenarios may happen during the registration phase in the local view: (i) If the peer is selected as the superpeer, the area information such as areaID and local area information such as the mapping of the Hilbert curve are retrieved from the global server. Despite this, the superpeer has to set up links with other superpeers at its parent and child level to enable cross-area routing; (ii) For normal peers, the registration process is straight forward. Once they successfully contact the local superpeer (i.e., local service manager), its areaID and peerID are assigned, where areaID is directly copied from the superpeer and peerID is determined by the local Hilbert curve. After that, it initializes local links with its left and right neighbor as well as peers in the finger table. Once the links are set up, its service profile can be indexed by respective peers in the local area. Service registration in the global view can be done concurrently with that in the local view. If the service peer is willing and is allowed to share services globally, its service profile is indexed by the corresponding class server at the middleware side through the local superpeer. In addition, service contexts are stored to support context reasoning.

When a service peer wants to leave the framework: If its role is the superpeer, it informs the global server by providing information about the candidate superpeer. After that, it shifts its routing states as well as local area information to the candidate superpeer. The local peers are notified by sending an update message along the Hilbert curve; meanwhile, the global server will inform the superpeers at its parent and child level for the updating of tree links. If the peer is of normal type, it only needs to inform the local superpeer and its left and right neighbor. Besides, all the indexed service keywords are shifted to its left neighbor which is the succeeding peer on the Hilbert curve. Finger tables of other peers are updated accordingly by using the stabilize protocol in Chord. In both cases, the service profile of the peer will be removed from the local and global view.

8.3.4 Service Discovery

Local Service Discovery

By arranging local service peers in a ring topology and with links constructed in Chord as shortcuts, local area service discovery can be done efficiently. We assume the user has some way to contact peers already in the framework. For instance, the server of the shop service may periodically broadcast its SSID through WiFi, mobile users in range can thus contact the server to access the framework. After gaining access, the user can specify his service discovery query by indicating a bag of keywords as well as the target service class. The query is then processed as follows: the bag of keywords are first hashed to the respective keys according to the hash function used in Section 8.3.1. Each key is looked up on the Hilbert curve by looking for the service peer storing it. The lookup process is exactly the same as the key search in Chord: For each step, the peer that is the highest predecessor of

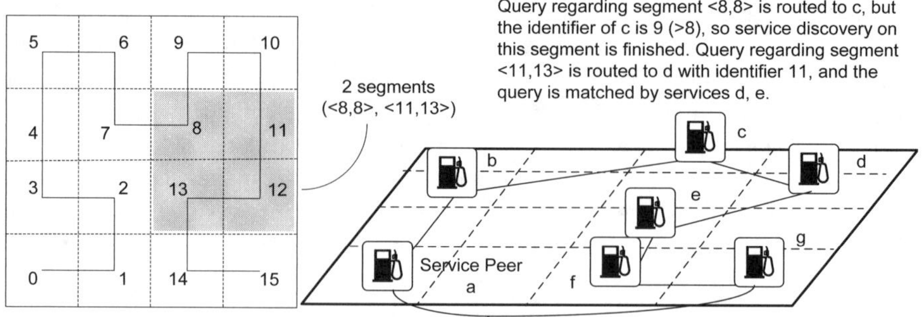

Figure 8.10: A sample range query and its routing behavior.

the key (in terms of its peerID) in the finger table is selected as the next hop. Note that the local ring topology guarantees the success of the query routing. Once the query reaches the peer storing the key, the indexed service peer information is returned. The union of all these results will form a candidate list for the user, and rankings can be determined among candidates according to their matching degree compared to the user's query.

A feature of the hierarchical organization of services in different areas is that the user may define his own local scope. For instance, if the user cannot find any satisfied service in the primary scope (lowest node in the geographic tree that contains the user's location), he may refine his scope by enlarging it up one level along the tree. The concept is realized by routing the queries to the local superpeer first and then to the superpeer at the parent level, or the next level up until the query reaches the superpeer of the defined scope. After that, it will route the query to all its decedent superpeers in its subtree, and local area service discovery will be triggered for each area.

Global Service Discovery

Global service discovery refers to those queries targeting areas outside of the primary scope. If the target area is specified by the user, normal peers will send queries to their respective local superpeers first. The local superpeer will then act as the mediator and route the query to the target area according to the tree structure defined in Section 8.3.1. If only service class is indicated, the query will be sent to our global server. The global server will redirect the query to the corresponding class server that does the matching. Note that the user's contexts as well as his preference can be catered to in this case. For any context information provided, the class server will parse them and pass to the reasoning engine for context reasoning. A list of candidate services based on the matching of keywords and contexts will be returned, and recommended candidates from the class rules defined can also be included.

Location-Based Range Search

The location-based range search is trying to solve common queries in service discovery, such as "looking for services within x meter range." The conventional DHT-based P2P systems is hard to support unless there is a centralized proxy to do the intermediate matching; whereas in our framework, both views can support this task easily. In the global view, a centralized context reasoning can be done for each service indexed in the class server. Because the concept is quite simple and intuitive, it is not further discussed here. In the local view, the physical locality of services is already embedded in the local organization by deploying the Hilbert SFC mapping. To discover a service within a certain range, the first step is to discover segments of the Hilbert curve involved in this range search, where the segments are due to the truncation of the linear curve by the range frame (Figure 8.10). Each segment is represented with *<startID, endID>*, where startID and endID are the identifiers of the starting and ending grid-cells of the segment on the Hilbert curve. The problem is then transformed to discover services on each of the segments. For the detailed segment discovery, the query is sent to discover the peer with $startID$, and routing strategy used in local area service discovery is applied here. However, it is not necessary that there must be a service peer whose peerID is equal to startID. As long as the first peer whose peerID is equal to or greater than startID is reached, the segment is discovered. The query is then matched for each service peer along the curve in a clockwise direction, and once the current peer has peerID that is greater than endID, the service discovery on this segment is finished. At last, matched services on all the segments are returned to the user.

8.3.5 Prototype Implementation

We have built a prototype for the above framework in our laboratory. The prototype implements a *CampusServiceHelper* application, which emulates the situation in an university environment where students or visitors try to acquire services within the campus, e.g., looking for bus services, carpark services, etc. The organization in the local view is based on the university map from Google Maps (35) and with a predefined geographic tree ontology. In the global view, services are differentiated by their application classes. On the application level, two GUIs and an application program based on Google Android (36) are developed: The two GUIs are for service registration at the service provider side and service monitoring at the middleware side (i.e., on the global server); while the application program is for service discovery at the user side. For the detailed scenario, the service is registered by filling the *ServiceRegistration* form, including *ServiceName*, *ServiceDescription*, *ServiceOpeningTime*, or *ServiceClosingTime*, etc. The request is sent to the local service manager according to the local ontology and once approved, the service provider can join the local topology and its service information is in-

Figure 8.11: Screenshots for service discovery.

dexed accordingly. On the GUI at the service provider side, the corresponding service icon will be displayed to indicate successful registration. If the service is not a private service or willing to share services globally, it can register with the global server, and the global server will store its service information as well as its contexts. After registration in the global view, the service icon will also be displayed on the GUI at the middleware side. For service discovery, the user has two options: (i) Local service discovery, which is targeting services in the same local area or according to the local scope defined by the user; and (ii) Global service discovery, which is targeting for services in the global picture. In both cases, location-based range search can be supported by indicating the range value in the user preferences. Screenshots for service discovery on the client side are shown in Figure 8.11.

8.4 A Process Matching Scheme for Web Services

In the framework of last section, service matching in both the views is done on the service description level and keyword-based; however, in the B2B environment, the description level matching may be not adequate as the notions of service adaptation and reusability are always highly desirable: In the former case, business developers for example, may adapt to the changing requirements by replacing existing services with compatible ones online; while in the latter case, multiple services can be composed to fulfill new business objectives. At this moment the process of incorporating services is largely reliant on human effort, and the matching of compatible or desired Web services is mostly done on the inputs/outputs level, which is insufficient for business developers to understand the behavioral properties or other constraints. To go beyond this limitation, services should be matched and integrated at the process level. In case there is no fully compatible service online, developers could look for the most similar one and use it as the starting point for adaptation. Therefore, in the second part of this paper, we further propose a structure behavioral approach for process matching of Web services.

8.4.1 Business Process Modeling

In real-life applications, business processes are usually described by markup languages (e.g., BPEL (37), PNML (38)) which define a set of business activity patterns. An activity pattern consists of objects, functions, structures, and is designed to specify the service behaviors and execution dependencies of the business process. However, for efficient design and verification, a process model is often required. It should capture information about how functions share data and how data flows from one function to another. Practical models adopted by business developers include Workflow Net (WF-net) (39), Petri Net (40) and Finite State Automata. In fact, all of them are kinds of graph models: They consist of nodes and edges, where nodes are for activity indication and labeled with function names; edges are for process flows or structure constraints to introduce process behaviors.

However, we argue that conventional graph models are not suitable for process matching, especially for efficient similarity calculation, since they consider process flows from the human perspective and the structure level information (e.g., structure type and ordering) is not obvious for machine processing; in addition, the Loop structure may introduce infinite cyclic problem during the matching process. We propose, therefore in this paper a Structure Series Tree (SST) model to further abstract conventional graph models (Definition 8.1). The process structures are abstracted to structure nodes in the SST. At this moment, the following four types of structures are considered: a. *Sequence (S)*; b. *Choice (C)*; c. *Loop (L)*; d. *Parallel (P)*.

DEFINITION 8.1 *The Structure Series Tree (SST) model is an abstraction of business processes. It consists of a set of structure nodes (or SST nodes) connected by SST edges.* $SST = (SN, SE)$, *in which*

- $SN = \{ST, LA, NUM, CON\}$ *is a structure node which indicates a process structure.* ST *represents the structure type, i.e.,* S, C, L, P*;* LA *represents the label of the structure node. It is derived from the structure type as well as the representing node in the graph models, such as the starting node of Choice and Parallel structures, or the ending node of Loop structure;* NUM *represents the total number of graph nodes (i.e., process functions) contained in the structure;* CON *represents for structure constraints such as branch number for Choice and Parallel structures;*

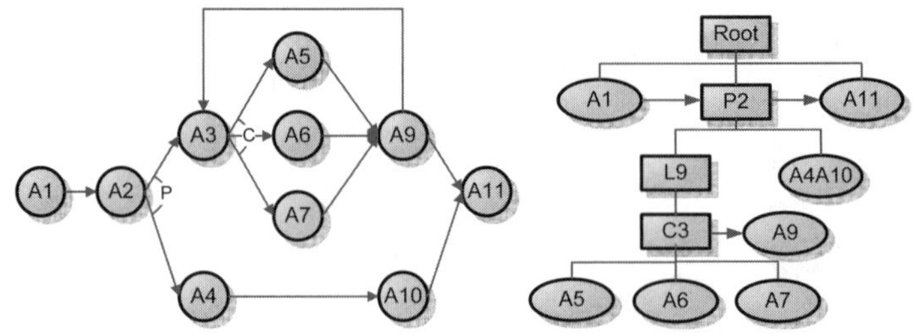

Figure 8.12: Transforming WF-net model (left) to SST model (right).

- *$SE = \{ET, SN_1, SN_2\}$ is an edge, where ET represents types of structure dependency, i.e., Parent-Child (PC) or Preceding-Succeeding (PS). The former one represents structure belongings, such as nested structures; while the latter one is for structure orderings, such as structure flows.*

Note that SST is not a rigid tree although represented in a tree form for intuitive views. Figure 8.12 shows an example of a process graph model and its transformed SST model. The function name of each graph node is abstracted to $A1$, $A2$, etc. In the transformed SST, Sequence node is labeled by its internal function sequence, e.g., $A4A10$; while the labels of Choice, Loop, and Parallel nodes are determined by their structure types as well as the representing nodes as defined in Definition 8.1, e.g., P (ST) + $A2$ (the starting node of the structure) $\Rightarrow P2$ (LA).

SST Model Construction

There are two steps to transform the process markup files to SST models: First, transform to conventional graph models, and then transform to SST models. There are already proposals regarding the first step, such as that in (41) and (42). In this paper, we focus on the second step, and to further simplify the construction process, we assume there is a 2-dimensional matrix P describing the graph model, in which $P[s,t] = 1$ represents a Sequence edge from node s to node t; $P[s,t] = 2$ or 3 represents a Choice or Parallel edge. In the SST construction process, the structure scope of each node in the graph model is captured by a so-called Node-Structure (NS) stack. Algorithm 1 sketches the whole construction process.

SST Model Discussions

The way to abstract business processes on the structure level allows us to better capture their behavioral properties as well as their dependency relationships. The structure-orientated design eases the process adaptation and reuse: Missing structures or behaviors in a process template can be replaced

Algorithm 1 *SST_Construction*

```
Require: Graph Matrix P
Root ⇐ Dummy SST root; sn ⇐ Starting node of the graph model;
NS ⇐ Node-Structure stack; V ⇐ SST node set;
Push sn − Root to NS;
while NS ≠ do
   ns ⇐ Pop from NS;
   n ⇐ ns's node information; s ⇐ ns's structure information;
   if n is the ending node of s then
      Create SST node of type s and add to V;
   NN ⇐ n's next node set; PN ⇐ n's previous node set;
   if Structure type of Choice, Parallel then
      Create the corresponding structure cp;
      Look for the ending node en of cp;
      Push en − cp to NS;
      Push all nodes in NN with cp to NS;
   else
      Create a Sequence structure seq;
      Create SST node of type Sequence and add to V;
      Push all nodes in NN with seq to NS;
   while PN ≠ do
      pn ⇐ Pop from PN;
      if pn → n is a Loop edge then
         Create a Loop structure lp;
         Insert pn − lp to NS according to lp's scope;
```

by existing ones that achieve the same goal. The weight for a specific structure can be imposed easily, so that the requestor may search processes with the most desirable behavior. Besides, by comparing SST models, any structure or function difference can be reflected and penalized in the similarity calculation. To summarize, SST allows us to do practical structure behavioral matching rather than node/edge subgraph matching as in the conventional graph matching methods.

8.4.2 SST Model Indexing

To support efficient process search, two kinds of indexing schemes are proposed, namely Function Sequence Indexing (FSI) and Structure Dependency Indexing (SDI). The FSI is used to support fast retrieval of relevant processes, as well as similarity calculation on the function level; while the SDI is to index structure dependencies among all the SST nodes and support similarity calculation on the structure level.

Function Sequence Indexing

We index SST nodes by using the Trie (43) data structure, which builds a prefix tree for function labels, with each leaf node in the Trie indicating a business function or node in the original graph model. The FSI scheme indexes SST nodes depending on their types: If it is of type Sequence, it is broken into separate functions and indexed to the Trie respectively. The Reference Pointer (RP) is introduced in the Trie to maintain the function order information, e.g., A4's RP is pointing to A10 in the example; If it is of type Choice, Loop, or Parallel, the function of its representing node is indexed, e.g., $A2$ is indexed for $P2$. Each indexed function identifies its belonging SST node by using SST ID and SST node ID derived from the SST construction process.

Structure Dependency Indexing

The Structure Dependency Matrix (SDM) is used to index structure dependencies, namely structure ordering and belonging. The dependency relationship is identified by the type of the edge between two SST nodes. The SDM is constructed by the following rules:

- Each SST is added with *Start* and *End* nodes to state the starting and ending states of the process. The *Start* node points to SST nodes containing nodes without incoming edge in the original graph model, and *End* node is pointed by SST nodes containing nodes without outgoing edge. There is no function in the two nodes;
- $SD(SN_0, SN_0) = 0$;
- $SD(SN_0, SN_1) = 1$ if SN_1 is SN_0's direct succeeding node, i.e., through PS edge;
- $SD(SN_0, SN_1) = 2$ if SN_1 is SN_0's direct child node, i.e., through PC edge;
- $SD(SN_0, SN_1) = 0$ otherwise.

Figure 8.13 shows the revised SST model and its corresponding SDM as for Figure 8.12. As there may be many entries with zero value, storage optimization techniques may be applied here, but this is not the scope of this paper.

8.4.3 Business Process Search

For real-time process search, the problem is defined as given a requested process with/without weight indicated for each structure or substructure, we look for all candidate processes that simulate partial or all the required behaviors, and rank candidates based on their similarity values. There are two

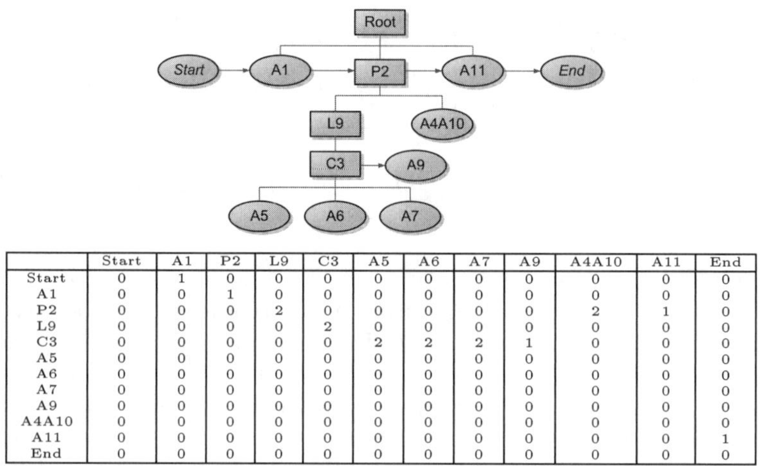

	Start	A1	P2	L9	C3	A5	A6	A7	A9	A4A10	A11	End
Start	0	1	0	0	0	0	0	0	0	0	0	0
A1	0	0	1	0	0	0	0	0	0	0	0	0
P2	0	0	0	2	0	0	0	0	0	2	1	0
L9	0	0	0	0	2	0	0	0	0	0	0	0
C3	0	0	0	0	0	2	2	2	1	0	0	0
A5	0	0	0	0	0	0	0	0	0	0	0	0
A6	0	0	0	0	0	0	0	0	0	0	0	0
A7	0	0	0	0	0	0	0	0	0	0	0	0
A9	0	0	0	0	0	0	0	0	0	0	0	0
A4A10	0	0	0	0	0	0	0	0	0	0	0	0
A11	0	0	0	0	0	0	0	0	0	0	0	1
End	0	0	0	0	0	0	0	0	0	0	0	0

Figure 8.13: Revised SST with *Start* and *End* nodes and its SDM.

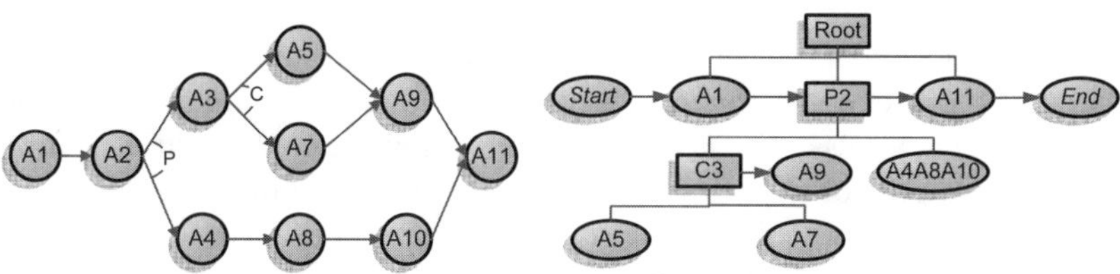

Figure 8.14: Requested business process (left) and its SST model (right).

steps involved in solving the problem: SST Node Search and SST Similarity Calculation. For this section, the process model in Figure 8.14 is used as the requested process and the model in Figure 8.13 is illustrated as a candidate.

SST Node Search

The first step to look for a candidate process is to find processes with relevant SST nodes. The "relevance" is mostly determined on the function level: The candidate SST should contain at least one function with the same label (or fulfilling the same task if different labeling rules are applied) as that in the requested SST. Since Sequence node is constructed differently from other three types of nodes; thus, there are two types of node searching schemes: Sequence node search; Choice, Loop and Parallel node search. For the former one, the details of the functions as well as the sequence order are considered; while for the latter one, differences of structure types and constraints are considered to derive edit cost on the structure level.

Sequence Node Search The search of candidate Sequence nodes relies on the Trie in the FSI. The edit operations on the function level include (i)

	A	E	B	C	D	Cost (FC)
ADBC	0_0^D	1	0_1^C	0_1^{Null}	0_2^B	$1+1\times 2=3$
ACDE	0_0^C	0_1^{Null}	1	0_2^D	0_2^E	$1+1\times 2=3$
AED	0_0^E	0_0^D	1	1	0_1^{Null}	$2+0=2$
EDF	1	0_0^D	1	1	0_1^F	$4+0=4$

Table 8.1: Illustration of the edit cost calculation on Sequence node

inserting a function of the requested process; (ii) deleting a function from the candidate process. The usual substitution and reordering can be considered by doing (ii) first and then doing (i). In the actual search, we modified the algorithm of Levenshtein distance (44) by allowing multiple candidates to be matched at the same time; that is, rather than retrieving all the candidates first and then doing the edit distance computation, we do matching while retrieving, so existing candidates may be filtered once they exceed the cost threshold set by the requestor. The basic cost unit is set to be 1 Function Cost (FC) for each insertion or deletion.

Table 13.1 illustrates the results for searching function sequence $AEBCD$ (a Sequence node) among several candidates in Column 1. In the table, 0 indicates the presence of the requested function in the candidate sequence; while 1 indicates a miss. In addition, the information of the matched function sequence (i.e. a part of the requested sequence) is shown in subscript and identified with a number; the expected next function is shown as superscript. The search process strictly follows the function order of the requested sequence, i.e., A to D. While computing the edit cost due to reordering, backward trace is applied and those functions with same subscript number is considered as a whole for insertion or deletion. The complexity of the modified algorithm is $O(kn^2)$, where k is the number of candidate sequences and n is the average length of function sequences.

Choice, Loop, and Parallel Node Search Compared to Sequence nodes, these nodes place more emphasis on the structure level constraints to reflect process behavioral properties. However, there is no standard way to measure structural differences. Some paper is using a predefined constant value for structure cost; while in this paper, we define it to be related to the structure size as well as structure constraints such as branch number. The structure cost unit is defined as $SC = \alpha \cdot |ST|$ FC, where α is a parameter adjusting the importance of structure level similarity compared to that of function level ($\alpha \geq 0$). If $\alpha = 0$ it approximates node/edge similarity without considering process structure differences; $|ST|$ indicates the total number of functions contained in the structure. The following steps are used to derive the cost due to structure differences.

1. Look for the representing function of the requested structure in the Trie. If the function is not found, which means no candidate is available, the node is considered missing. The edit cost for the node is equal to SC;

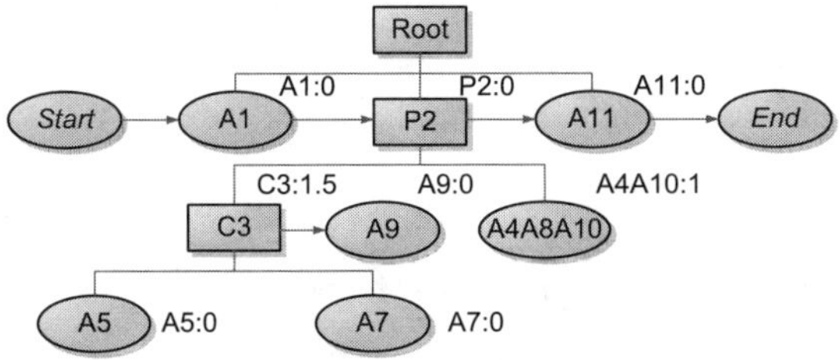

Figure 8.15: Candidate node with cost indicated for each requested node.

2. If the function is found. The corresponding candidate SST node is retrieved. The two nodes are matched based on their structure types and number of branches (for Loop structure, number of branches is 1). If the types are different, the cost is SC; If the types are the same, but branch numbers are different, the cost is defined as $min(SC, \frac{|branch_r - branch_c|}{branch_r} \cdot SC)$, where $branch_r$ and $branch_c$ are the number of branches in the requested and candidate node respectively.

Following the two types of node search schemes, candidate nodes are found for each requested SST node (Figure 8.15). The extra labels show candidate nodes in Figure 8.13 as well as their edit cost when $\alpha = 1$.

SST Similarity Calculation

Once candidate processes are found, the overall edit cost is computed. Note that for a requested structure node, there may be multiple candidate nodes from the same SST. We adopted a heuristic approach by assuming each requested SST node is mapped to the candidate node with minimum cost as derived in last section. In addition, the additional cost on the structure level can be due to (i) inserting a PC/PS edge of the requested SST; (ii) deleting a PC/PS from the candidate SST. These operations are realized by the candidate SDMs, and the cost for each operation is set to be α as defined. The whole algorithm works in a recursive way: It computes edit cost for candidate SST in a bottom-up, right-left fashion according to the requested SST. In the end, it adds in additional cost due to deletion of any function in the candidate process that is not encountered in SST node search phase. Algorithm 2 shows part of the pseudocode that computes distance cost for Choice, Loop, or Parallel nodes. For the illustrated example, the edit cost for the candidate SST is 6.5 when $\alpha = 1$.

The distance cost derived reflects the effort involved to modify the candidate process to the requested one, and thus to evaluate their "closeness." Based on it, the Business Process Similarity (BPS) can be calculated by the formula:

$$1 - \frac{Dist\_Cost}{|SST_{candidate\_request}|} \tag{8.1}$$

Algorithm 2 *SST_Dist_Calculation(node_r)*

```
Require: Candidate SST Node Map candidate_map, Candidate SDM
    candidate_sdm
dist_cost ⇐ 0
if node_r.type.equals("Choice" or "Loop" or "Parallel") then
  candidate ⇐ candidate_map.get(node_r);
  for index = 0 to node_r.children_size do
    dist_cost ⇐ dist_cost+SST_Dist_Calculation(node_r.children[index])
  if candidate ≠ null then
    for index = 0 to candidate_sdm.length do
      if candidate_sdm[candidate.id][index] > 0 then
        dist_cost ⇐ dist_cost + α;
    for index = 0 to node_r.children_size do
      candidate_child ⇐ candidate_map.get(node_r.children[index]);
      if candidate_child ≠ null then
        if candidate_sdm[candidate.id][candidate_child.id] ≠ 2 then
          dist_cost ⇐ dist_cost + α;
        else
          dist_cost ⇐ dist_cost − α;
      else
        dist_cost ⇐ dist_cost + α;
return dist_cost;
```

where $|SST_{candidate\_request}|$ is the maximum cost that may occur for transforming the candidate process to the request one. It includes all the possible costs for edit operations in the SST node search phase, as well as that on structure dependencies. In the example, $|SST_{candidate\_request}| = 48$; thus, the BPS value for the candidate process is 0.865.

Structure Weight & Similarity

The SST model also allows structure weight ω to be imposed by requestors. Initially, each SST node is assigned with an equal weight (i.e., $\omega = 1$), where *Root* and *End* nodes are not considered. Requestors then may adjust weight of each SST node, if certain behaviors or structures are highly desirable. The adjustment increases the weight of the current node; while decreasing others. $\sum \omega$ is a constant that is equal to $|SN| - 2$, where $|SN|$ is the total number of structure nodes in the SST. Suppose each SST node is with weight ω_i, by specifying the current node SN_c with weight ω_c, the weights of other unspecified nodes will become $\omega_i - (\omega - \omega_c) \cdot \frac{\omega_i}{\sum \omega_{unspecified}}$, where $\omega_{unspecified}$ represents the weight of unspecified SST nodes. The fraction $\frac{\omega_i}{\sum \omega_{unspecified}}$ keeps the ratio of weights of specified nodes to those of unspecified ones. With the weight modified, the new edit cost derived from that structure node will be $Cost(SN_c) \cdot \omega_c$, where $Cost(SN_c)$ is the edit cost derived in Section 8.4.3.

Node Type	Mutate Operation	Mutate Type
Sequence node	Change node label	Function level
	Add a new node behind	Function level
Loop representing node	Change node label	Structure level
	Add a new node behind	Function level
	Remove loop edge	Structure level
Choice/Parallel representing node	Change node label	Structure level
	Add a new branch	Structure level

Table 8.2: Process mutation operations

The weight of a SST edge is determined by the weights of its two connected nodes: The smaller one is chosen as its value, and it affects the structure dependency cost for the edge in the same way as that for SST nodes. In other words, ω imposes a penalty coefficient to the process structure: the larger (smaller) it is; the more (less) penalty will get for each edit operation on the structure. In addition, the subprocess search may be supported by removing the Start and End nodes in the requested SST model so that the search is looking for all candidate processes that contain the requested structure.

8.4.4 Experiments & Verification

To our knowledge, currently there is no benchmark available in evaluating business process matching schemes. We did some preliminary experiments on a simulated process database created by *BusinessProcessGenerator*. The generator randomly generates processes according to the node number specified by the user. At this moment, we allow each structure type (S, C, L, P) to appear with equal chance (i.e., 0.25) in the generated processes, and each node is randomly labeled with a function name extracted from a pool of Web service WSDLs. In addition, nested structures are allowed in the generation. In order to test the correctness of our proposed approach, we add in "noise" to the generated process. The idea behind this is that by comparing with the original process, the more noise we add, the more variation there will be, and thus the mutated process will be more and more dissimilar with the original one. The noise is added according to a ratio $(0 - 1)$, which measures how much percentage of the original process nodes will be mutated. The mutation operations involved are listed in Table 8.2, and it will be randomly picked up according to the node type. Besides, we have implemented Dependency Distance Measure (DDM) in (27) and Graph Edit Distance (GED) in (45) for verification and comparison purpose. We have generated 10 business processes with 50 nodes per process. For each process, we mutate it 10 times at each noise ratio (0.2, 0.4, 0.6, 0.8, 1). We issue the original process as the requested one and mutated processes as candidates. The prototype is implemented on a $P4$ ($2.6GHz$, $1GB$ RAM) machine.

Figure 8.16 shows the node and structure similarity in the mutated processes as compared to the original one. The mutation process works as expected: it

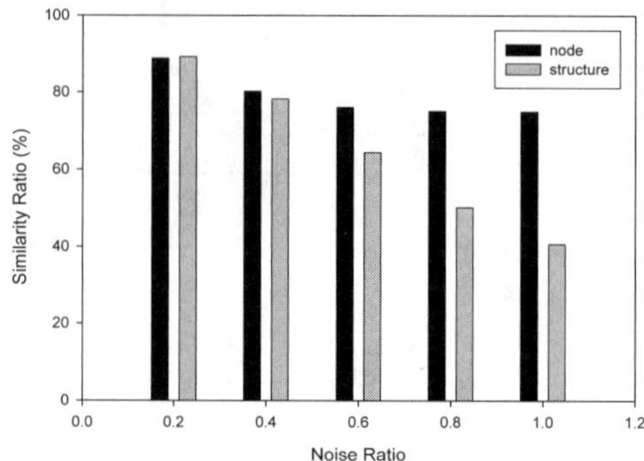

Figure 8.16: Overview of mutation process.

decreases the node similarity to 75%, and structure similarity to 40%.

Figure 8.17 illustrates the patterns of edit cost and BPS along with the increasing noise ratio. Note that for a fair comparison with DDM and GED, we allow $\alpha = 0$ to measure the edit cost on the node/edge level, and the edit cost due to structure level operations is set to constant 1 in Section 8.4.3 and 8.4.3. Figure 8.17 shows edit costs for all approaches are increasing while more noise are added. We observe that DDM and our approach ($\alpha = 0$) have a lower cost value compared to GED, since DDM only considers operations on the edges (i.e., node dependencies) but not nodes; while ours does not consider edge operations, although some of them can be realized by function/structure reordering. We also observe that our edit cost increases exponentially when $\alpha > 1$, which is due to the dominance of structure level cost and the way we compute it. However, the cost is normalized when computing the BPS (Equation 8.1); thus, α does not have too much impact on the overall trend for the BPS as shown in Figure 8.18. The BPS chart also demonstrates that our approach better reflects the structure similarity presented in Figure 8.16, i.e., BPS declines from 0.87 to 0.41 when $\alpha = 1$. Note that DDM has lower BPS values as it imposes rather strict requirements on the similarity metric: It only considers edge similarity but not node similarity.

The quantitative BPS value has the potential for process categorization or filtering; however, in some applications, only the result order makes sense. We have verified our approach with DDM and GED by comparing the top two candidates in the result list. When $\alpha = 0$, 90% of our candidates are the same as that in GED, and 88% to DDM, which shares almost the same result for the case GED to DDM (i.e., 90%). However, when $\alpha = 1$ and 2 the result similarity becomes 84% and 80% respectively, which is due to the structure level cost as defined in this paper.

We have also tested the efficiency of process indexing (including SST construction) and online search for the collection of 500 processes. The average

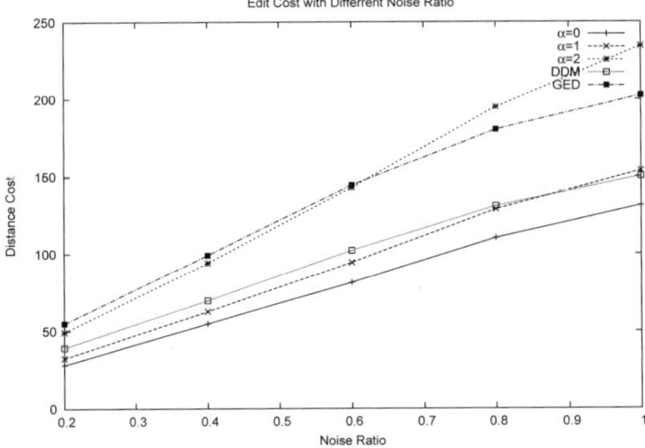

Figure 8.17: Edit cost patterns for 50-node processes.

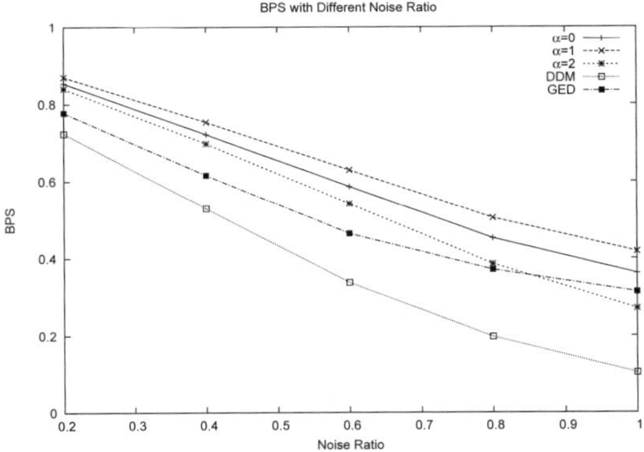

Figure 8.18: BPS patterns for 50-node processes.

indexing time for one 50-node process is 182.69ms, and the average retrieval time for one candidate process is 1.24ms. If we increase the process size, i.e., from 50-node to 100-node, we find the average indexing time increases to 403.74ms, but the retrieval time does not increase much, only by 0.62ms for our test cases.

8.5 Conclusion and Future Work

In this paper, we have presented a pragmatic approach to context-aware service organization and discovery. By "pragmatic," we mean the local view organizes services in a distributed manner and caters for easy local administration as well as privacy protection. The discovery of service is location-aware and location-based range search can be supported effectively; whereas the global view supports class-specific context reasoning for services selection, ranking, and recommendation. Both views are extensible, such as adding new areas/classes to the local/global ontology. We have implemented a prototype in our laboratory for a proof of concept. In the second part, we identified description level matching of Web services may be not adequate for service adaptation and reuse in B2B applications; thus, a process matching scheme is specifically proposed for Web service discovery. The designed SST model captures the structure information of the process and provides a way of computing process similarity based on structure differences. Preliminary experiments are carried out for scheme verification. In the near future, we would like to focus on the development of real applications by using the proposed framework as well as the process matching scheme and test their effectiveness in a practical environment.

References

[1] UDDI, http://www.uddi.org/pubs/uddi_v3.htm, 2004.

[2] Jini, http://java.sun.com/products/jini/2_1index.html, 2005.

[3] SES: Local Search Marketing, http://videos.webpronews.com/2007/02/20/ses-local-search-marketing/, 2007.

[4] Bluetooth, http://www.bluetooth.com/bluetooth/, 2004.

[5] UPnP, http://www.upnp.org/, 2009.

[6] Service Location Protocol, http://tools.ietf.org/html/rfc2608, 1999.

[7] D. Bachlechner, K. Siorpaes, H. Lausen, and Dieter Fensel, "Web Service Discovery — A Reality Check," in Proceedings of the 3rd European Semantic Web Conference (ESWC), 2006.

[8] Gnutella Specification, http://rfc-gnutella.sourceforge.net/src/rfc-0_6-draft.html, 2002.

[9] I. Stoica, R. Morris, D. Karger, M. Frans Kaashoek, and H. Balakrishnan, "Chord: A Scalable Peer-to-Peer Lookup Service for Internet Applications," in Proceedings of ACM SIGCOMM, pages 140–160, 2001.

[10] S. Ratnasamy, P. Francis, S. Shenker, and M. Handley, "A Scalable Content-Addressable Network," in Proceedings of ACM SIGCOMM, pages 161–172, 2001.

[11] F. Banaei-Kashani, C.C. Chen, and C. Shahabi, "WSPDS: Web Services Peer-to-Peer Discovery Service," in International Symposium on Web Services and Applications, 2004.

[12] K. Verma, K. Sivashanmugam, A. Sheth, A. Patil, S. Oundhakar, and J. Miller, "METEOR-S WSDI: A Scalable P2P Infrastructure of Registries for Semantic Publication and Discovery of Web Services," *Journal of Inf. Tech. and Management*, 6(1):17–39, 2005.

[13] JXTA, https://jxta.dev.java.net/, 2008.

[14] M. Balazinska, H. Balakrishnan, and D. Karger, "INS/Twine: A Scalable Peer-to-Peer Architecture for Intentional Resource Discovery," in Proceedings of the 1st International Conference on Pervasive Computing, 2002.

[15] C. Schmidt and M. Parashar, "A Peer-to-Peer Approach to Web Service Discovery," *Journal of World Wide Web*, 7(2):211–229, 2004.

[16] C. Tang, Z. Xu, and M. Mahalingam, "pSearch: Information Retrieval in Structured Overlays," in Proceedings of ACM HorNets-I, 2002.

[17] Y. Zhu and Y. Hu, "Semantic Search in Peer-to-Peer Systems," in *Handbook of Theoretical and Algorithmic Aspects of Ad Hoc, Sensor, and Peer-to-Peer Networks*. J. Wu (ed.), Auerbach Publications, pages 643–664, 2006.

[18] G. Pirro, D. Talia, P. Trunfio, P. Missier, and C. Goble, "ERGOT: Combining DHTs and SONs for Semantic-Based Service Discovery on the Grid," in Technical report, TR-0177, Institute on Knowledge and Data Management, CoreGRID — Network of Excellence, 2008.

[19] K. Arabshian and H. Schulzrinne, "GloServ: Global Service Discovery Architecture," in Proceeding of MobiQuitous, 2004.

[20] L. Huang, "A P2P Service Discovery Strategy Based on Content Catalogues," in Proceeding of the 20th CODATA International Conference, 2006.

[21] N.J.A. Harvey, M.B. Jones, S. Saroiu, M. Theimer, and A. Wolman, "SkipNet: A scalable overlay network with practical locality properties," in Proceedings of the 4th Conference on USENIX Symposium on Internet Technologies and Systems, 2003.

[22] A. Bernstein and M. Klein, "Towards High-precision Service Retrieval," in Proceedings of International Semantic Web Conference (ISWC), 2002.

[23] S.S. Bansal and J.M. Vidal, "Matchmaking of Web Services Based on teh DAML-S Service Model," in International Joint Conference on Autonomous Agents and Multiagent Systems, 2003.

[24] J. Pathak, N. Koul, D. Caragea, and V.G. Honavar, "A Framework for Semantic Web Services Discovery," in Proceedings of the 7th Workshop on Web Information and Data Management (WIDM), 2005.

[25] D. Grigori, V. Peralta, and M. Bouzeghoub, "Service Retrieval Based on Behavioral Specifications and Quality Requirements," in Proceedings of BPM, 2005.

[26] A. Wombacher, "Evaluation of Technical Measures for Workflow Similarity Based on a Pilot Study," in Proceedings of the 14th International Conference on Cooperative Information Systems, 2006.

[27] J. Bae, L. Liu, J. Caverlee, and W.B. Rouse, "Process Mining, Discovery, and Integration using Distance Measures," in Proceedings of IEEE International Conference on Web Services (ICWS), 2006.

[28] M. Minor, A. Tartakovski, and R. Bergmann, "Representation and Structure-Based Similarity Assessment for Agile Workflows," in Proceedings of the 7th International Conference on Case-Based Reasoning, pages 224–238, 2007.

[29] Y.H. Xiao, H. Dong, W.T. Wu, M.M. Xiong, W. Wang, and B.L. Shi, "Structure-Based Graph Distance Measures of High Degree of Precision," *Pattern Recognition*, 41(12):3547–3561, 2008.

[30] J. Bae, J. Caverlee, L. Liu, and H. Yan, "Process Mining by Measuring Process Block Similarity," in BPM Workshops, pages 141–152, 2006.

[31] R. Yang, P. Kalnis, A.K.H. Tung, "Similarity Evaluation on Tree-structured Data," in Proceedings of SIGMOD, 2005.

[32] A.M. Houyou, A. Stenzer, and H.Meer, "Performance Evaluation of Overlay-Based Range Queries in Mobile Systems," in Wireless Systems and Mobility in Next Generation Internet: 4th International Workshop of the EuroNGI/EuroFGI Network of Excellence, 2008.

[33] Open Directory Project, http://www.dmoz.org/, 2009.

[34] T. Gu, H. K. Pung, and D.Q. Zhang, "A Service-Oriented Middleware for Building Context-Aware Services," *Journal of Network and Computer Applications (JNCA)*, 28(1):1–18, 2005.

[35] Google Maps, http://maps.google.com/, 2009.

[36] Google Android, http://code.google.com/android/, 2009.

[37] BPEL, http://docs.oasis-open.org/wsbpel/2.0/OS/wsbpel-v2.0-OS.html, 2007.

[38] PNML, http://www.pnml.org/, 2009.

[39] W.M.P. van der Aalst, "The Application of Petri Nets to Workflow Management," *Journal of Circuits, Systems, and Computers*, 8(1):21–66, 1998.

[40] Petri Nets, http://www.informatik.uni-hamburg.de/TGI/PetriNets/, 2009.

[41] A. Wombacher, P. Fankhauser, and E. Neuhold "Transforming BPEL into Annotated Deterministic Finite State Automata for Service Discovery," in Proceedings of IEEE International Conference on Web Services (ICWS), 2004.

[42] S. Hinz, K. Schmidt, and C. Stahl "Transforming BPEL to Petri Nets," in Proceedings of BPM, pages 220–235, 2005.

[43] E. Fredkin, "Trie Memory," in Communications of the ACM, 1960.

[44] L.I. Levenshtein, "Binary codes capable of correcting deletions, insertions, and reversals," *Soviet Physics-Doklady*, 10(8):707–710, 1966.

[45] H. Bunke and B.T. Messmer, "Similarity measures for structured representations," in *Topics in Case-Based Reasoning*, 1994.

Chapter 9

A Context Model to Support Business-to-Business (B2B) Collaboration

Puay Siew Tan, Angela Eck Soong Goh, and Stephen Siang-Guan Lee

Abstract

There has been a lot of context-aware computing research in recent years, especially in ubiquitous computing. However, there are very few context-aware work for Business-to-Business (B2B) collaboration. This chapter puts forth a *B2B Context Model* to support context-aware approaches in B2B collaboration, in particular to support supply chain applications. The model is built on the twin foundations of typical *context* elements and *domain specific contexts* from supplier selection. This model was validated for completeness against an industry de facto standard, RosettaNet. As an example of how to use this model, an implementation for partner selection to support B2B collaboration is presented, incorporating a hybrid Multicriteria Decision Making (MCDM) technique, *Fuzzy Deviation Measure.*

9.1 Introduction

Businesses no longer operate simply. They form and operate in value chains, pursuing a common goal and benefitting from a competitive advantage through shorter time-to-market, reduced cycle times, and enhanced customer service (21). The present-day globalized and networked economy increases the complexity of manufacturing operations, as manufacturing shifts from North America and Western Europe to Asia, in particular (9). As businesses now operate across geographical regions, such value chains are becoming even more pervasive.

One such value chain is the ubiquitous supply chain, often characterized by Business-to-Business (B2B) collaboration activities, ranging from supplier selection to transaction-type activities like purchase orders to inventory information sharing. Supply chains typically have many partners. Companies now scour the world for the best partners that can produce high quality goods/ services at the lowest cost (16). As a consequence of this globalization (27), the number of supply chain partners has increased, partners are more geographically dispersed and operate in different time-zones, rendering traditional collaboration methods inefficient. Traditional methods of collaboration through phone calls, faxes, and face-to-face communication are giving way to electronic or virtual forms of communication such as e-mails, electronic exchanges, and video conferencing. Thanks to the Internet, easy information access and effective remote communication have facilitated real-time business transactions. Geographically dispersed corporations can now collaborate anytime and integrate their business processes. However, such integration poses a major challenge, especially for supranational corporations or multinational corporations (MNCs) that have hundreds of trading partners (5). Such MNCs drive and dictate B2B collaboration as supply chain masters (10). In Singapore for example, these MNCs typically have suppliers that are local enterprises consisting of both big and small enterprises.

Effective B2B collaboration in a supply chain depends on B2B connectivity across the entire value chain. Such connectivity is needed not only to expedite transactions and thereby reduce costs, but also to seize opportunities in a rapidly changing business environment. In a 2007 report by Aberdeen for example (27), the foremost motivator for companies to globally invest in supply chain applications has been "the need to change and adapt business processes over time." In another report (14), the two main drivers were "improve the response to market demand" and "better manage variability in the supply chain." These trends are forcing companies to improve their supply chain responsiveness, to "quickly identify and react to changes in supply, demand, and execution threats/opportunities" (28). Furthermore, there is greater volatility in demand, resulting in uncertainty of demand forecasts (26). Consequently, *sense and respond manufacturing* is gaining recognition (11), underpinned by outsourcing (22). In *sense and respond manufacturing*, instead of predicting demand (which is often inaccurate), a company or enterprise focuses on responding to changes to demand in a shortest time possible. Real-time collaboration with partners is critical to empower sense and respond manufacturing (11). It is crucial to identify potential partners and form the partnerships quickly (PNP being desirable) (23).

9.1.1 Challenges in Forming Virtual Enterprises for B2B Collaboration

In the *sense and respond manufacturing* environment, partners can be identified either by a restricted search (e.g., those who have been prequalified earlier) or by an open search. In both cases, the notion of virtual enterprise has been investigated extensively (6). A *virtual enterprise* is an *ad hoc* alliance of enterprises that come together to exploit core competencies and resources in order to seize business opportunities.

Companies forming such virtual enterprises can discover potential partners by searching profiles of services (e.g., Web services). Such services could be made available publicly (facilitating open search) or to limited partners only (restricted search). Current capabilities, such as the Web Service Description Language (WSDL), being syntactic and nonsemantic by nature, are not able to provide enough useful business information to match potential partners well. Semantics is necessary because services (through the service profiles representing partners' products and services) with the same meaning but different keywords cannot be identified. Much research into the semantic approach to overcome this problem has been reported, one such being a survey done by (17). Semantics presumes the existence of ontology, a formal representation of the relationships between a set of concepts within a domain. It is unrealistic to assume that every organization or partner has an ontology, given the diverse nature of partners in a supply chain.

Another challenge is the acquisition of timely information to dynamically identify and match potential partners. Most B2B activities are process-governed but dominated by the exchange of business documents. In a typical supply chain, examples of such documents include purchase orders (POs), goods receive notes (GRNs), and advance shipment notes (ASNs). These documents are rich in contextual information, for example the company's name, address (billing, delivery, etc.), date of delivery (or receipt), location of the goods, the status of payment or invoice, and other information like inventory level and expected demand of particular items. Such contextual information (e.g., that derived from customer PO) can be used as a basis to support *context-aware B2B collaboration*, including searching and identifying the relevant partner(s) for outsourcing for example. The contextual information is expected to be encapsulated as context-aware profiles, complementing existing WSDL and OWL-S profiles. This is further discussed in section 9.3.3.

9.1.2 Motivation

In order to apply context-information in support of B2B collaboration, it is important to understand the meaning of "context" and how context (or context information) has been modeled or represented in earlier research. Section 9.2.1 discusses a classification of context across different application domains. A review of supplier selection criteria from existing literature is presented in section 9.2.2. Supplier selection criteria is important for several reasons: (1) partner selection is typically the starting point for any virtual enterprise formation, notably *sense and respond manufacturing;* (2) much work has been reported on supplier selection criteria.

The *B2B Context Model* formulation is discussed in Section 9.3. Section 9.4 introduces a hybrid Multicriteria Decision Making (MCDM) technique that is used to support the context-aware approach in B2B collaboration, with partner selection as a specific example. This section also includes the experimental setup and test results obtained using the context-aware approach for partner selection. Section 9.5 concludes this chapter, with several pointers for further investigation.

9.2 Foundation for the B2B Context Model

This section provides an overview of the foundations used for formulating the *B2B Context Model*; namely context elements used in context-aware computing across various domains and supplier selection criteria. A detailed treatment on the understanding of context, classification of context-aware elements, suppliers selection, and also the appropriateness of context elements identified for inclusion into the model are discussed in a separate publication (24).

9.2.1 Classification of Existing Context-Aware Related Elements

The Cambridge dictionary (7) defines context as "the situation within which something exists or happens, and that can help explain it." The common understanding of context in context-aware computing is aligned with this definition, but adapted for specific applications. This led to widely diverse definitions and interpretations. Bazire and Brézillon (4) analyzed 150 definitions of context and tried to find relationships between the various definitions using latent semantic analysis (LSA) and STONE techniques. They concluded that context definition is varied, and depends very much on the discipline involved. Consequently, the use of context information (namely the type or elements of context) also differs significantly.

The varied and diverse facets of context information across ubiquitous computing, information, and knowledge management and other related fields of work were analyzed to identify the dominant categories or elements for building a context model to support B2B collaboration. Arising from this exercise, these context elements can be broadly categorized into *Physical*, *Cognitive* and *Information* aspects. The *Physical* aspect is derived mainly from ubiquitous

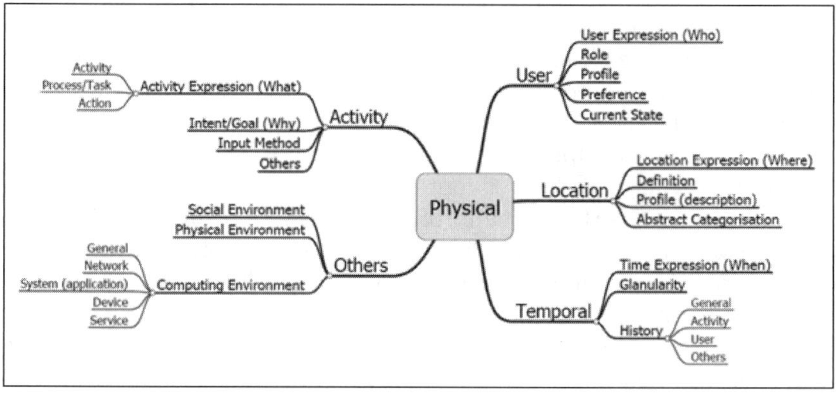

Figure 9.1: Elements in the *Physical* Aspect.

computing research, referring to context pertaining to the physical environment. In contrast, the *Cognitive* aspect views context from a cognitive or decision-making viewpoint. In the *Information* aspect, context is essentially used to handle meta-data related to information processing such as query processing, for example. (A detailed treatment of the classification process is not possible within this chapter.) Of these three aspects, the *Physical* aspect showed the most promise in providing a starting point in developing a *B2B Context Model* to support B2B collaboration. Figure 9.1 shows a mind-map of the five distinct categories within the *Physical* aspect, namely *User*, *Location*, *Temporal*, *Activity*, and *Others*. Each of the categories is further subgrouped into elements and subelements. An *element* typically represents a *context concept*, and where relevant, is further broken down into subelements.

The *User* category contains elements related to a user of a system or service, such as *User_Expression* (direct user information), *Role* (user's role), *Profile* (skillset or personalization information), *Preference* (personal preference), and *Current_State* (e.g., current location or role). The *Location* category holds context related to the "place or where" of the user. It includes the *Location_Expression* (location identifier), *Definition* (type of location identifier, e.g., *symbolic* for room or *geometric* for GPS), *Profile* (meta-data to describe location, e.g., nearby services or shops), and *Abstract_Categorization* (user defined categorization, e.g., *external* or *internal* and *office* or *home*). The *Temporal* category consists of elements related to time. Such elements include *Time_Expression* to provide the current time stamp, duration, and/or range of time context, *Granularity* (measure of time) and *History* to capture past activities with time-related information. The fourth category, *Activity*, encompasses contexts that describe an activity or sequence of activities. The *Activity_Expression* captures what a user is doing (action), with sequential information where needed. The intent of the activity is captured through the

Intent/Goal element, whereas *Input_Method* encapsulates the way information is captured. The *Others* element is a placeholder for other artifacts that might arise or are related to activities. The *Others* category captures contexts that describe the setting where a system or service operates in, including social, physical, and computing environments.

Interestingly, there is a lack of context application in the business domain. One possible reason is the difficulty in defining and modelling business information as context information. Business information is often interrelated and is best portrayed in a complex hierarchical model. The work of Kumar (13) is an indication of such complexity.

From a B2B perspective, *any information that characterizes the situation of a participant in an interaction* is considered *context*, including the user and the application themselves. For example, the needs of a requestor for a service or product are considered as context. The ability of partners (providers) to satisfy requestors' requirements is also considered as context. This is best illustrated through an example. Say a company is looking to purchase a `Product-A`. It has a quotation for the product with the following information:

Price is $3.50/unit for *order quantity* of 2000 units or more. However, the price is $5.00/unit for *order quantity* that is less than 2000 units. The quotation has a *validity* of 4 weeks and an *order lead-time* of 2 weeks. For *order lead time* of 1 week, an additional $0.20/unit will be charged. Price is inclusive of local *deliveries* only, overseas orders need to pay for freight.

Here, it is obvious that *temporal* (duration in weeks) and *location* (local, overseas) contexts are needed. However, in deciding whether to place an order for `Product-A`, there are other contexts that must be considered; *cost* (price and freight charges if not in the same country), which in turn is related to other contexts, *order quantity* and *order lead-time*. Furthermore, the *prices* quoted has a *temporal* aspect too (validity of 4 weeks). This example illustrates the complex linkages of contexts in B2B domain and why elements such as *price*, *transportation costs*, and *lead-time* are needed to provide a holistic representation of B2B context. It is also apparent that a *B2B Context Model* for supporting B2B collaboration cannot be derived from existing literature of context-aware research only (namely the *Physical* aspect), thereby needing other inspiration. This is discussed next.

9.2.2 Identification of Relevant B2B Contexts

In order to understand and derive relevant *B2B context* for the model, inspiration is drawn from supplier selection criteria. Supplier selection is a key activity in B2B collaboration as it is often the starting point for any virtual enterprise formation. Furthermore, there are many published works on supplier selection criteria, providing a good foundation to uncover *B2B context*.

This section reports on published supplier (or vendor) selection criteria, including supplier prequalification and selection. From a literature review of

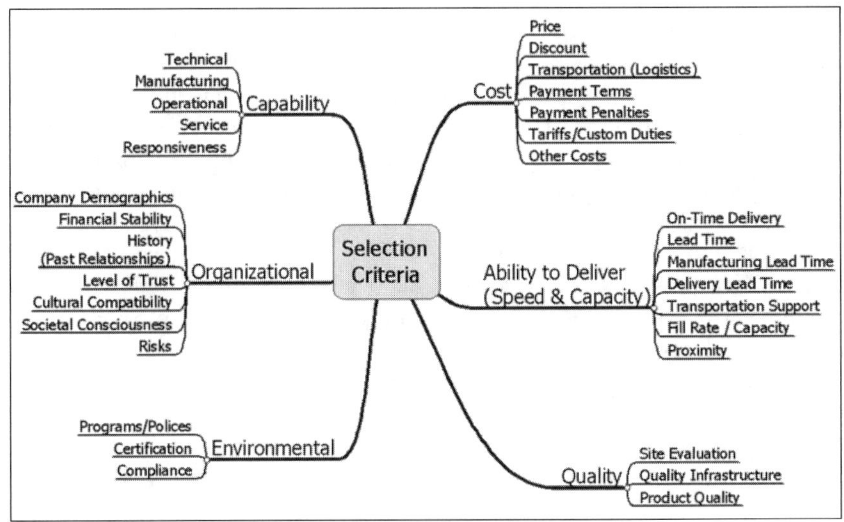

Figure 9.2: Categorization of Supplier Selection Criteria.

73 pertinent publications, 6 major categories, namely *Cost, Ability-to-Deliver, Quality, Capability, Organizational*, and *Environmental* were ascertained. In each category, different selection criteria were found (see Figure 9.2). The criteria in each category are not exhaustive but provide a reflection of factors unearthed during the literature review. A detailed treatment is out of scope of this chapter.

The evaluation criteria in the *Cost* category are based on the price of a product or service (*Price*), applicable discounts (*Discount*) and other associated costs of the product or service such as transportation (*Logistics*), tariffs and custom duties (*Tariffs*), and warranty or maintenance cost and foreign exchange fluctuations (grouped under *Other Costs*). Another important facet is the *Payment Terms* and *Payment Penalties* of the product or service. In retail commerce, *payment terms* often refer to credit or some other form of payment. However, in B2B transactions, deferred payment (e.g., 60 days) and letter of credit (LC) are the typical modes. *Payment Penalties* is usually described as a certain percentage of the total contracted value of a purchase.

The *Ability-to-Deliver* category deals with criteria such as *On-Time Delivery*, the length of time taken to deliver a product or service (*Lead-Time* including *Manufacturing* and *Delivery*) and the level of *Transportation Support* (including modes, frequency, and capacity of transportation). Also, the capacity to fulfill orders (*Fill Rate/Capacity*) and the distance between the manufacturing sites and customers' locations (*Proximity*) are important considerations, too.

In the *Quality* category, *Site Evaluation*, on-site evaluation of a candidate supplier's physical infrastructure and processes, is a common activity, particularly during supplier prequalification. As such, the willingness of a company to accommodate site evaluations (in particular its production facilities) is important. Another important measure here is the *Quality Infrastructure* that the candidate supplier has instituted, for example ISO 9000 certification. The last factor is *Product Quality* track record; for instance defect rate, rejects rate, and reliability.

The *Capability* category outlines the ability of a company to deliver a product or service. Here factors like technical competency (*Technical*), *Manufacturing* know-how and the ability to integrate and share information (*Operational*) and pre- and postsale support (*Service*) are important concerns. Another important consideration is swiftness of action in response to a request (*Responsiveness*).

At times, in particular for potential partners with no prior business liaison or from a different country/region, the *Organizational* aspects are key considerations. In this category, the overall standing of the company may be assessed, including *Company Demographics, Financial Stability*, and the *Risks* that the company is exposed to. Where past business relationships exist, then past business experience (*History*) and the *Level of Trust* that can be ascertained are useful. In this day and age, *Cultural Compatibility* and *Societal Consciousness* are gaining importance, especially when there is a need to work closely (e.g., co-design to roll out a new product). Environmental (or green) considerations are also fast gaining importance, with new awareness and legislations, in particular for products that are targeted for the European Union market. Under the *Environmental* category, the *Policies and Programs* of a company relating to issues like recycling and other environmental conservation efforts, the type of *Certification* (e.g., ISO 14001) achieved and *Compliance* with regulatory requirements could be crucial, especially for certain industries like medical devices manufacturing.

Technically, the evaluation criteria themed in the selection criteria mindmap should underpin a *B2B Context Model*. However, some of the information is not easily processed or modeled; they may be qualitative or aggregate in nature (making it harder to process) and/or not easily obtained (as the sources of information are either not within the control of collaborating parties or inaccessible because of confidentiality). Also, different types of companies attach different importance to the selection criteria (1, 29). For example, if a company is extremely cost conscious, such as small companies with very low profit margins, the main selection criterion could be just *cost*. In support, the delivery *lead-time* could be another consideration. In contrast, for bigger companies, cost is not the sole or key criterion. Factors like *timely delivery* based on planned schedule and *product quality* could be key considerations. Product *price*, could be relegated to secondary importance. In addition, companies typically consider the *cost* aspect in a holistic manner, not just the product price alone (8). Other related factors like *transportation cost* and prevailing *taxes or tariffs* are also included in the *total cost* calculations of an order. Thus, the authors have chosen to incorporate only key information that are easily accessible into the *B2B Context Model.* A detailed analysis of this is not possible within this chapter.

9.3 Proposed B2B Context Model

Using the B2B context perspective, and the foundations from the context classification and supplier selection criteria reviews, an initial *B2B Context Model* was formulated, centered around the three context categories identified in section 9.2.1; *User, Temporal,* and *Location.* The mind-map of this initial formulation is as shown in Figure 9.3. The next section discussed how the model was verified for completeness.

9.3.1 Verification of Model Completeness

The initial *B2B Context Model* was compared to relevant Partner-Interface-Processes (PIPs) of the RosettaNet standard. The RosettaNet standard (3) is a set of XML-based specifications that define message guidelines, business process interfaces, and implementation frameworks that enable businesses to communicate efficiently across multiple platforms, applications, and networks. Founded in 1988, RosettaNet is supported by a nonprofit consortium of industry leaders. Today, RosettaNet has become a de facto B2B collaboration standard, in particular for high-tech manufacturing.

The **Partner Interface Processes (PIPs) & Dictionaries** specifications define a common set of properties for PIPs®, associated with basic business activities. PIPs specify the structure and format of business documents, activities, actions, and roles for each trading partner. PIPs are grouped into 7 distinct clusters, supporting different aspects of the trading network for B2B collaborations (18). Each cluster can be decomposed further into segments and then finally into the PIPs themselves. As part of the PIP definition, the payload information defines the type and structure of business information to be exchanged between collaborating parties. This payload information is a good reference against which to benchmark the *B2B Context Model,* as it contains information (including contextual) commonly used and exchanged in B2B collaboration.

Cluster 3 and Cluster 4 PIPs were identified as being relevant to B2B collaborations in a supply chain. Cluster 3 is related to order management, from quotations to transportation and distribution to finance matters, whereas Cluster 4 is about inventory management, including tasks like forecasting, inventory, and sales reports. Several Cluster 7 PIPs that relate to distribution of work orders and work-in-progress information were also considered as

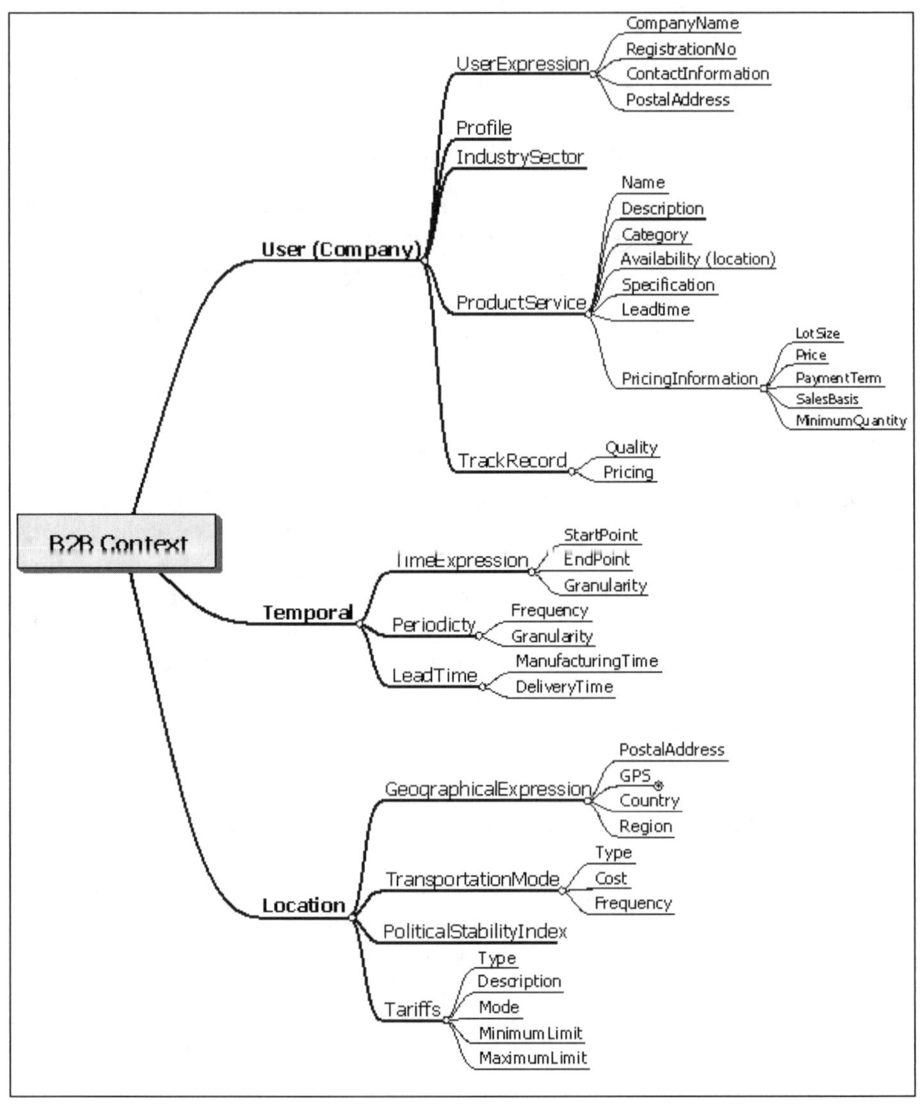

Figure 9.3: Initial mind-map of the B2B Context Model.

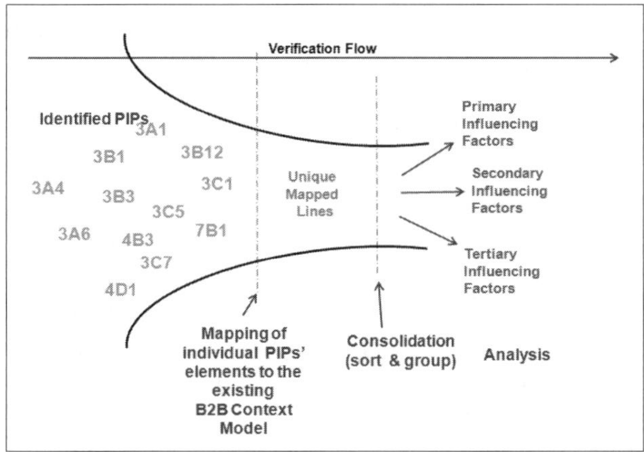

Figure 9.4: Schematic representation of the verification methodology.

being relevant to this verification. 27 relevant PIPs from a total of 46 PIPs were selected for this verification.

Verification Methodology

The high level steps in the verification process (shown graphically in Figure 9.4) are as follows:

(1) Select and map contextual PIP elements onto the initial *B2B Context Model*

(2) Identify potential PIP context elements to add to the current *B2B Context Model*

(3) Repeat steps (1) to (2) for all selected PIPs

(4) Consolidate the results by sorting and grouping the identified candidate additions into influencing factors

In Step (1), to determine whether a given element is contextual or otherwise, the definition, constraints, and examples provided by the RosettaNet PIP documentation (Guideline Information) are used. Every contextual element identified and represented in the *B2B Context Model* is denoted by a "**Y**" in the model. Otherwise a "**C**" is used instead, to indicate that the leaf node element in the PIP is a candidate addition to the model. In Step (4), the potential PIP context elements, marked as "C," are sorted and grouped into 3 types of influencing factors:

- **Primary Influencing Factor** – Factors that are crucial for B2B collaboration, and omitted from B2B Context Model.

- **Secondary Influencing Factor** – Factors that are crucial for B2B collaboration, present in B2B Context Model but require refinement.
- **Tertiary Influencing Factor** – Factors that are not crucial for B2B collaboration (nice to have), present in *B2B Context Model* but must be refined.

Mapping Results

From the verification, 10 PIP context elements were identified, i.e., elements marked with "C." They are further grouped into 3 types of factors with different degrees of influence; **0** Primary Influencing Factors, **5** Secondary Influencing Factors, and **5** Tertiary Influencing Factors. As there were no primary influencing factors, it can be inferred that the current *B2B Context Model* contains all the essential elements needed to support B2B collaboration. However, the model can be improved by inclusion of the Secondary and Tertiary Influencing Factors.

The relevance of these PIP context elements for inclusion were further analyzed. Table 9.3.1 provides a detailed listing of the justification for either inclusion or exclusion into the *B2B Context Model.* In summary, four influencing factors (three secondary and one tertiary) are included, as they can enhance the model's capability to better mimic real-life B2B scenarios. The PIP context elements are *Delivery Exception, Step Pricing, Interest Information,* and *Product Identification.* These elements are mapped into different B2B elements according to Table 9.3.1.

9.3.2 Proposed B2B Context Model

The verification exercise has confirmed that the initial *B2B Context Model* embodies all critical contextual elements needed for B2B collaboration. Several noncritical elements were also identified, as discussed earlier. Furthermore, several shortcomings were also identified during the verification exercise; in particular the way context elements are structured:

- The initial model is context centric, centered around 3 categories of context, namely *User, Location,* and *Temporal.* Such a categorization cannot capture the semantics between context elements. For example, most companies practice price differentiation that is time and quantity sensitive. The current model is unable to encapsulate such complexity.
- Also, associativity of elements from different context categories are not captured. For example, cost is related to both product price and other factors like transportation.
- Resulting from the context centric focus, some elements are actually meaningless without the relevant linkages. For example, the *Tempo-*

Influencing Factor		Description	Include?	Reason	B2B Element Name (new/changed)
Secondary	Delivery Exception	Allows providers to advise requestor of exception reason and time.	Yes	Allow providers to state the coverage, time of delivery, and price validity period.	- Market Availability - Validity
	Step Pricing	Allow product's variance in pricing in regards to specified date ranges and quantities.	Yes	Step pricing is the implementation of different pricing for different period of time and for different quantity of product ordered. Required in model to allow companies price differentiation in similar products.	Restructure the model to allow price variation with quantity (as Pricing Information)
	Interest Information	Describe interest information.	Yes	RosettaNet adopted this to provide information on interest due to late payment. Late payment penalties Influence purchase decisions.	- Payment Penalties
	Terms and Conditions Information	Describe the terms and conditions agreed between two parties under which a product/service is provided.	No	Most are standard, nonstandard terms and conditions require manual intervention. Not relevant in the context of B2B collaboration.	- Not applicable
	Handling Charges	The monetary amount to be paid for product handling for pre-, during and post-transportation. Typically	No	Such charges are usually included in the price, as declassification creates too much complexity.	- Not applicable
Tertiary	Contact Information	Provide communication and address information for contacting a person, organization, or business.	No	RosettaNet way of classifying contact information is more complete. But current B2B context model's "Contact Name" is enough for representing the user expression. Adding more complete information will just make the B2B context model more complex but give no advantage in partner selection.	- Not applicable
	Storage Description	The collection of business properties that describe properties of goods or packages being held in	No	Related to packaging information. Thus not particularly relevant to B2B collaboration.	- Not applicable
	Mass Physical Dimension	The physical dimension of unit of mass used for measuring the masses of weights, volumes, etc.	No	Related to packaging information. Thus not particularly relevant to B2B collaboration.	- Not applicable
	Proprietary Information	Description of information, relating to a product.	No	RosettaNet uses it as a free-form-text to describe proprietary Information. Not useful for contextual B2B collaboration.	- Not applicable
	Product Indentification	Describes proprietary and global identifier information regarding a product.	Yes	RosettaNet provides an unique identfier for products globally, related to GS1 system. Adopted into model for similar reason.	- Global Product Identifier (GTIN)

ral::TimeExpression is meaningless unless linked to other context elements like *Temporal::Leadtime* and *User::ProductService::Leadtime.*

Arising, a new *B2B Context Model* containing two main categories; *OrganizationalProfile* and *ProductService* is proposed (Figure 9.5), embedding the context aspects of *User, Location* and *Temporal. OrganizationalProfile* embodies all context information related to an organization (user); *Organizational_Expression* for storing direct organization information (e.g., name and industry sector), *Geographical Expression* to capture the location context; both direct and indirect and *Current* to capture a snapshot of the current state of the company in terms of *financial, quality,* and *turnover.* It also has other location related context, *Ease_of_Doing_Business*, based on the country the company is located in.

ProductService is a hierarchical structure that captures the semantic relationship of a product or service. In *PricingInformation, Costing* is scoped by other elements like *Payment, Validity,* and *Quantity* (minimum or order) for instance. It is important to note that the generic context elements from *Location* and *Temporal* are still present, except that they are embedded within the B2B context elements. Examples include *PricingInformation::Costing::SalesTerritory::SourceCountry,* and *PricingInformation::Validity::TimeScale.*

By rationalizing the model, the *B2B Context Model* is now able to address the limitations highlighted earlier. The model has a mix of both quantitative and qualitative criteria (e.g., *Turnover* is High). The *OrganizationalProfile* category in particular, has many qualitative elements, for example *Current::Turnover*[1] and *History::Years _of_Operation.* Under the *ProductService* category, *PricingInformation::Payment::PaymentTerm* is also a qualitative criterion. These criteria typically require special processing, and will be discussed further in section 9.4.2.

9.3.3 Information Sources and Usage of the Model

The frequency of change and validity of context information within the model is expected to vary. For instance, information pertaining to a company, such as *OrganizationalProfile::OrganizationalExpression and ProductService::Product_Identifier* is not expected to change very often. Others like *ProductService::PricingInformation* would change fairly often, depending on the products and services. *Tariffs* for the product/service on the other hand may stay fairly constant, unless there are changes in the tax structure of a particular country for example.

The information sources for the elements are expected to be varied. For example, company related information is expected to be published by the company itself or could be obtained from business documents that are commonly

[1]The notation used is in the form of *Element::Subelement,* omitting the category.

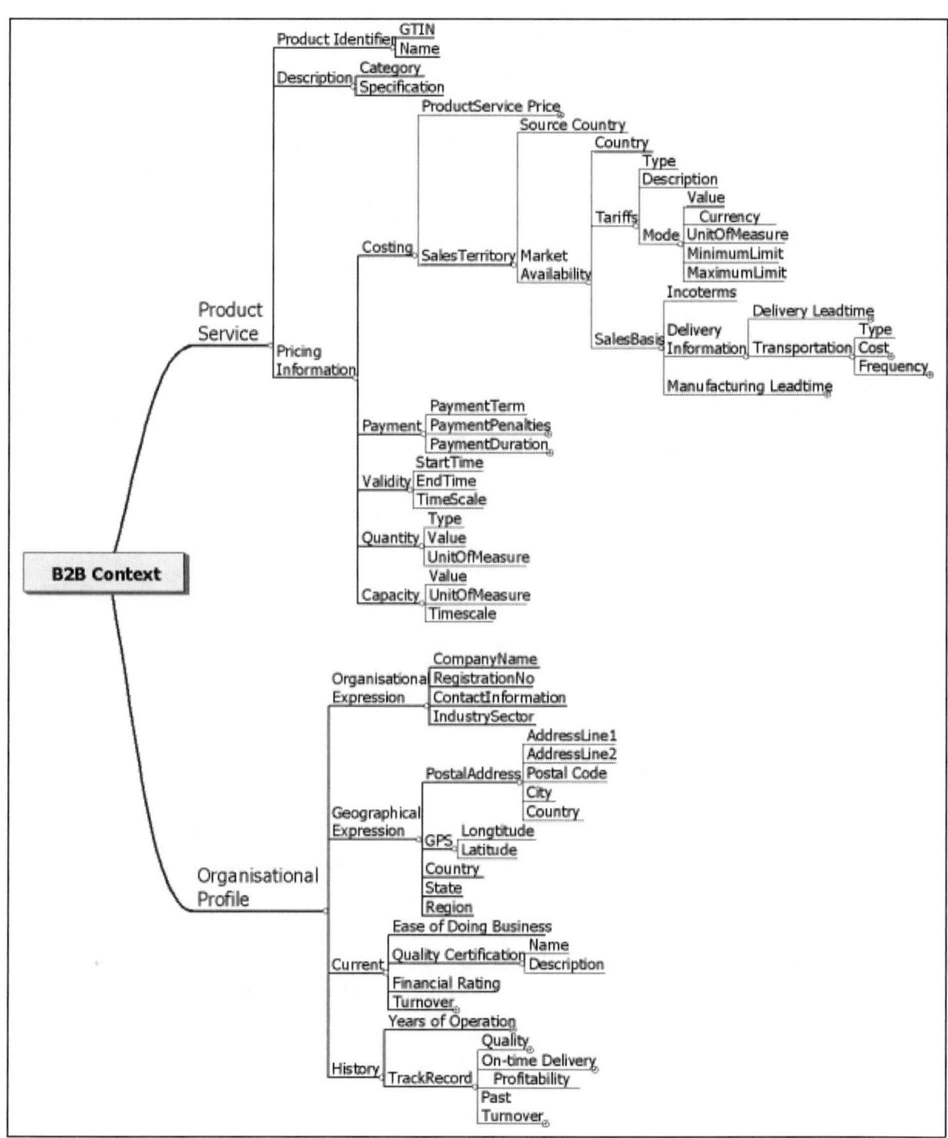

Figure 9.5: Proposed *B2B Context Model.*

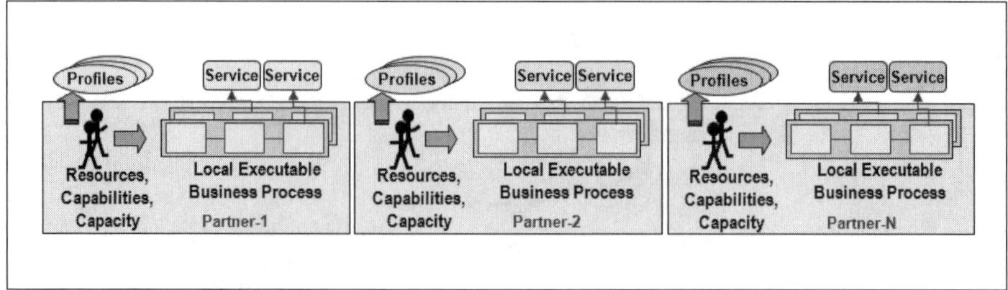

Figure 9.6: Realization of the model as *context-aware profiles.*

exchanged during B2B transactions. Information on products/services would have to be provided by the company itself, such as product catalogues and quotations. Again, such information is exchanged as business documents, too. Other information like the GPS (Global Positioning System) coordinates can be obtained automatically from external sources. Web sites readily provide such information (e.g., http://www.gps-data-team.com/map/). Information on ease of doing business could also be obtained externally, such as from The World Bank Group (http://www.doing business.org/economyrankings/).

There are numerous methods for sensing and capturing the relevant information into the context model, such as data mining and auto-updates for example. This is not the focus and scope of this chapter. The assumption is the context information can be formulated into *context-aware profiles* to support context-aware B2B collaboration, in particular the formation of virtual enterprises (see Figure 9.6). The partners are assumed to be operating in a Service Oriented Architecture (SOA) environment, offering *services* that can execute a formed business process as part of a B2B collaboration between partners (23). These profiles function like the "yellow pages" of these services. In the Web Services environment, the equivalent would be WSDL, and OWL-S (Semantic Markup for Web Services) profiles for the semantic version of Web Services. Here the context-aware profiles are oriented towards capturing business information (as reflected by the *B2B Context Model*) that is typically not available in either the WSDL or OWL-S profiles, and is expected to complement such profiles.

9.4 Application and Evaluation of the B2B Context Model

The use of the *B2B Context Model* to support the context-aware approach in B2B collaboration (partner selection as a specific usage example) has been validated using quantitative criteria (25). In this chapter, the focus is on validating the qualitative criteria, through a hybrid fuzzy approach that is able to handle both types of criteria.

9.4.1 Hybrid Multicriteria Decision Making (MCDM) Technique for Context-Aware Partner Selection

In partner selection, the importance of the various selection criteria (1) depends on a company's business strategy, goal, or objective in a collaboration, and other prevailing business conditions such as distance, available lead-time, and costs, for example. Given a set of selection (matching) criteria and corresponding weights to denote relative importance, the challenge is then to devise an automated context-aware methodology to identify and rank suitable partners. A hybrid multicriteria decision making (MCDM) technique incorporating fuzzy logic is proposed. In MCDM techniques, a generalized set of steps based on the following is typically adopted (2):

a) Generation of ideas
b) Structuring of ideas
c) Model building
d) Determining the relative importance of the criteria
e) Determining the impact of the alternatives on the criteria (scoring)
f) Processing the value to arrive at a ranking for alternatives
g) Final decision making and review

Here, Step (a) and (b) are already predetermined and generated through the context-aware approach, while Step (c) is partially automated by choosing from the criteria inherent in the *B2B Context Model.* Once the criteria are determined, MCDM techniques can be employed in Step (d) through Step (f). In Step (g), the user may accept or reject this short list of ranked candidates, or to modify the ranking by changing the weights of some criteria.

In Step (d), the Analytic Hierarchy Process (AHP) provides an intuitive yet easily quantifiable method of determining the relative importance of weights using the Saaty scale (20, 19). Put simply, AHP makes pair-wise comparisons of criteria/alternatives and then scores them on a scale ranging from 1 to 9 (9 being the widest disparity of relative importance). For example, consider two criteria, price and lead-time. If price is just as important as lead-time, then the score assigned to this pair-wise comparison is 1. If price is only marginally more important than lead-time, then a score of 3 is apt. However, if it is significantly much more important, a score of 7 or 9 could be assigned. On the other hand, when lead-time is compared to price, a score of less than 1 is expected (for instance $1/3$, $1/7$, or $1/9$). These scores are then normalized against the Saaty scale, where the summation of all weights, $\sum_{i=1}^{n} w_i = 1$, for n-criteria.

For Step (e) a hybrid technique, combining the *Deviation Measure* (25) method and *fuzzy* logic is used. The *Fuzzy Deviation Measure* method is based on the concept that a requestor has *preferred values* (collaboration context).

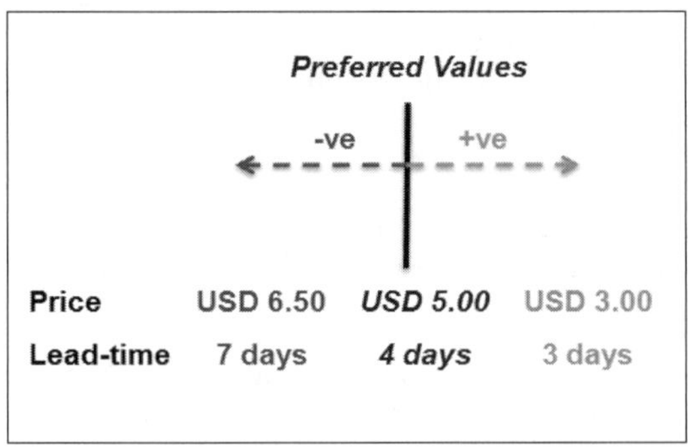

Figure 9.7: A deviation measure from preferred values.

Potential partners are assessed for the deviation from these *preferred values*; positive (e.g., lower price) or negative (e.g., longer lead-time), as illustrated in Figure 9.7. The steps to compute the *Deviation Measure* are as follows:

1. *Calculate the normalized performance matrix*

   By applying deviation measure, the normalized performance value of the i-criterion in the j-alternative, v_{ij} can be computed depending on whether the objective is to minimize or maximize a particular criterion.

   (a) For a criterion where the objective is to *minimize* the value,

   $$v_{ij} = (\frac{r_i - a_{ij}}{r_i}) \tag{9.1}$$

   (b) For a criterion where the objective is to *maximize* the value,

   $$v_{ij} = (\frac{a_{ij} - r_i}{r_i}) \tag{9.2}$$

   where r_i= requestor's preferred value, a_{ij}= alternative's value

2. *Calculate the weighted normalized performance matrix*

   The weighted normalized individual criterion score,

   $$ICS_{ij} = w_i v_{ij} \tag{9.3}$$

   where w_i= weight of i-criterion

3. *Calculate the overall suitability score OSS_j* for all the criteria for each j-alternative,

   $$OSS_j = \sum_{i=1}^{n} ICS_{ij} \tag{9.4}$$

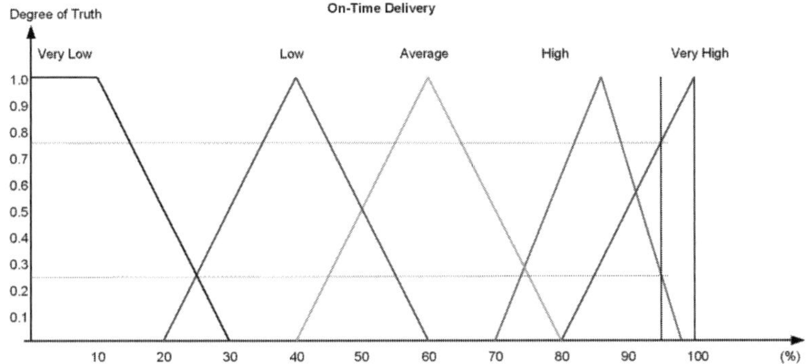

Figure 9.8: Fuzzy Model *of On-Time_Delivery.*

A positive score indicates that the alternative is a viable alternative, whereas a negative score indicates that it is not viable, when compared to the ideal or preferred values.

4. *Rank all alternatives* by their overall suitability scores (OSSs) in descending order. The highest ranked alternative is the best candidate.

For step (1), where a criterion is qualitative in nature (e.g., *Organizational-Profile::Current::Ease_of_Doing_Business*) a separate algorithm, based on Fuzzy Set Theory (30), is used to determine the *crisp values* that can be used for the ranking process. The next section will illustrate this in greater detail.

9.4.2 Fuzzy Processing of Qualitative Criteria

Membership Functions Design of Qualitative Criteria

In fuzzy logic, the membership function represents the *degree-of-truth* of a statement like "The turnover of a company is high." The membership functions for each criterion are designed to be either 3 or 5 categories. As an example, the membership functions for *On-Time_Delivery* (*Very Low*, *Low*, *Average*, *High*, and *Very High*) are shown in Figure 9.8.

Fuzzy Controller

Using the Fuzzy Controller that is based on the Mamdani concept (15), the stages involved in handling fuzzy criteria can be broken into (12):

- ***Input Stage*** – Mapping of performance values of each criterion to appropriate membership functions (based on predesigned membership models)

  Using the example of a *On-Time_Delivery* of 95% and the fuzzy model in 9.8, the related fuzzy conditions are $(OnTime\ Delivery = High)_{0.25}$

Lingustic Value	Rank Values	
	5-Category	3-Category
Very High	1.0	1.0
High	0.8	-
Average	0.6	0.6
Low	0.4	0.3
Very Low	0.2	-

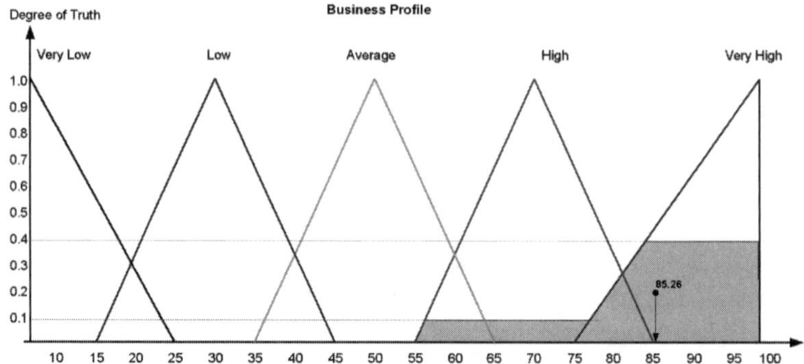

Figure 9.9: Fuzzy Model of Business Profile.

and $(OnTime\ Delivery = Very\ High)_{0.75}$. 0.25 and 0.75 represents the *membership values* or *degree-of-truth* for each of the linguistic statement. In order to automatically generate fuzzy rules, a *rank value* is defined. *Rank value* is a number between a 0 and 1 that is assigned to a *linguistic value* like *high, average*, or *low*. The rank values for 3- or 5-category models are shown in Table 9.4.2.

- ***Processing Stage*** – Invocation of relevant fuzzy rules and generating the result for each rule, including aggregating the results into a single output

  A fuzzy rule typically has the form of IF-THEN statements, where the IF part is the *antecedent* and the THEN part is the *consequent*. *Antecedents* are combined using fuzzy operators such as AND, OR, and NOT. The AND operator usually uses the minimum weight of all operators, while OR uses the maximum weight. The NOT operator results in a complementary function to the *antecedents*. For the qualitative criteria, the *consequent* of the rules (the THEN part) is aggregated into a single outcome (or output) known as the *Business Profile* (see Figure 9.9).

  There are altogether 10 criteria with either 5-category or 3-category membership functions. Instead of predefining the fuzzy rules, an algorithm to automatically generate the relevant fuzzy rules was designed and implemented. The details are:

1. Determine the *rank value* of each input fuzzy condition arising out a criterion, using Table 9.4.2
2. Calculate the average *rank value* from all the fuzzy input criteria for each possible combination, $AveInRV_n$ for n*th*-combination
3. Determine the outcome of all the input conditions (*consequent* portion of each rule), $OutRV_n$ based on the following:
   - If $AveInRV_n > 0.83$ then $OutRV_n = Very\ High$
   - If $AveInRV_n < 0.83$ and $AveInRV_n > 0.65$ then $OutRV_n = High$
   - If $AveInRV_n < 0.65$ and $AveInRV_n > 0.50$ then $OutRV_n = Average$
   - If $AveInRV_n < 0.50$ and $AveInRV_n > 0.30$ then $OutRV_n = Low$
   - If $AveInRV_n < 0.30$ then $OutRV_n = Very\ Low$

In other words, these are the "generated rules." Using an example for two fuzzy criteria of *On-Time_ Delivery* (input value 95%) and *Quality* (input value 0.12%) would yield the following fuzzy conditions:

- $(OnTime\ Delivery = High)_{0.25}$ and $(OnTime\ Delivery = Very\ High)_{0.75}$
- $(Quality = High)_{0.10}$ and $(Quality = Very\ High)_{0.40}$

The 4 possible combinations, and their respective $OutRV_n$ are:

(1) $(OnTime\ Delivery = High)_{0.25}$ and $(Quality = High)_{0.10}$, then $AveInRV_1 = \frac{0.8+0.8}{2} = 0.8$ and $OutRV_1 = High$

(2) $(OnTime\ Delivery = High)_{0.25}$ and $(Quality = Very\ High)_{0.40}$, then $AveInRV_2 = \frac{0.8+1.0}{2} = 0.9$ and $OutRV_2 = Very\ High$

(3) $(OnTime\ Delivery = Very\ High)_{0.75}$ and $(Quality = High)_{0.10}$, then $AveInRV_3 = \frac{1.0+0.8}{2} = 0.9$ and $OutRV_1 = Very\ High$

(4) $(OnTime\ Delivery = Very\ High)_{0.75}$ and $(Quality = Very\ High)_{0.40}$, then $AveInRV_4 = \frac{1.0+1.0}{2} = 1.0$ and $OutRV_1 = Very\ High$

- ***Output Stage (Defuzzification)*** – Conversion of the output into a specific control output value (crisp value).

  Here the centroid method of defuzzification is adopted. In determining the areas to calculate a centroid, the typical treatment of AND or OR operators are adopted to determine the relevant membership values. Using the same example:

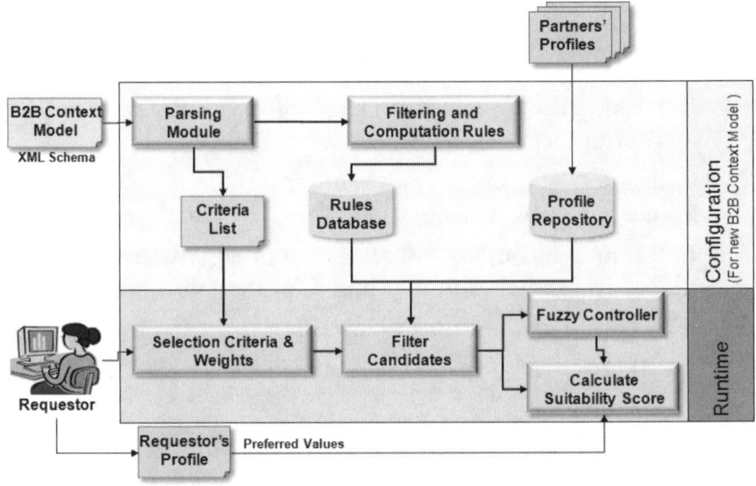

Figure 9.10: Cognitive Advisor's Usage Flow.

(1) Membership value $= min(0.25, 0.10) = 0.10$ of *High*

(2) Membership value $= min(0.25, 0.40) = 0.25$ of *Very High*

(3) Membership value $= min(0.75, 0.10) = 0.10$ of *Very High*

(4) Membership value $= min(0.75, 0.40) = 0.40$ of *Very High*

Using the centroid calculation method for defuzzification, $C = \frac{\int xf(x)dx}{\int f(x)dx}$, the crisp value obtained is *85.26*, as shown in Figure 9.9.

9.4.3 Prototype Implementation

The *B2B Context Model* and the hybrid *Fuzzy Deviation Measure* were implemented in a prototype called *Cognitive Advisor.* It is a Web-based application based on components and implemented using Java. It has been tested for a small number of quantitative criteria (25). The usage flow is shown in Figure 9.10. A request and its associated criteria is initiated through a Graphical User Interface (GUI). These criteria are then weighted by the the AHP pair-wise comparison technique. Thereafter, the system takes over the processing, including filtering of incompatible candidates, activating the fuzzy controller for qualitative criteria, and finally, calculating the *OSS* for each potential candidate and ranking them according to the *OSS* values.

Criterion\Candidate	Requestor	Tomoto	Matreng	Hamind	Indisar Puri	Deltazer	Chenpo
Quality (%)	0.5	0.01	0.03	0.1	0.08	0.15	0.12
On-time Delivery (%)	80	99	98	99	96	99	95
Profitability (%)	10	25	22	24	20	24	19
Past Turnover ($ mil)	20	183	230	148	122	151	125

Test Criteria	Criterion Type*	Test-1	Test-2	Test-3
Price (cost)	T	-	0.60	-
Lead-time	T	-	-	0.50
Capacity	T	0.50	-	-
Past Turnover	L	0.17	-	0.17
Quality	L	0.17	0.20	0.17
On-Time Delivery	L	0.17	0.20	0.17

* T=quantitative criterion, L=qualitative criterion

9.4.4 Setup of Test Runs

In verifying the viability of the context-aware approach for partner selection using the hybrid *Fuzzy Deviation Measure* approach, test scenarios that encompasses both qualitative and quantitative criteria were employed. A sample of the qualitative data of six candidate providers, are shown in Table 9.4.4. Altogether, 20 candidate providers were used in the tests. Table 9.4.4 shows the criteria and corresponding weights used in the four tests. The weights shown are derived from the AHP pair-wise comparison (see 9.4.1). The tests (`Test-1`, `Test-2`, and `Test-3`) were designed to understand the effect of fuzzification on the rankings (OSS values), by using a dominant quantitative criterion (capacity, price, and lead-time) that has a high weight of at least 0.5, and several qualitative criteria.

9.4.5 Test Results

Figure 9.11 shows the effect of fuzzification on the OSS values of candidate providers, as compared to when no fuzzy implementation for the qualitative criteria for `Test-1`. For no-fuzzy implementation, the performance values (v_{ij}) of the qualitative criteria are baselined at `-1`. Thus, the improvement shown is against this baseline, factoring in the respective criterion weight. Through fuzzification, a more accurate OSS of candidate providers is achieved. Similar improvements are observed for `Test-2` and `Test-3`. Figure 9.12 shows the difference in OSS values for all three tests. The average improvements recorded are 0.54, 0.45, and 0.54 for `Test-1`, `Test-2`, and `Test-3`, respectively. The standard deviation for the tests are also at an acceptable range of between 0.08 to 0.14. It is important to note that since the qualitative criteria in `Test-1` and `Test-3` are the same and have the same weights, the

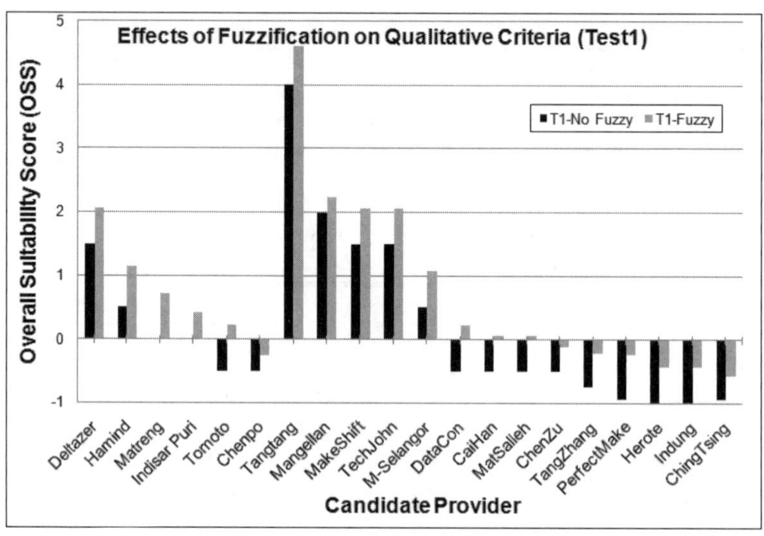

Figure 9.11: Effects of Fuzzification (Test-1).

OSS difference for all candidate providers should be the same, regardless of the quantitative criterion. This is confirmed, as shown in the graph, with the same mean and standard deviation values.

Figure 9.12 also shows that the OSS difference is significantly higher for some companies (e.g., `Tomoto`) as compared to others (e.g., `ChenPo`), for all tests. This indicates that the fuzzy processing was correctly implemented, as `ChenPo` has significantly lower performance values (see Table 9.4.4).

9.5 Conclusion

This chapter discusses a *B2B Context Model* for supporting B2B collaboration. In this model, the main categories from the *Physical* aspect of context-aware computing (largely from ubiquitous computing), namely *User*, *Location*, and *Temporal* are adopted. However, instead of being context-centric, the model is B2B centric, with main categories such as *ProductService* and *OrganizationalProfile*. This is mainly due to the complex nature of *B2B context*; creating many semantic relationships and some context elements are meaningless (e.g., time) unless linked to other context elements from the B2B domain. The model is expected to be instantiated into *context-aware profiles* to complement typical Web services profiles like WSDL (Web Service Description Language) and OWL-S (Semantic Markup for Web Services)

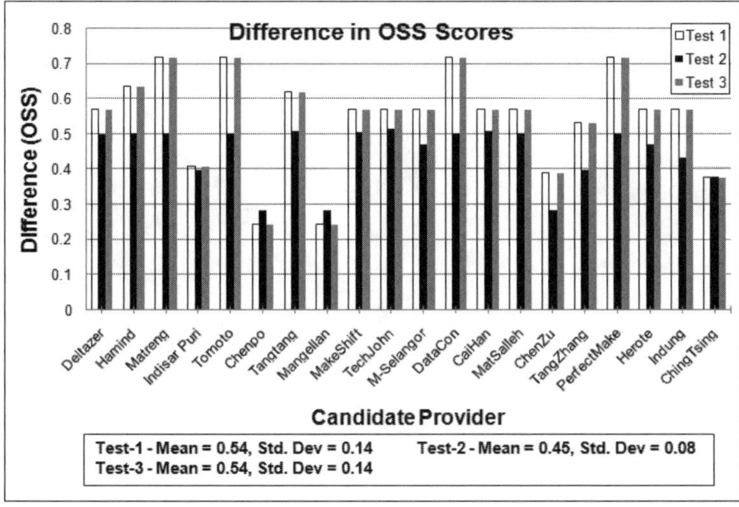

Figure 9.12: Effects of Fuzzification on the OSS.

with more business-related information. The model contains both qualitative (e.g., *Quality* and *Financial Rating*) and quantitative elements (e.g., *Price* and *Cost*). This model has also been verified for completeness against an industry de facto standard for B2B collaboration, RosettaNet.

The *B2B Context Model* was then implemented to support context-aware partner selection, as an example of context-aware B2B collaboration. A hybrid technique called *Fuzzy Deviation Measure* was introduced to rank candidate partners. This technique is able to handle both qualitative and quantitative elements within the model. Through the use of fuzzy logic, the influence of qualitative criteria are now more accurately represented.

In future, this model is also expected to support other B2B collaborations in general, such as *business process formation* based on an articulated goal (collaboration context) and *exception management* in formed business collaboration for supply chain applications using context-aware adaptation based on newly detected context conditions. Investigations into this *B2B Context Model* are also ongoing, including using an alternate representation format (currently it uses XML Schema) such as ontology. The fuzzy method adopted for handling qualitative criteria will also be investigated further, to further enhance its accuracy and ability to mimic industry norms.

References

[1] Umit Akinc. Selecting a set of vendors in a manufacturing environment. *Journal of Operations Management*, 11(2):107–122, 1993.

[2] Kolera Radhika Apaiah. *Chapter 5: An Overview of Multi Criteria Decision Making (MCDM) of Designing Food Supply Chains – A structured methodology: A case on Novel Protein Foods*. PhD thesis, 2006.

[3] Hussein Badakhchani. Introduction to RosettaNet, 20 Sep 2004.

[4] Mary Bazire and Patrick Brézillon. Understanding context before using it. In A. Dey, editor, *CONTEXT 2005*, volume LNAI 3554 of *LNAI*, pages 29–40, Paris, France, 2005. Springer.

[5] L. M. Camarinha-Matos. Virtual organizations in manufacturing: Trends and challenges, 2002.

[6] L. M. Camarinha-Matos, editor. *Collaborative Networked Organizations: A Research Agenda for Emerging Business Models*. Springer, 2004.

[7] Cambridge. Cambridge advanced learner's dictionary, retrieved on 17 November 2007.

[8] Luitzen de Boer, Eva Labro, and Pierangela Morlacchi. A review of methods supporting supplier selection. *European Journal of Purchasing & Supply Management*, 7(2):75–89, 2001.

[9] Deloitte. Mastering complexity in global manufacturing – powering profits and growth through value chain synchronization, 2003.

[10] C. M. Goh, P. S. Tan, S. G. Lee, and S. P. Lee. Key issues in manufacturing enterprise integration - the singapore perspective. In *3rd IEEE International Conference on Industrial Informatics, 2005 (INDIN '05)*, 2005 3rd IEEE International Conference on Industrial Informatics (INDIN), pages 390–395, Perth, WA, Australia, 2005. IEEE.

[11] Kees Jacobs, Peter Jordan, Ruud van der Pluijm, Ard Jan Vethman, and Sabine Ritter. 2016: The future value chain. *Executive Outlook*, 6(4):46–63, 2006.

[12] Jan Jantzen. Tutorial on fuzzy logic, 1998.

[13] P. Kumar, S. Gopalan, and V. Sridhar. Context enabled multi-cbr based recommendation engine for e-commerce. In *IEEE International Conference on e-Business Engineering*, pages 237–244, Beijing, China, 2005. IEEE Comput. Soc.

[14] Matthew Littlefield. Demand driven manufacturing: Synchronizing demand with production, Nov 2007.

[15] E.H. Mamdani. Application of fuzzy logic to approximate reasoning using linguistic synthesis. *Computers, IEEE Transactions on*, C-26(12):1182–1191, Dec. 1977.

[16] Ram Muthukrishnan and Jefferey A. Shulman. Understanding supply chain risk: A McKinsey Global Survey. Technical report, Sep 2006.

[17] Le Duy Ngan, Angela Goh, and Cao Hoang Tru. A survey of web service discovery systems. *Int. J. of Information Technology and Web Engineering*, 2(2)(65-80), 2007.

[18] RosettaNet. Overview clusters, segments, and pips - version 02.06.00, Jan 2009.

[19] Thomas Saaty. Decision making - the analytic hierarchy and network processes (ahp/anp). *Journal of Systems Science and Systems Engineering*, 13(1):1–35, 2004.

[20] Thomas L. Saaty and L. Vargas. *Fundamentals of Decision Making and Priority Theory With the Analytic Hierarchy Process*, volume Vol. 6 of *Analytic Hierarchy Process Series*. RWS Publications, 2000.

[21] Gunjan Samtani and Dimple Sadhwani. B2bi and web services: An intimidating task? In P. Fletcher and M. Waterhouse, editors, *Web Services Business Strategies and Architectures*, pages 57–59. Wrox Press, 2002.

[22] Robert Shecterle, Nari Viswanathan, and Melissa Spinks. The supply chain executive's strategic agenda 2008: Managing global supply chain transformation, Feb 2008.

[23] P. S. Tan, A. E. S. Goh, S. S. G. Lee, and E. W. Lee. Issues and approaches to dynamic, service-oriented multi-enterprise collaboration. In *Industrial Informatics, 2006 IEEE International Conference on*, pages 399–404, 2006.

[24] P.S. Tan, A.E.S. Goh, and S.S.G. Lee. Context information support for B2B collaboration. *International Journal of Web Engineering and Technology (IJWET)*, 5(2):214–245, 2009.

[25] P.S. Tan, E.W. Lee, K. Mous, S.S.G. Lee, and A.E.S. Goh. Multi-criteria, context-enabled b2b partner selection. pages 1699–1706, Oct. 2008.

[26] Nari Viswanathan. Demand Management in Discrete Industries: Order to Delivery Exceleence, Sep 2007.

[27] Nari Viswanathan. On-demand applications in supply chain: Enable flexible business processes. Technical report, Aberdeen Group, Aug 2007 2007.

[28] Nari Viswanathan. Supply chain innovator's technology footprint 2008: Technology enablers for driving supply chain transformation, May 2008.

[29] I. H. Yigin, H. Taskin, I. H. Cedimoglu, and B. Topal. Supplier selection: An expert system approach. *Production Planning and Control*, 18(1):16–24, 2007.

[30] L. A. Zadeh. Fuzzy sets. *Information and Control*, 8:338–353, 1965.

Chapter 10

Context-Aware Mobile Grids

Stefan Wesner, Antonio Sanchez-Esguevillas, Victor Villagra, and Babak Farshchian

Abstract In this article an approach for the integration of Session Initiation Protocol (SIP)-based session and context management, the Virtual Organization concepts from the Grid community and SOAP-based service interactions is presented. The technical approach is motivated by two simplified scenarios that have been realized in the frame of a large scale European collaborative research project called Akogrimo. Furthermore, different context gathering approaches are discussed and how in such a distributed application setting a potential adaptation to these different contexts, such as location or device capabilities can be used for realizing mobility-aware service-oriented Grids.

10.1 Introduction and Scenarios

The impact of the increasing capabilities of mobile devices and the almost ubiquitous availability of Internet connections with high bandwidth allow new kind of applications where mobile users and workers aim to participate in distributed collaborative applications at any time everywhere. Service-oriented Architecture (SOA) has achieved the status of a de facto standard approach for the realization of Web-based distributed applications. The most prominent realization technology of the SOA concept are Web Services. However, distributed applications often need additional functionality beyond Web Service interactions such as audio and video communication or the streaming of very large data sets. Applications beyond Web-based interactions are much more sensitive to changes of network or device capabilities and need to be dynamically adapted to the changing context of the user. Gathering context information ranging from location over device capabilities up to user profiles and contracts requires a more broad view on the SOA concept beyond pure Web Services using advances from research on the Future Internet.

In order to motivate the need for context aware mobile grids two selected scenarios representing abstractions of realized demonstrators in the European research projects Akogrimo (25) and CoSpaces (13) are described in this section from different application domains ranging from personal systems for a large user community down to scenario in a commercial environment for engineers using partially highly specialized environments and resources.

The following two scenarios aim to demonstrate how context awareness provide the necessary information about the user's environment that can be applied transparently for the user to the way services are provided to her/him. Location is the most common example, but there exist many other types of context information, like those provided by sensors like accelerometers (acceleration), gyroscopes (orientation), weather-related parameters like temperature, humidity, light, wind speed, etc. whose values, if talking about large extensions with several measurement points can be communicated through Wireless Sensor Networks (WSN); health-related parameters like blood pressure, heartbeat rate, insulin level and whose values are communicated through so-called Body Area Networks (BAN). Furthermore, context also apply to the things and people that surround a user in a given place and time. For identifying things, the most extended technology used is RFID (Radio Frequency Identification) tags. For identifying people, apart from biometric technologies (e.g. face recognition) a more simple non intrusive way is through communication with personal devices, like mobile phones, through Personal Area Networks (PAN) radio technologies like Bluetooth or Zigbee. Additionally, context covers also the current situation such as "free time or work time," "available versus occupied," as well as the changing situation within a collaboration e.g., from "participant to presenter."

The scenarios assume the existence of a Grid and Cloud environment ensuring the availability of powerful resources accessible from anywhere on demand as utilities. Putting context awareness and the availability of remote services and resources available for mobile users (mobile grids), we can imagine a myriad of scenarios that would simplify our daily routines enormously. The model would be as follows: local sensors (e.g., like the ones in the mobile phone) provide context information on the go as we move. Additional context information can be captured from general networks of sensors (e.g., city temperature) and can be combined with situation/collaboration status context in order to have a highly dynamic and configurable environment that adapts to the changing environment but similarly to the change in the collaboration.

10.1.1 Personal Services Scenario

The first scenario aims to illustrate from a consumer's viewpoint how such an adaptive environment can be used in private and in worker contexts. It is assumed that the person carries a digital companion or Personal Assistant

(PA) able to use available local services and gather context information as well as to communicate with external services provided. This scenario is partially driven by work presented in (22, 21).

The day starts and Maria is sleeping, the PA has regularly sent requests to the agenda Grid service (service in short from now on) for updated schedule information for the day (meetings, tasks,etc.) and another alarm service calculates the appropriate wake up time based on this information and personal preferences (e.g., "rush-and-go-breakfast"). When the time arrives, a local sensor together with a remote sleep resource detects whether Maria is still asleep and if the wake-up alarm or notification alarm should be triggered. Fortunately, this morning no physical but a video conference meeting is scheduled and the agenda service has calculated this extra sleeping time taking into account the time saving. Once awake (the sensor makes sure Maria is awake as a condition to stop the alarm, a so-called annoying persistent waking up system), a domotics service commands the blinds to open and opens the window (Marias profile indicates that she wants fresh and clean air in the room). If Maria is not at home but in a hotel room the device uses the local facilities if available (and performs some similar tasks at home to make the impression someone is at home to fool potential thieves).

Next Maria moves to the home office (e.g., room with a laptop, dock station, screen, keyboard, mouse and Webcam). A sensor in the laptop detects that she is approaching and turns the computer system on avoiding the initial waiting time. Once more, this location service could be requested to an external location resource that can make more complex calculation to accurately estimate the time it takes to arrive in the room. The laptop connects to a virtual environment in the office, which again is a remote desktop resource (that might be hosted for example by an external service provider, that has outsourced all the desktop environments of the company in its Internet Data Center).

Next, Maria has the multivideoconference, in which her suppliers, partners, customers, and colleagues (boss and team) are involved, each of them in a different location (even country). No matter where each of the participants are, because the conferencing resource connects the people automatically, not the offices where they are usually located (their location and device capabilities context is gathered). Her neighbor rings the door to ask for salt, but since Maria is in the meeting the bell does not even ring in order not to disturb. This is possible as the bell is not a disconnected facility in the house but is part of the personal communications service merging all types of voice communications, from the mobile, the fixed, the IP softphone, or the real world. As the meeting is taking longer as expected the agenda service is updated automatically and a corresponding replanning process is triggered. This triggers a notification for Maria that the plan for todays lunch break to buy a gift for her parent's wedding anniversary is at risk. Just after the video conference Maria orders online for her parents a last-minute gift. The selection of the gift is supported by the profile of her parents in a social network and her locally stored "wish list." As a result of the video conference several action points had been assigned to Maria. All work items are associated with the necessary services and resources (e.g., documents, database queries, etc.). Upon completion of an action point the consumers of the results are automatically notified.

10.1.2 Engineering Scenario

In order to further motivate this view the following scenario outlines the close interaction between different types of applications and components in a distributed engineering application:

Jeff is just on his way to a project partner and has some time waiting for his connecting flight to check out recent posts in the company internal secured Twitter with his mobile. He is delighted to see the post from the job scheduler that the long-awaited simulation results are available and that Sally, his co-worker, has already picked up the next tasks in the workflow to prepare the post-processing and to book the windtunnel for comparison of the results with the real experiment. Some minutes later the post-processing application posted a new entry in the corporate Twitter that the new material is available. Jeff starts a session for browsing these results. Based on the limited capabilities of his mobile device and the rather slow network connection, the provided data is mostly meta-data and some associated low-quality pictures. Additionally, the public environment blocks the delivery of any confidential content. Some hours later Jeff continues the data analysis session by transferring it from the mobile to his powerful laptop and asks Tom (the partner is currently visiting) and Sally to join this session. The available local resources such as the stereo-display for 3D visualization is detected and authorized by Jeff to be included in the session. The increased capabilities of his devices, the local resources, and the much better network connection allow a more sophisticated setting. The services supporting the collaboration session automatically adapt to the different capabilities of the available resources and analyze if the collaboration goal set by Jeff "validate the results and decide if the milestone have been achieved," and its corresponding minimal requirements on the client side are met or could have been provided by back-end services. During the session where an audio/video conference is combined with a shared view of the simulation results and links to reference documentation, e.g., with the original metrics assigned for achieving the milestone, they participants agree that the milestone has been achieved and Sally confirms her booking of the wind tunnel for the next step. In order to make the team aware of this decision and provide a convenient access to the key results, they reconfigure the session from analysis to publication mode and ask Sandra to join the session. They jointly agree on text, links, and pictures to be placed within the corporate blog and which parts should go into the news section on the public Web page. The underlying infrastructure supports them in setting the appropriate access rights to the data and results linked in the blog entry.

10.1.3 Summary of Key Challenges

The scenarios in the previous section shows that the application context is changing not only due to changing connectivity and devices used in the collaboration but also due to a change in the business process, current activity, collaboration or interaction with other persons using different kinds of devices, depending on the necessary purpose. Furthermore, private and commercial activities have to be seen in an integrated way as many dependencies need to be considered.

The derived key challenges are:

1. The SOA concept cannot be limited to Web Services but cross-layer cooperation is necessary from network over middleware up to the application layer in order to allow information about device capabilities, current location, etc., to be communicated to the provisioning services and similarly to allow the services to communicate their requirements on the communication infrastructure. This covers the need to integrate different elements ranging from doorbells, Twitter entries over Web Services up to video communication, or visualization services.

2. The collection of a wide number of raw context information is not sufficient but need to be propagated to a complete context view and requires an automated evaluation of the information and an automated adaptation (controlled and steered by the user as well as company policies).

3. The standardized and rich description of external service providers, as well as the context information, is essential for the realization of such a scenario.

4. The collaboration phase (e.g., expressed in business processes) has a major impact on how context information is interpreted and how the reaction needs to be. For example, the context "low bandwidth, small device" is no problem in the notification phase of the second example but is a showstopper for the collaboration phase.

10.2 What Is in Context, and What Is Out: The Need for Adaptation

Context is the information that is needed to give meaning to any collaborative act. In day-to-day collaboration scenarios we can distinguish among three types of information exchange among collaborating parties (7):

- Intentional communication: Information contained in explicit utterances that are exchanged between the participants. This includes both verbal and non verbal (e.g., gesture) communication among humans or exchange of communication messages among IT systems.

- Feedthrough: Information that is mediated through artifacts involved in collaboration. For instance, rearranging documents on a table might give a specific signal to collaborating people.

- Consequential communication: Information that is generated due to the fact that our bodies and our body language are all visible to other parties.

The quality and the effectiveness of collaboration depend on proper access to this information by all parties involved. Context information, in order to have value, needs to be shared among collaborating parties, be it humans or computing systems.

In real-world scenarios, a large amount of context information exists in the environment, but not all this information is of importance. The importance of context information can vary based on time, user attention, and the ongoing collaboration. The human brain is highly adaptive regarding what context information it prioritizes above other. Computer systems are often very limited in that they have a fixed (hard-coded) model of context, and degree of adaptation is low. This results in computer systems that are not perceived by humans as intelligent or flexible. Adaptation to the situation and correct prioritization of context information is one of the biggest challenges in context-aware computing.

10.2.1 Context Gathering Approaches: The Need for Standards

Context information can be collected from the physical world (through physical sensors) or it can be collected from various information systems. At the physical world level, a lot of attention is paid to communication standards that allow signals from the physical world to be represented in digital forms. At the information systems level, research is focused on creating models (ontology) of context information so that interoperability among computer systems can be guaranteed. In general standards are needed at different levels:

- Context-sensor discovery standards: Context information, especially from the physical world, is spread around. Users are mobile and move from one context to another. Discovering sources of context in various locations is the first step towards context-aware computing. For instance, when a user enters a room with a temperature sensor, the sensor needs to be discovered by a user's application before the application can make any use of the temperature data provided by that sensor. Discovery systems and standards are developed for various purposes and by various vendors and standardization bodies (4).

- Data standards. In order for the temperature data to be understandable by an application, the data needs to be represented and exchanged in a standard format. Standards such as Universal Plug and Play (UPnP) (2) for home appliances, and Continua (1) for medical devices are being developed in order to guarantee this type of data interoperability.

- Context models: Besides the shared understanding of what bits of context mean, relations among these pieces of context data are of importance not only to provide a better understanding of the real context, but also to enable reasoning based on the data. A number of context models are developed that support relationships among context data (9). In the next section we will look at a number of models that can be useful in supporting human collaboration.

10.2.2 Sharing Context: The Need for Articulation of Shared Models

There are a number of human collaboration theories that propose to structure collaboration according to different criteria. This kind of structuring means that parts of the shared context become more relevant based on which theory one is considering and what the structure means to the users. Examples of such theories include:

- Activity theory (12, 14): According to this theory activities are the center of attention for collaborating individuals. An activity is an analytical collective phenomenon that consists of a subject (a person or a group with a motivation), an object (being produced or manipulated by the activity), and tools (used to mediate the relation between the subject and the object). Based on this model, shared context should emphasize the activities that the users are involved in, the users' relations to tools and objects, and the temporal development of these relations.

- Social worlds (6): Social worlds is a grounded theory initially developed by the sociologist Anselm Strauss. The locale framework was later developed to bridge the social worlds theory of Strauss into a more informative theory for technology development (5). According to these theories, collaboration is structured within social worlds, or locales, and is carried on in a series of actions embedded in interactions. Interactions are carried out by interactants (individuals or groups or societies) who have their own identity, biography, and perspective. Collections of interactions are the basis for trajectories that guide the development of social worlds in various ways.

- Distributed cognition (10): A theory focusing on the representation of knowledge "*both inside the heads of the individuals and in the world*" (16). This theory is a further development of cognitive theories that mainly focused on knowledge representation in brain. The unit of analysis here is not an activity or a social world, but a cognitive system composed of individuals and artifacts. In this perspective, distributed cognition treats collaboration from a systems theory perspective: all pieces in the cognitive system are equally important for successful completion of collaboration.

- Situated action (24): This theory is probably the one that imposes the least structure on shared context. The assumption is that any structuring or planning of a collaborative effort and related context is a rationalization that can be done afterwards and not a priory to the effort. The unit of analysis in situated action theory is a setting that is contained in an arena. A setting is the relationship between an individual and his/her surroundings (the arena). This means that relevant context is totally dependent on the individual (or group of individuals) who is currently relating to available context. Context that is important in one setting might be of no value in other settings.

The above theories are not the only ones, but illustrate in a satisfactory way the various focus on shared context that is demonstrated in the literature. These theories, through defining a system of concepts such as activity, social world, cognitive system, and setting, put a different emphasis on what is considered important context. Although the sources of shared context are the same (e.g., intentional, feedthrough, consequential, collected from the physical, and the virtual world), the way this information is grouped, presented, and emphasized is different. In addition, humans have the ability to switch from one model to another. For instance, while engaged in a structured activity, a user might choose to switch to a flexible and non structured mode of operation, and vice versa.

In day-to-day interactions, users will not follow strict models of collaboration. It is therefore not wise to have a strict and rigid model of context underlying any computer system. On the other hand, most of the activities of users are based on a number of common features that can be a common denominator for a context model. For instance:

- Most context is representing an action or activity or can be subsidized as plain data describing a setting or situation, etc. Context of this type can in its most flexible form be implemented in the form of a bag, or a container of data. This idea has been utilized in creating blackboard models of context (27).

- Another common denominator is the users. Context is often related to users who need to cooperate. In some cases these users are human beings, in other cases they might be computer systems or agents. Models of context should provide a flexible way to connect context information to users who are represented by that context information (through intentional, feedthrough, and consequential information exchange).

- A third element common to the models discussed above is the existence of artifacts. Any context model should support a flexible notion of artifacts that can be defined and annotated by users.

We believe the three elements above can constitute the basis for a context model that can be shared among various systems. Of course for advanced context management we will need much more (specialized) information in a context model. But for making context awareness ubiquitous and naturally supported by computers, a minimum model, such as the one proposed here, can play a central role.

10.3 Service Grids in Mobile Environments

The scenarios described above share the common need of accessing resources and services that are delivered by different companies and need to differentiate between the resources shared in the private context (for example the wish list for the parents gift) and the information in the commercial context (e.g., availability for a meeting). Similarly, the context communicated is different for different interactions. For the neighbor ringing the bell the communicated status is "not available" whereas the status for the ongoing video-conference is "available." Within the Grid community a concept called Virtual Organization (VO) evolved from definitions in the field of economics as *a temporary cooperation of independent companies (suppliers, customers, even erstwhile rivals) linked by information technology to share skills, costs, and access to one another's markets ...* towards a more sophisticated model defined as *a temporary or permanent coalition of geographically dispersed individuals, groups, organizational units, or entire organizations that pool resources, capabilities and information to achieve common objectives* (26). Virtual Organizations can provide services and thus participate as a single entity in the formation of further Virtual Organizations. This enables the creation of recursive structures with multiple layers of *virtual* value-added service providers.

As shown in Figure 10.1 such Virtual Organizations might cover across several real organizations and some resources or services (shown as shapes within the company clouds) are provided exclusively to one Virtual Organizations while some others are offered as a shared resource. Typically the relationships within such Virtual Organizations are based on either static or dynamically established Service Level Agreements (SLA) (3) describing the conditions on how the services are provided or how the consumer needs to

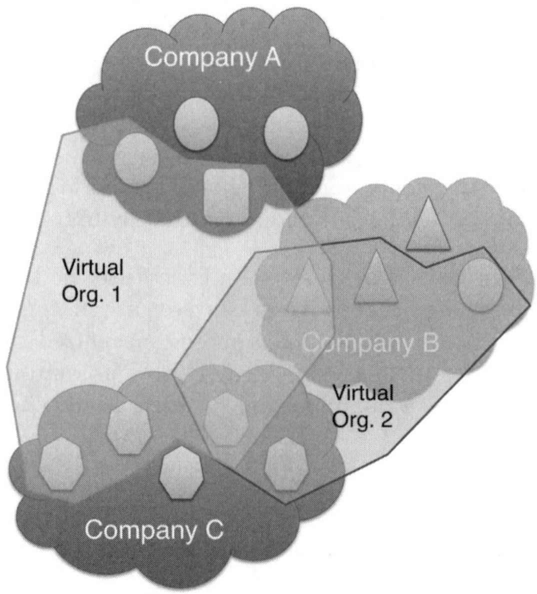

Figure 10.1: Virtual Organization Concept.

behave to receive the agreed service quality. As an example the calendar service provider guarantees the treatment of the data with an appropriate level of security, ensures reliability of the data access, etc., and promise to be able to respond to up to 1000 requests per minute in average in less than 500 ms. If the consumer queries the service 3000 times a minute the response time is no longer guaranteed to be within the boundary.

However, this model had been designed for Business-to-Business scenarios where the coalition between the companies building the Virtual Organizations persist for days, weeks, or even several months (e.g. to realize a complex product such as an airplane). This model is not sufficient for the scenarios described above where a very dynamic change and adaption of the service mix is necessary and additionally local resources (e.g., sensors, equipment such as displays) and an integration with communication sessions is necessary. So, the classical Grid concept of a Virtual Organization has to evolve to the concept of a Mobile Dynamic Virtual Organization (MDVO) (15). It is mobile because some of the elements of the Virtual Organization can be moving and have to use or provide the services while moving, with the consequences that this implies at a network level (different bandwidths, networks handovers, nomadicity of the element, etc.). It is dynamic because the elements can be changed due to a variety of reasons: the element is suddenly disconnected and then reconnected (e.g., it enters a tunnel while being in a train), or it switches from one terminal to another terminal, or even the Grid management services can choose to look for another element more suitable (better bandwidth, better context, etc.).

The combination of a mobile network infrastructure and mobile network services with Grid service is the main issue that has to be addressed in order to provide Context-aware Mobile Grid Services. In this environment, it can be assumed that service consumers and service providers can be mobile and nomad, so the Grid architecture has to take into account this requirement while trying to establish a Virtual Organization.

10.3.1 Network Mobility

The evolution from current telecommunication systems to the future Next Generation Networks is an ongoing process towards a global service convergence, where networks can provide a high variety of services, and telecommunication services like voice and video can coexist with advanced SOA-based services. This evolution has to be achieved regardless of the underlying QoS-enabled transport technology being used, offering access to different service providers and supporting generalized mobility that will allow consistent and ubiquitous provision of services to users, anywhere and at any time.

Generalized mobility means, within this context, that the concept of mobility itself has to be understood in a broad sense, since several aspects and characteristics of the infrastructure can be mobile. The traditional Network Layer has to evolve towards a "Mobile Network Layer," where different types of mobility can be considered:

- Device mobility, meaning that a terminal can remain connected to the network across different network access points and/or technologies;
- User mobility, which refers to the capability that a user can use different devices to access the same set of services;
- Session mobility, which focuses on the capability of transferring running service sessions between terminals;
- Service mobility, enabling the invocation of services running on nomadic or mobile machines.

The mobile network infrastructure has to provide support for all the mobility types defined above, which basically relies on a powerful signaling framework, supported by the cooperative work of a variety of protocols. In this way, while device mobility is supported by the Mobile IP version 6 (MIPv6), user, session and service mobility relies on the Session Initiation Protocol (SIP) infrastructure.

Device mobility using MIPv6 is completely transparent to the applications and services, so there is no need to customize the Grid architecture in order to include this type of mobility. Usage of specific network modules and services, like a Home Agent, is needed but it does not make any impact to applications. So, Service Grid scenarios where there exist devices (either service provider or service consumers) that are connecting from a different network than their home network or are moving to another network infrastructure (handover) can be realized with the same existing Grid architecture, with just changes at the network level in order to support MIPv6 and related modules.

But user, service, and session mobility is not provided in a transparent way to applications and services. This type of mobility is supported with the usage of a specific signaling network protocol (SIP) that has to be used and managed by the applications. So, there is a need to combine traditional Grid architectures with network signaling protocols for managing several type of mobility of the Grid elements. This is the case of user mobility or nomadicity, where a user can connect and use or provide a service from different terminals.

As a result depending on the available underlying network infrastructure different type of mobility can be supported. While for a full support Mobile IPv6 is needed the mobility supported by SIP can also be realized on top of IPv4. But using SIP is not straight forward: it is needed to use SIP in order to manage Grid sessions, while the usual SIP application is management of multimedia sessions. In other words, it is needed to use SIP for managing SOAP-based sessions. This combination of SIP and SOAP is one of the key challenges of a Mobile Grid solution.

10.3.2 Session Initiation Protocol

SIP (20) is a protocol targeted to the establishment, modification, and termination of sessions. It has been designed by the IETF to be used as a signaling framework in IP networks. In the context of SIP, a session is understood as an association created between two or more participants to exchange some kind of data. SIP's specification only defines the signaling messages and the operation of the protocol, decoupling it from the actual type of data to be exchanged once a session has been successfully established. Despite this decoupling, SIP has been traditionally used to manage multimedia sessions.

SIP functionality makes it especially useful in mobile and ubiquitous environments. In such environments, users can move and change the terminal they are using to access the services. The SIP servers tracks each user in order to keep the information about its (logical) location up to date. As part of SIP's specification, the different entities that form part of the signaling infrastructure are defined. The most important ones are:

- User Agents (UA): Hardware and/or software applications that the user uses to handle SIP signaling.

- Proxies: Message routers. They use the headers present on each message to route them accordingly.

- Registrar: Handles the Location Database. This database contains the logical location information (e.g., IP address) of each user present in the network.

Of course, SIP's specification also defines the format of the messages that are used to manage the sessions. SIP messages are plain-text, human-readable messages that contain a set of headers and, in some cases, a payload. The specification of SIP defines the core set of messages, which are the minimum ones that every implementation must support. Several new messages have been defined to extend the basic functionality of the core set, allowing the use of SIP for supporting new signaling services. The basic SIP messages are shown in the table 10.1.

Table 10.1: SIP basic message set

Message	Meaning
REGISTER	User registration
INVITE	Session establishment
ACK	Session establishment acknowledge
BYE	Terminates an ongoing session
CANCEL	Cancels a previous request
OPTIONS	Capability consultation

A SIP-based signaling infrastructure provides seamless mobility and ubiquity for users. This infrastructure will be able to route signaling messages to every user with independence from the access technology, as long as the user has IP connectivity. This is done through the registration mechanism. In more detail, when the user gets connected to the network, he/she issues a REGISTER message to let the system know his/her logical location (i.e., the IP address of the device being used). Every user is designed by a non changing, globally unique identifier called SIP Uniform Resource Identifier (URI). A REGISTER message creates a binding between a given SIP URI and the IP address where the user is reachable at a given moment. If this logical location should change, the user registration must be updated accordingly.

Figure 10.2 describes the basic signaling that is exchanged in a typical SIP interaction. In the figure, Alice establishes a session with Bob and uses it to send and receive some data to and from Bob. Once the session is not needed, Bob terminates it. During the establishment phase, INVITE requests and responses to them contain session descriptions. The Session Description Protocol (SDP) is used to describe and characterize the session that the endpoints are willing to establish. Session Description Protocol's (SDP) specification defines a format to describe streaming media sessions, focused

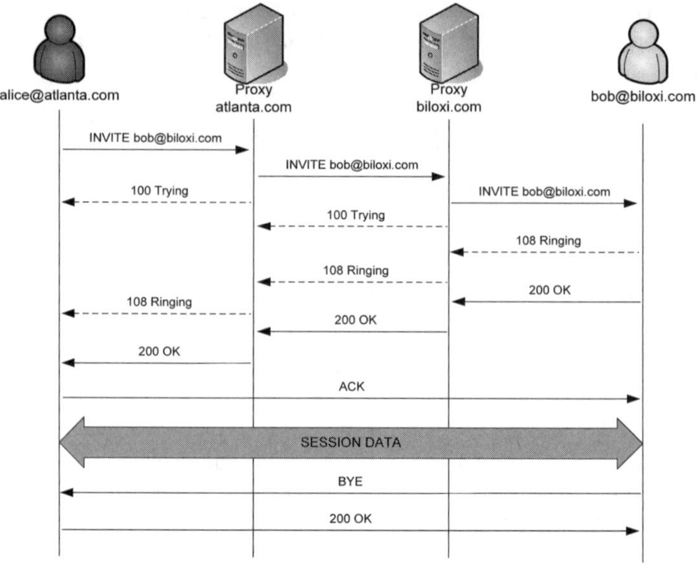

Figure 10.2: Basic SIP signaling for session establishment.

on session announcement, session invitation, and other forms of multimedia session initiation. SDP specification defines a session as a *set of multimedia senders and receivers and the data streams flowing from senders to receivers.* As an example, we can consider a multimedia conference, including audio and video streaming. Since it does not provide the media content itself, but facilitates an agreement regarding media type and formats between two endpoints, forward compatibility of upcoming media types and formats is granted.

An SDP (8) session description is entirely textual and usually includes the following information:

- Session name and purpose.
- Time(s) the session is active.
- The media comprising the session, and the information needed to receive those media (addresses, ports, formats, etc.).

Optionally, and considering that resources necessary to participate in a session may be limited, additional information may be also included, such as the session bandwidth consumption, the contact information for the responsible person, etc.

From a formal point of view, an SDP session description consists of a number of lines of text of the form:

`<type>=<value>`

where `<type>` is exactly one case-significant character and `<value>` is a structured text whose format depends on `<type>`. An example of an SDP session description is described in the following figure:

```
v=0
o=jdoe 2890844526 2890842807 IN IP4 10.47.16.5
s=SDP Seminar
i=A Seminar on the Session Description Protocol
e=j.doe@example.com (Jane Doe)
c=IN IP4 224.2.17.12/127
t=2873397496 2873404696
a=recvonly
m=audio 49170 RTP/AVP 0
m=video 51372 RTP/AVP 99
a=rtpmap:99 h263-1998/90000
```

Roughly speaking, this SDP session description defines a receive-only Real Time Protocol (RTP)-based audio and video session named "SDP seminar," whose responsible person is Jane Doe. The user wanted to receive the audio and video RTP streams at IP address 224.2.17.12, concretely at ports 49170 and 51372 respectively. Chosen formats are audio G.711 and video H.263, identified by the RTP payload type number.

The IETF also defines the way in which two entities can use the SDP protocol in order to reach an agreement on the parameters of the session to be established between them. This information, that comprises actions like the generation of the initial offer, how to process it in order to generate the answer, how to add, remove or modify media information, etc., is described in detail in RFC 3246 (19).

As RFC 4566 (8) mentions, *"SDP is intended to be general purpose so that it can be used in a wide range of network environments and applications."* This work is leveraging this idea, using SDP to describe sessions for Web Services in a Grid environment.

10.3.3 Combining Mobility and Service Grids

As said before, Service Grids may require mobility and nomadicity of their elements. Adding mobile services and resources to the Grid poses a lot of challenges. Of course, taking advantage of the mobility support already present in the network by means of a SIP-based signaling infrastructure would reduce the effort that re-implementing it at the Grid layer would require. Dealing with nomadic nodes and services does not mean that every entity at the Grid layer must be mobility-aware.

The proposed approach (see also (11, 18, 17)), which this section presents in detail, suggests the utilization of the network signaling infrastructures to locate where the services are running. Once the system is aware of the nodes the services are running in, the traditional interaction can take place.

The basic element of the proposed solution is the use of SIP to control SOAP sessions. The objective is to use the capabilities of tracking and locating users of the network's SIP-based signaling infrastructure to locate not only the users (as shown in Figure 10.3), but also the services. Once the service is located, the usual SOAP messages can be exchanged.

According to this concept, the Grid Management layer will use SIP signaling to establish a SIP session with the required services. This SIP session establishment will be used to exchange the information needed for the subsequent SOAP interactions. In particular, the IP address of the machine where the service is actually running and the URI of the WSDL file that describes the service. The following figure shows the messages exchanged in the proposed scenario.

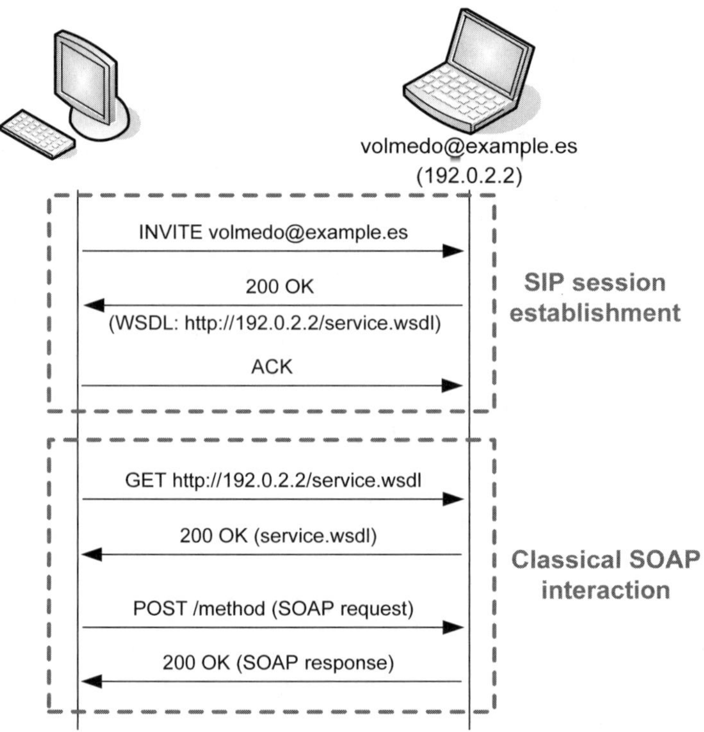

Figure 10.3: Using SIP to locate a mobile service.

Of course, this approach implies the usage of SIP URIs in order to identify the services. Using SIP URIs, the underlying SIP infrastructure can be used to locate the service and route signaling messages to its current location. Additionally, every server willing to be SIP-reachable must be registered in the SIP registration entities. This registration must be updated every time the location of a service changes to allow the SIP registrar to keep track of the service logical location. When a request for a SIP URI arrives to the proxy, it will query the registrar to know where to route the request.

The utilization of the "SIP with SOAP" approach not only allows locating mobile services, but also opens the door to all the functionalities provided by SIP. Some of the most interesting capabilities the Web Services layer can take advantage of are:

- SIP's support for presence and context information propagation. Thanks to this, it is possible for a Web Service client to know (in real-time) whether a service is available or not, letting the user decide to look for another service or be notified when the service is available again.The usage of SIP's subscription and notification features allows any module to track any registered information, like presence or context information.
- The use of Third Party Call Control (3PCC) techniques to establish Web Service sessions. This allows the inclusion of the consumption of a service by a client as part of a business process workflow.
- The support for session mobility (19). The properties of a SIP session can be changed while in use, without the need to terminate it and establish a new one. This allows an ongoing session to be transferred from one terminal to another or to recover a session after an eventual loss of connectivity (18).

The integration of the SOAP-based Grid Services and the SIP signaling can be done in two ways: a **built-in SIP aware applications approach** (provide a SIP API to the Grid applications, which will handle SIP interactions themselves); or a **delegates/intermediaries approach**, which consists on the provision of some intermediate modules, dealing with the SIP interactions on behalf of the Grid applications.

The first approach is "cleaner" in terms of architecture, but implies that existing Grid applications have to be modified in order to use the interfaces offered by some SIP stack for SIP protocol handling. The second approach minimizes the implementation efforts on the applications, since they will need to use only SOAP messages even for SIP-related methods.

At least, two intermediaries are envisaged: a specialized application running on the terminal where the services to be invoked are expected to be present, dealing with the SIP signaling on behalf of the services; and an accessible Grid Service located in some server (the Mobile Grid Service Gateway), capable to convert an incoming request for services information into the underlying SIP process supporting it, which was described previously, and including SIP context-awareness capabilities, in order to inform about the availability of the targeted resources.

This is the solution adopted in the Akogrimo project (25, 11). As shown in Figure 10.4 while Mobile Grid Services providers and clients will use SIP messages directly to contact each other, standard Grid Services providers and clients will relay on Mobile Grid Service Gateways. These gateways will offer a standard Grid Service interface to Grid Service clients and providers, producing the corresponding SIP signaling when they want to communicate with a Mobile Grid Service provider or client.

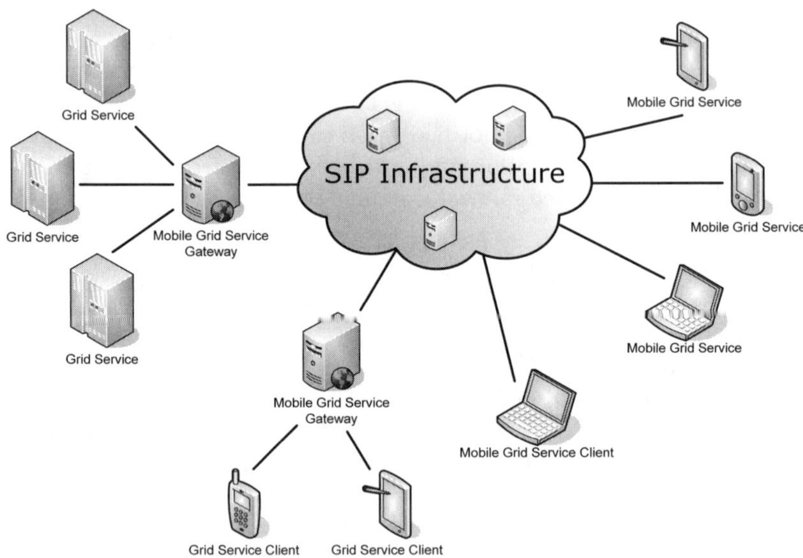

Figure 10.4: Generic architecture service.

As described earlier, every Mobile Grid Service provider and client must be registered in the SIP registration entities. Usually, this registration will be performed at startup. As registration is the mechanism that SIP uses to track and locate users, this registration will need to be updated any time a change in the location occurs.

When a Mobile Grid Service client decides to use a Mobile Grid Service, the client will send a SIP INVITE message targeted to the URI that identifies the service. The SIP entities will then locate the service and route the message accordingly. This INVITE message will create a new SIP session between the Mobile Web Service client and the service. The SDP payload in this INVITE message will characterize the session as a Grid Service session, i.e., a session oriented to the consumption of a Grid Service through the exchange of SOAP messages.

In the answer to the session establishment request, the service will also provide a SDP payload. This payload will contain all the details that the client needs to know in order to use the service, i.e. the IP address of the machine where the service is running in and the WSDL describing the service. Figure 10.5 shows how one of these SDP payloads could look like.

```
v=0
s=-
c=IN IP4 192.0.2.2
t=0 0
m=application 8080 TCP/http soap+xml
a=setup:passive
a=wsdl:http://192.0.2.2/service.wsdl
```

Figure 10.5: SDP answer with WSDL URI service

This SDP example informs the client about a service listening for requests at TCP port 8080 of a machine with an IP address of 192.0.2.2. Furthermore, it provides the URI to the WSDL file that defines the methods offered by the service. As it can be seen in the example, a new SDP attribute is used to inform the client about the location of the WSDL file that describes the service. From the viewpoint of SDP standardization, only this new attribute, the *wsdl* attribute, needs to be defined. The first steps on this direction have already been undertaken. This SDP moficiation only affect to the SIP-enabled applications in the Mobile Grid clients, but is transparent to all the rest of SIP entities that can be used without any modification.

Once the answer has been received by the client and the SIP session has been correctly established, the usual Grid Service interactions can take place. Every client downloads the WSDL file to know which methods are offered by the service and how to invoke them, and issues SOAP messages to invoke the needed methods. When the client consumes the service and the session is not needed anymore, it is terminated using the corresponding SIP messages.

10.4 Adaptation Approaches

The previous section discussed how SIP can be used to support for the Web Service Layer context information such as presence and device capabilities as well as how SIP session mobility can be integrated with Web Service interactions. In order to realize with this context information an environment for context-aware mobility-supporting Grids, the information needs to be propagated within the Virtual Organization that corresponding adaptations can be realized. The Virtual Organization concept is supporting this propagation process as all service providers and consumer of services need to perform certain steps to join or leave a Virtual Organization. In this process, also described as VO Lifecycle, a new VO participant needs to sign up to certain rules and policies for example via a dynamically negotiated Service Level Agreement (SLA). In this step of joining the VO consumers or providers can be automatically subscribed to certain notification topics. Additionally subscriptions to notifications topics within a Virtual Organization can be driven by workflow engines based on the current status within a business process or can be driven by the consumers or providers themself. Service-oriented Grids relies for the realization of such a notification system typically either on WS-Notification or WS-Eventing. So for the considerations in the following sections one can assume that the events about changing context (together with other events communicated as part of the VO operation not related to context changes) are delivered to all necessary entities using a VO-wide notification system.

10.4.1 Direct Adaptation of the Service

The first and easiest way to react to a changing context is to have a service provider as a direct subscriber for context changes. So if a profile changes (e.g., reduced network bandwidth) the service can adapt e.g., by producing a lower quality video stream. Similarly a change in the device with different capabilities such as available video codecs can be covered. Similarly changes in the presence status, e.g., to "offline" might temporarily disable the provision of the service with a resume based on a corresponding "online" notification.

While this solution clearly allows a very efficient adaptation to different context changes, it comes along with two major drawbacks:

1. the resource demand for delivering the service varies significantly
2. a service cannot easily change its behavior once deployed or different behaviors for different customers requires the provision of different versions of the same service

A change of the consuming device might require a different encoding or a complex rendering process to a small screen with significantly increased computing and memory demand for the service delivery. On the other hand, if the consumer changes its status to "offline" the service provision is paused. This problem can be solved by hosting the service within a flexible hosting environment, as offered by many cloud computing service providers (23).

The second problem, the lack of flexibility, is more challenging. If the service behavior to different context situations is an inherent part of the service implementation, all potential cases for all type of customers cannot be divided from the implementation to deliver the service. For this reason the approach cannot be used for services that require personalization or show a volatile behavior change depending on small context changes.

10.4.2 Adaptation on Workflow Level

While changing the service behavior can cover certain cases for context adaptation the lack of flexibility as outlined above is obvious. A more flexible approach is to rely on aggregated services that are built from several core components. The aggregation can be driven by a workflow combining different services in order to deliver a complex service. Adaptation to context changes can be realized by replacing certain components within the workflow or by replacing or modifying the workflow itself. Beside the possibility to have such an aggregation process as a service provider in order to offer more complex and more personalized services consumers also can realize their own workflows e.g., following a mash-up style approach.

As an example consider the case where a client device cannot do the rendering of the shared data anymore itself due to its limited computing and graphics power. The solution for this problem could be that the data provider service does not deliver the raw data anymore but delivers a video stream or static pictures instead. However, a more efficient and flexible approach would be to change the workflow for the result delivery. Instead of delivering the data to be rendered directly to the client device a different workflow involving a rendering service in between delivering the results, e.g., as pictures or a movie to the client device can be realized. Such solutions assume that for the conceptually unlimited space of different context profiles different options in the workflow had been premodeled and a context change received by a workflow manager triggers a change in the branch of the workflow or replaces a certain part of workflow with another workflow snippet.

This approach is already quite flexible and allows personalized services and a wide range of functionality (using different workflows) *within* a service provider domain. However, this approach fails to adapt to a situation where a certain service provider is unable to deliver the service anymore due to an overload situation or critical failure of the underlying infrastructure. This includes a context change demanding for a significantly increased demand on the resources that cannot be covered by the available resources.

10.4.3 Adaption on VO Level

Slightly more complex is the situation if the change of the workflow requires additional participants in the collaboration. In the scenario introduced above, where with a context change an additional rendering service became necessary, a better adaptation can be reached leaving organizational boundaries (in the current virtual organization no such provider is present). Consequently, such a provider needs to be discovered and integrated into the virtual organization. So a change in context might trigger an evolution of the Virtual Organization adding a new organization or service provider into the collaboration group. Service Grids enable such functionality with standardized semantically enriched service descriptions (including Service Level Agreements that can be offered).

Beside a changing context from a single user viewpoint the collaboration itself undergoes certain changes. A collaboration session might start as a regular information exchange where certain problems are discovered. In order to address the problems immediately additional users and resources might be needed and added ad hoc to the collaboration. Such changes might require participants to change their sessions to more powerful devices, move to trusted areas and so on. This means that certain operations need to be paused and resumed based on certain conditions, and business processes might have synchronization points where all participants are on hold until certain conditions are met.

Such a change requires a wide range of additional adaptations. For example, if the reaction to the context change is to change the collaboration model, add new providers and resources as further context change events are triggered. The change of the collaboration might require an update of the access right of certain VO members or some other members are no longer needed within the VO. While such an approach clearly offers the largest potential for reacting to changing context it also leads to a non deterministic system as changes at provider A to adapt to a context change event might trigger further events impacting other providers, etc. Assuming a site autonomy allowing each provider to realize the adaptation in a transparent way to the other VO members the number and type of potential further context change events is unpredictable. The lack of a central control instance allows the realization of a scalable and massively distributed approach at the expense of unpredictable behavior of the overall system. This problem can be partially solved by binding all cross-provider interactions to certain Service Level Agreements or enforce the providers to realize a certain black box behavior as proposed in the Web Service Choreography Description Language (WS-CDL).

10.5 Conclusions and Future Work

This article has shown how a combination of SIP-based session and context management can be combined with a concept from the Grid community, namely Virtual Organizations and SOAP-based service interactions in order to realize context aware and adaptive distributed applications. The concepts outlined are partially based on results of the Integrated Project Akogrimo co-funded by the European commission under contract number IST-2003-004293. The engineering scenario is based on results achieved within the Integrated Project CoSpaces co-funded by the European commission under contract number IST-5-034245.

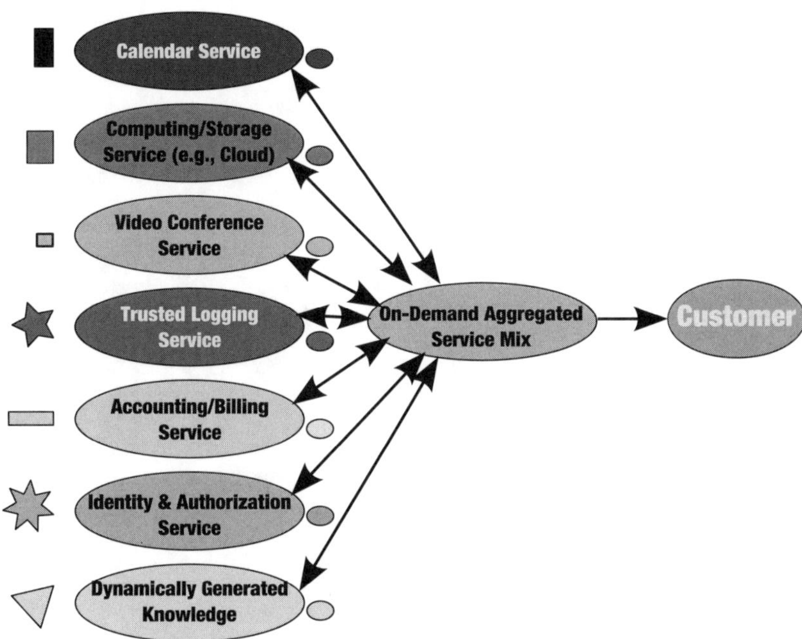

Figure 10.6: Potential Future Service Aggregation Model.

Current developments in Grids and the Internet of Services will further increase the complexity of context adaptation as the proposed workflow-based service aggregation in order to react on context changes, assuming that all services that can be combined are somehow on the same level and do not have any inherent dependencies. This layered model is currently under discussion and the aggregation model shown in Figure 10.6 combining service "bricks" from different layers is a potential replacement.

This model combines services from different layers ranging from connectivity services such as transport over utility services such as identity or accounting & billing up to knowledge services. Such a model allowing a service mix to be dynamically realized from services offered by different providers will allow even more dynamic and flexible reactions to context changes, in particular if the mix can be changed by the consumer itself. As the mix cannot be controlled anymore by a single service provider, the adaption of the different elements can only be realized if all components are described using rich semantics and based on generally agreed information models.

In conclusion, the current existing approaches for adapting to context change are able to react on a limited number of predefined context situations within a controlled environment such as a service provider domain or a virtual organization consisting of a comparably small number of participants. The lack of semantically rich description models allowing an automated reasoning on the context state and the capabilities of different service components is the current showstopper for more flexible solutions.

References

[1] Continua health alliance. http://www.continuaalliance.org/.

[2] UPnP forum. http://www.upnp.org/.

[3] Francesco D'Andria, Josep Martrat, Giuseppe Laria, Pierluigi Ritrovato, and Stefan Wesner. An enhanced strategy for sla management in the business context of new mobile dynamic vo. In Paul Cunningham and Miriam Cunningham, editors, *Exploiting the Knowledge Economy: Issues, Applications, Case Studies*, eChallenges 2006. IOS Press Amsterdam, 2006.

[4] W. Keith Edwards. Discovery systems in ubiquitous computing. *IEEE Pervasive Computing*, 5(2):70–77, 2006.

[5] Geraldine Fitzpatrick. *The Locales Framework: Understanding and designing for Wicked Problems.* Kluwer Academic Publishers, 2003.

[6] Geraldine Fitzpatrick, William J. Tolone, and Simon M. Kaplan. Work, locales and distributed social worlds. In *Proceedings ECSCW*, Stockholm, Sweden, 1995. Springer.

[7] Carl Gutwin and Saul Greenberg. A descriptive framework of workspace awareness for real-time groupware. *CSCW*, 11(3–4):411–446, 2002.

[8] M. Handley, V. Jacobson, and C. Perkins. Sdp: Session description protocol. Internet RFC 4566, July 2006.

[9] Karen Henricksen, Jadwiga Indulska, and Andry Rakotonirainy. Modeling context information in pervasive computing systems. In *Proceedings of Pervasive '02*, pages 167–180, Z?žrich, Switzerland, 2002. Springer.

[10] Edwin Hutchins. *Cognition in the Wild.* The MIT Press, Cambridge, MA, 1995.

[11] Juergen Jaehnert, Antonio Cuevas, Jose I. Moreno, Victor Villagra, Stefan Wesner, Vicente Olmedo, and Hans Einsiedler. The akogrimo way towards an extended ims architecture. In *Proceedings of the 11th International Conference on Intelligence in Networks (ICIN 2007)*, October 2007.

[12] Victor Kaptelinin and Bonnie A. Nardi. *Acting with Technology: Activity Theory and Interaction Design.* The MIT Press, October 2006.

[13] Alexander Kipp, Lutz Schubert, and Matthias Assel. Supporting dynamism and security in ad-hoc collaborative working environments.

In *Proc. 12th World Multi-Conference on Systemics, Cybernetics and Informatics*, pages 259–263, 2008.

[14] Kari Kuutti. The concept of activity as a basic unit of analysis for CSCW research. In *Proceeding ECSCW*, Amsterdam, The Netherlands, 1991. Springer.

[15] Christian Loos, Stefan Wesner, and Juergen M. Jaehnert. Specific challenges of mobile dynamic virtual organizations. In Paul Cunningham and Miriam Cunningham, editors, *Innovation and the Knowledge Economy: Issues, Applications, Case Studies.* IOS Press Amsterdam, 2005.

[16] Bonnie A. Nardi. Studying context: A comparison of activity theory, situated action models, and distributed cognition. In Bonnie A. Nardi, editor, *Context and Consciousness*, pages 69–102. The MIT Press, Cambridge, Massachusetts, 1995.

[17] Vicente Olmedo, Antonio Cuevas, Victor Villagra, and Jose I. Moreno. *IP Multimedia Subsystem (IMS) Handbook*, chapter Next Generation Grid Support over the SIP/IMS Platform, pages 133–157. CRC Press, 2009.

[18] Vicente Olmedo, Victor A. Villagra, Kleopatra Konstanteli, and Julio Berrocal Juan E. Burgos. Network mobility support for web service based grids through the session initiation protocol. *Future Generation Computer Systems*, November 2008. ISSN 0167-739X.

[19] J. Rosenberg and H. Schulzrinne. An offer/answer model with session description protocol (sdp). Internet RFC 3264, June 2002.

[20] J. Rosenberg, H. Schulzrinne, et al. Sip: Session initiation protocol. Internet RFC 3261, June 2002.

[21] Antonio Sanchez, Belen Carro, Pedro Romo, and Carolina Pinart. Telco services for end customers within Spanish programmes. *IEEE COMMUNICATIONS MAGAZINE*, 46(6):24+, JUN 2008.

[22] Antonio Sanchez, Belten Carro, and Stefan Wesner. Telco services for end customers: European perspective. *IEEE COMMUNICATIONS MAGAZINE*, 46(2):14+, FEB 2008.

[23] Lutz Schubert, Alexander Kipp, and Stefan Wesner. *Above the Clouds: From Grids to Resource Fabrics*, pages 238–249. IOS Press, 2009.

[24] Lucille Alice Suchman. *Plans and Situated Actions: The Problem of Human-machine Communication.* Cambridge University Press, Cambridge, 1987.

[25] Stefan Wesner. Towards a mobile grid architecture. *IT Information Technology*, 2005.

[26] Stefan Wesner, Lutz Schubert, and Theo Dimitrakos. Dynamic virtual organisations in engineering. *Notes on Numerical Fluid Mechanics and Multidisciplinary Design*, 2005.

[27] Terry Winograd. Architectures for context. *Hum.-Comput. Interact.*, 16(2):401–419, 2001.

Chapter 11

Leveraging Context-Awareness for Personalization in a User Generated Services Platform

Laurent-Walter Goix, Luca Lamorte, Paolo Falcarin, Carlos Baladron, Jian Yu, Isabel Ordas, Alvaro Martinez Reol, Ruben Trapero, Jose M. del Alamo, Michele Stecca, and Massimo Maresca

11.1 Introduction

With the wide spread of the Web 2.0 philosophy, end-users are being empowered with tools for proactively creating and sharing their own content and services according to their needs, mainly in the Internet arena, thus creating expectations also in the telecommunication world. The preferred approach is to develop platforms for User Generated Services (UGS) that enable users to author and compose ("mash-up") their own adaptive and personalized services, to capitalize user community's collective intelligence to create new knowledge and value, foster creativity and opt for the most useful ones without waiting for operators to offer them in their service portfolios. Personalization is perceived as an added value factor as it drives the service usage and as a consequence the flow of revenues for service providers and operators. Service personalization is the way to provide users with services tailored to their needs, preferences, interests and expertise.

While personalization has always been a desirable feature of applications and services, in the past it had to be operated manually by the user, modifying parameters, options and preferences by hand. The applications simply lacked enough information about the status (i.e.: context) of the user in order to be able to automatically adapt to it. However, in recent times, the increasing amount of mobile devices allows data collection for real time characterization of users' context. Now, services have enough information available in order to interpret the intentions of the user and react accordingly without the need for a human controller. This opens the door for a huge amount of useful service adaptations that were impossible or impractical before, due to the lack of specific knowledge of the user or difficult usability. Think for instance in the change of portrait to panoramic visualization style in mobile devices according to the orientation. Most users did not have enough knowledge of the operating system in order to do it by hand. But even more, if they knew how to, they will probably not use that feature because it is simply not very useful to change it manually every time the orientation of the device changes. However, modern devices which pack accelerometers are able to do it automatically, becoming a very desirable functionality.

This automatic adaptation and personalization may be performed at all levels of a service, and in this path towards personalization context awareness is a fundamental enabler: the emergence of mobile devices and embedded systems allow users to be connected to a variety of electronic devices, from very different environments and in short time periods. Mobility introduces factors such as freedom of movement, ubiquity, localization, convenience, instant connectivity and personalization, which are perceived by end-users as extremely valuable. Device-based content adaptation, location based services and content filtering based on user preferences are some examples of how personalization is understood as *context-aware* in the service provision.

In some cases, such personalization can be explicitly defined at creation time making the service flow depend upon the context of the user. However services that do not embed context-based conditions still need to be adapted and personalized to achieve a good user experience, requiring the platform itself to perform adaptation implicitly based on the user context information (amongst others).

The OPUCE project has been providing an open service infrastructure to enable end-users for easy service creation and deployment in heterogeneous environments, allowing services to be accessed in a seamless way by a variety of devices and networks.

The purpose of this chapter is to describe the approach followed by OPUCEfor enabling user generated services in general and more precisely how context awareness is used to empower personalization and fine-grained adaptation of services for each single user. In particular, after a short presentation of the OPUCE project, section 2 addresses the modeling of the user context to cope with a highly volatile, yet heterogeneous and infinitely expansible set of information. Our reference architecture for embedding context wareness features in the core of our UGS platform is then described in section 3. Section 4 focuses on the solutions adopted to achieve the right level/trade off of explicit and implicit adaptation of services based on context information, both at service creation and service execution time, and also discusses the scalability, performance and privacy-related issues raised throughout the platform specification and development.

11.2 The OPUCE Project

OPUCE (1, 11, 9, 16) is a project aimed at designing a platform for the creation and execution of converged services merging capabilities from the Internet and Telecommunications world, but oriented to non-expert end-users. The idea behind OPUCE is to take the Web 2.0 and User Generated Content (UGC) paradigms and expand them into the telco world and the creation of services (in addition to contents), giving birth to Telco 2.0 and User Generated Services (UGS). In the Web 2.0, the users are in the centre of their own experience in the internet, acting not only as consumers of information, but also as creators, as popularized by sites like Wikipedia [1] or YouTube [2]. OPUCE allows end-users to create their own highly personalized services using graphical tools to define the service workflow, and including not only applications in the Web, but also Telecommunication features through links to operators' networks, thus resulting in what is called Telco 2.0.

The OPUCE platform includes:

- Tools for the specification of a service logic using graphical abstractions and drag-and-drop interfaces;
- Tools for publication, discovery, recommendation and sharing of the created services with other users;
- A service execution environment which links the Web and the operator domains;
- Repositories to store user profiles and context data;
- And a Web Portal acting as the user interface towards the platform.

The OPUCE service model sees services as a composition (and thus are called composed services) of basic building blocks which implement atomic functionalities, like "Retrieve_RSS_Feed" or "Send_SMS". These basic building blocks are called Base Services in the OPUCE terminology. In order to build a composed service, the end-user acting as a service creator drags and drops those building blocks into an editing canvas and links them to form a workflow.

For that, each Base Service exposes a set of actions and events. Events are linked to actions, so when a Base Service fires an event, it could be linked to an action of another Base Service, which is then executed. Figure 13.1 presents an example of a composed service that monitors a Gmail account and forwards emails received to a mobile phone as SMSs. When the Gmail

[1] http://en.wikipedia.org/

[2] http://www.youtube.com/

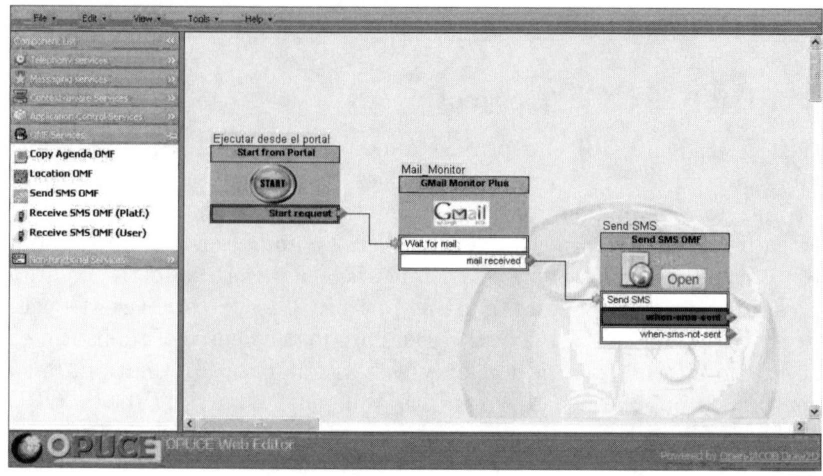

Figure 11.1: Snapshot of the OPUCE service editor

monitor detects a new incoming message, fires the event "When-target-mail-received". This event is linked to the "Send SMS" action of the Send SMS Base Service, and thus this action is executed.

Additionally, these Base Services expose a set of properties to configure them and store output information. In the example of the Figure 13.1, the Gmail Monitor Base Service will have a property "email_title" that will be forwarded to the "SMS_content" property of the Send SMS Base Service, so the title of the email received will be the content of the SMS. Additionally, the Send SMS Base Service has a "Target_Phone_Number" that has to be specified.

OPUCE composed services are therefore a service workflow together with all the information associated with it necessary for its proper operation inside the OPUCE platform, like constant values, AAA information, graphical representation, etc. This information is stored in a XML document following the OPUCE service description model (9). Therefore, while the service workflow is written in BPEL language and in theory could run under any BPEL engine, OPUCE composed services are designed to be executed only inside the OPUCE Service Execution Engine. The reason for this is that OPUCE services may access operations provided by a network owned by an operator and are potentially subject to tarification, so execution should be done under a controlled environment.

Base Services are also represented by XML documents following the OPUCE Base Service Specification (9). Information about graphical representation, deployment conditions, icons, events and actions, AAA and access control lists is also present. However, instead of a BPEL workflow, their executable logic is a Web Service.

Base Services must run inside one of the Web Service containers of the OPUCE Service Execution Environment, because due to security, accounting and tarification reasons the operator of the platform always needs to keep track of the operation of Base Services. To access functionalities external to the OPUCE platform, Base Services could be implemented to act as proxies accessing remote capabilities through the Internet, but it is necessary to keep this access point controlled inside the OPUCE Execution Environment for the reasons mentioned.

To build a Base Service, it is necessary to write the code packing the functionality outside the OPUCE environment, using traditional development methods and languages like Java or C# (The OPUCE Service Execution Environment is able to handle several different Web Service containers, so any technology could be employed as long as a proper container is installed on the platform). However, in order to facilitate the creation of new OPUCE Base Services, a tool known as Base Service Manager is provided by the OPUCE platform to automatically produce the Base Service description wrapping the Web Service code and deploy it in the appropriate container depending on the language in which the Web Service is written.

The operations of the Web Service implementing a Base Service functionality follow a fixed structure specified in the OPUCE service model (9). The *invokeAction* operation is used to manage all the action invocations instead of having a different one for each action exposed, so in the end all Base Services follow the same prototype regardless their actual interface.

During the execution, the orchestrator lying inside the OPUCE Service Execution Environment runs the BPEL scripts of composed services. This orchestrator invokes the appropriate Base Services according the BPEL script, using a SOAP message sent to the Web Service implementing that Base Service. If this Web Service produces events, they are routed back towards the orchestrator invoking the Event Gateway, which is a Web Service acting as an input stage for the orchestrator.

In the end, the event-action composition model is completely implemented using invocations of Web Services.

The User Information Management (UIM) is the OPUCE subsystem that stores user information comprising dynamic data, defined as context, and static personal and preferences information, named as profile. Context information is made available to all OPUCE modules throughout the whole lifecycle of services, i.e. from creation (to create context-aware shareable services) to execution (to adapt services dynamically based on run-time user context), and drives the different adaptation operations that are presented in section 4.

One of the main challenges of User Generated Services is to have them *adapted* to each user profile and context. Adaptation here is understood as modifying the service workflow and contents to better fit the customer requirements. End-users taking the creator role of User Generated Services are not able to put together complex adaptation logics due to their lack of specific expertise, so it should be as automatic as possible, with tools to facilitate the specification of adaptation conditions.

Therefore, a Context Aware service platform needs mechanisms to describe how user context drives the service execution. Such mechanisms must firstly support creators in defining their services' features so that these services can be easily tailored and adapted at subscription or execution time according to specific consumers' preferences and characteristics. The definition of such features requires the service model and description itself to natively support adaptability. And secondly, at execution time these mechanisms should take the necessary low-level actions in order to carry out the adaptation defined by the service creator at a higher abstraction level.

11.3 Modeling Context Information

Nowadays in the area of service personalization, some service providers are coping with context data acquisition, interpretation, and aggregation, but there are different opinions on what should be considered as "context". In order to set a boundary between what is context and what is not, we follow the viewpoint of Schilit et al. (10) by taking context as the user's location, the social situation and the nearby resources.

11.3.1 Identifying Context Information

Users' terminals (either fixed or mobile) can be used by network operators and service providers to learn more about user's environment, habits and preferences and to take advantage of this information for service personalization.

Such information can be gathered in two ways, either explicitly, based on direct feedback or input, no matter the user is aware of it (e.g. it could be some application monitoring in the background, versus a web-based portal asking for food preferences), or implicitly, based on some analysis and post-processing that deduce some information out of other explicit information.

Another orthogonal dimension to be considered for gathering user information and in particular user context is related to the technological approach to be used for retrieving this information, which is either device-based, i.e. through an application in the end-user device (PC, mobile) capturing user-related information (e.g. application usage, GPS location, nearby Bluetooth devices, and so on), or network-based, thanks to a system in the network holding information about the user. Such approaches are complementary, since some information resides on the end-user device only, other is available from the network.

In addition, some data only available on the device can be aggregated with information centrally available to mobile operators and service providers, for example to extract a higher-level description of the user context: in such way this higher-level context information can be directly treated by applications.

On this wide user-related information, a generic categorization can be the following:

- User Profile: information inserted by the user that is persistent in time, e.g. agenda, buddy-list, service settings/profile;
- User Context: information not directly inserted by user but gathered or inferred from devices; this is usually made of more volatile/transient data, e.g. presence, location, current device capabilities, bandwidth, and so on;
- Service Usage: which service is used by which user, considering the context parameters; it represents the history of service usage;
- Device Usage: what is the user doing on the own device, linked somehow to service usage, but for example related to applications used on PC or on the mobile phone.

We can also classify user information with respect to its persistence in time:

- Static Data: persistent data, i.e. their duration in time is quite long (like preferences and rules) or fixed (birthday);
- Dynamic Data: information is quick like position, mood, presence, device info, etc.

11.3.2 Context Information Modeling

One of the objectives of research is also to develop a uniform context model representation language in order to facilitate context sharing and interoperability of context-related data among applications. Bolchini et al. (3) surveyed different data-representation of context information, while Strang et al. (12) listed six different ways of modeling context, namely:

- Key-Value Pairs: is the simplest data structure for modeling contextual information; they are easy to manage, but lack capabilities for complex structuring for enabling efficient context analysis algorithms.
- Graphical Models: are mainly used during the analysis phase to define a shared model among developers, and they are good to describe the structure context information and derive some code from the model, but they are usually not machine-readable.

- Object Oriented Models: are inspired from the OO paradigm and details of context processing are encapsulated on an object level and hence hidden to other components. Access to contextual information is provided by means of specified interfaces.

- Logic Based Models: In a logic based context model, the context is defined as a set of facts, which may be expressions evaluated to true in a particular time frame, and rules whose activation allows deriving new facts. Usually contextual information is added to and deleted from a logic based system in terms of facts or inferred from the rules in the system.

- Ontology Based Models: Ontologies are an emerging way to formally specify concepts and interrelations typical of human language. One of the first approaches of modeling the context with ontologies has been proposed by Otzturk and Aamodt (8) who argued the necessity of normalizing and combining the knowledge from different domains. More recently many OWL/based approaches have emerged (13, 15) describing the knowledge base with OWL-DL . Main issues with ontology-based models are related to reasoning on situations with a limited temporal validity.

- XML- Markup Scheme Models: they are hierarchical data structure consisting of XML tags with attributes and content. There are many other context modeling approaches mainly focused on extending the Composite Capabilities / Preferences Profile (CC/PP) (5) and User Agent Profile (UAProf) (14) standards, These kinds of context modeling approaches usually extend the basic CC/PP and UAProf terminology, trying to cover the high dynamicity of contextual information.

As these languages are often not publicly available or limited to few contextual data (4), we defined ContextML to fill this gap.

11.3.3 ContextML

In this work we have used a common language for context representation named "ContextML" (7). It is an XML-based language used by all elements of the context enabler within the UIM.

ContextML is based on the idea of "scopes", which are abstract definition of a coherent set of information related to user context. One or more scopes are associated to each specific domain of information. A scope is a simple table of concepts, grouped together and identified by a name, which is used as parameter name in ContextML. Such names could be easily mapped to concepts in an ontology. Scopes can be atomic or aggregated, when composed of different atomic scopes. The scopes currently defined are related to: location, calendar and device information.

For example, the scope named "position" groups latitude, longitude and range regarding a certain entity's location. Scopes can be atomic or aggregated, as union of different atomic scopes. Any user information given by a provider is characterized by an entity and a specific scope. When a provider is queried, it returns the required data in a XML document, which contains the following elements:

- contextProvider: a unique identifier for the provider of the data.
- entity: the identifier of the entity which the data are related to.
- scope: the scope which the context data belongs to.
- timestamp and expires: respectively, the time in which the response was created, and the expiration time of the data part.
- dataPart: part of the document which contains actual user information data which are represented by a list of a features and relative values through the <par> element ("parameter"). They can be grouped through the <parS> ("parameter struct") or <parA> ("parameter array") elements if necessary.

For example, the getCivilAddress method of the Location Provider for latitude 45.112 and Longitude 7.67 for user "elio" is invoked through:

```
GET
http://cark2.cselt.it/LP/LocInfo/getCivilAddress?
username=elio&lat=45.112&lon=7.67 HTTP/1.0
```

And returns the following XML content:


```
<?xml version="1.0" encoding="UTF-8" ?>
<contextML xmlns="http://ContextML/1.1"
xmlns:xsi="http://www.w3.org/2001/XMLSchema-instance"
xsi:schemaLocation="http://cark3.cselt.it/schemas/ContextML-1.1.xsd">
   <ctxEls>
      <ctxEl>
         <contextProvider id="LP" v="1.0.2" />
         <entity id="elio" type="username" />
         <scope>civilAddress</scope>
         <timestamp>2007-01-24T16:05:19+01:00</timestamp>
         <expires>2007-01-24T17:05:19+01:00</expires>
         <dataPart>
            <parS n="civilAddress">
               <par n="street">Via Arrigo Olivetti 5</par>
               <par n="postalCode">10148</par>
               <par n="city">Torino</par>
```

```
                <par n="subdivision">TO</par>
                <par n="country">Italy</par>
            </parS>
        </dataPart>
    </ctxEl>
  </ctxEls>
</contextML>
```

Location information is defined in three scopes: `cell` for GSM-based cell information, `position` for GPS-related information, and `civilAddress` for precise information about user address in a town.

The `cell` scope is typically provided by the CUF, whilst `position` and `civilAddress` are provided by the Location Provider part of the CER and detailed later on. In particular a `cell` scope contains data like the CGI (Cell Global Identifier), the NMR (Power level data), SGN (Service Cell Signal power), net type (GSM or UMTS), signal strength, extended name of the network operator, roaming (true or false).

`Position` scope includes data like latitude, longitude, accuracy (i.e. the precision of the previous measure), localization mode (i.e. based on GPS, cell-based or Wi-Fi based), and altitude.

`CivilAddress` scope contains information about country, city, postal code, street, building, room, aisle, floor.

Apart from location scopes, ContextML also defines Calendar-related scopes and Device-related scopes. Calendar-related scope includes information on agenda entries, and each entry is characterized by attributes like start/end time, event name and location, participants list, category (i.e. public or private). Device-related scopes are `DeviceInfo`, `DeviceSettings`, `Connectivity`, `commStats` (Communication Status), and `deviceApp` (i.e. the list of running applications on a device):

- DeviceInfo contains information about the Amount of total RAM in bytes, Phone IMEI code, Operating System short name, Number of CPUs, Screen height and width (in pixels), Color Deep, Java VM version, presence of Flash Plugin. DeviceSettings scope contains static information related to the user configuration, like languages used by the device.

- Connectivity scope includes Name, IP Address and SubNet Mask of the network adapter, usage of DHCP, description of the kind of Adapter (wlan, Ethernet).

- CommStats lists the available communication statistics, like the Number of SMS and e-mail unread, number of Calls Missed/Received, number of SMS/MMS sent/received , Last Call Received, Last SMS sent Number, Total time of communication.

11.4 Context Management Architecture

11.4.1 User Information Management: Architecture Overview

Designing a functional architecture for the User Information Management requires considering the different nature of the data, as described before: Profile, Context and Controls. This categorization suggests the implementation of three different modules, each of them in charge of handling a specific kind of information. Traditionally in the telecommunications world, these three types of information have been addressed separately by standard bodies, each of them providing a partial answer to a global question of gathering user information, which has been jealously isolated and spread over heterogeneous systems throughout the years. At the same time, this approach is still valid as it allows for fine tuning the needs and peculiarities of each single type of information as explained above. Furthermore, it allows integrating or re-using available implementation of components that address a single type of information and are already deployed and running in the operator's network, typically better exploiting standards through well-defined protocols, data descriptions and the methodology for handling the information. Of course not always what has been already specified can be reused as is, but it has been the selected approach for choosing the reference technologies that could have a significant exploitable impact when slightly improved or extended.

The outcome of such design for the User Information Management system resulted in the following components:

- User Profile Repository (UPR), for handling all static information;
- Context Enabler and Reasoning (CER), for feeding, elaborating and providing context data;
- AAA Module, for Authentication, Authorization and Accounting information management.

The whole User Information Manager, and so the composition of these three modules, act as the mediator between underlying enablers and the application/Service Layer where the User Generated Services platform lies.

The so-called enablers are the elements that provide low level User Information (e.g. context or profile) and can be categorized into four groups:

- End-user device: PDA, mobile phone;
- Intelligent Sensors: Body Area sensors (ECG monitors), Environmental sensors (humidity, temperature, light, imaging), Movement sensors (based on GPS receivers, compass, accelerometers, gyroscopes etc.);

- Network enablers: Address Book providers, Location providers, Presence providers, Calendar providers;

- 3rd party enablers.

The User Profile Repository, as described above, has been designed to store static data, or rather the kind of information that doesn't change often in time. Simple examples of these are preferences, personal details (e.g. gender, country of birth), buddy list and so on. Starting from that last type of information, the buddy list, the OMA/3GPP/IETF standard schema specification (sometimes called Group Management) has been reused (see resource-list references), whilst the profile schema was defined as an extension of the OMA user profile, that could embed specific information related to the User Generated Services information (e.g. type of user, service preferences, etc.) Considering that this standard has also defined the access, storage and manipulation of such information and of XML documents in general, the repository itself relies on an XML Document Management (XDM) server. The XDM specification is based on the IETF XML Configuration Access Protocol (XCAP) which enables manipulation of individual XML elements and attributes instead of the whole XML document. This reduces the bandwidth and processing power requirements significantly. XDM is also fully integrated with the IMS SIP implementation, which allows the powerful push mechanism for being notified whenever something changes in a document.

Regarding Context information, another approach has been followed given the requirements for major flexibility in the interfaces and in the processing of information: in fact the set of functionalities provided by XCAP protocol is rather limited and better suited for periodic editions of full or large portions of documents, and does not support any further intelligence in the server node to elaborate the updated information, which task is fully demanded to external entities. Instead, the CER embeds a Brokering Layer that is an infrastructure for storing and sharing user information in a modular fashion and specifically designed for handling dynamic data.

Looking at standard solutions, the OMA SIMPLE presence specification in the IMS extended world is known to address this type of information when seeing presence (and location extensions) as an important set of context information. Although it has been considered as a source (or a potential consumer) of context information within our platform, it has been discarded as main technology to process and exchange context information for three reasons. The first one is because of the Presence schema format and philosophy that requires the definition of new tags or extensions (and related schemas) for each piece of information that needs to be added or modified. This less flexible approach has conditioned the choice for another more abstract format, especially when considering how context information is growing, and thinking about the complexity of defining and managing several XML schemas and versions to accommodate the intrinsic versatility of context information as the technology enables new devices and sensors to provide additional information almost every month. The second reason for not choosing SIMPLE is due to the SIP protocol, which still has compatibility problems and a poor diffusion so far especially when coming to devices. We understand however that the protocol was also standardized for being used in mobile networks and expect a wider support and stability in the next years. Our current infrastructure is however flexible to accommodate this evolution towards IMS through gateways that would convert information back and forth between the two systems. Indeed the third reason for not using SIMPLE as generic infrastructure is related to the limited flexibility in processing/elaborating information on the server-side, for example managing dependencies between information and their sources or providers. For example, a source could publish a user GPS location and another service could provide the related reverse geocoding information to retrieve the civic location given GPS coordinates as input. The SIMPLE infrastructure would typically require the Presence User Agent to access the service directly and publish as well the outputted civic location. However, such approach is limited in requiring such direct access from each device whilst the operator may have a contract with external GIS providers to perform this context enrichment centrally, synchronizing the validity of the civic location information with the GPS one. Other such examples can be made that show the limits of the SIMPLE infrastructure in handling dependencies in a proper way, avoiding complex exchanges that anyway do not prevent discrepancy of information.

The last part of user information has been demanded to the AAA module, which has to be linked deeply with already existing systems and standards used in the operator's network. It deals with several kinds of user information such as identity, credentials and privileges (for Authentication), usage records (for Accounting), and the relation of the user with the platform, its services and their controlled access (for Authorization) . AAA (Authentication, Authorization and Accounting) infrastructures are critical in a service delivery platform, where users must be identified and trusted to let them access services, and where records on platform and service usage (e.g. for creating, subscribing or executing a service) must be adequately saved in order to later apply charging policies to the users. This module was based on several standards, such as Diameter and Liberty for example. Its architecture was inspired by other existing reference models for AAA, such as the ones defined by the IETF in its Generic AAA architecture (especially regarding the layered architecture), in its AAA Authorization Framework, for the separation of policy retrieval, evaluation and enforcement,and in the Pluggable authentication Modules of the XSSO standard.

The core of the Authentication functions is based on the use of a Liberty Alliance based Identity Provider (IdP), backed by a Diameter server (Diameter is an open, robust, standardized, AAA protocol family, which is being employed by 3GPP). The Liberty infrastructure has been the primary choice for federated identity because of being an open standard with open source available implementations, besides being robust and Web Services oriented (against web based Open-ID) for an easy integration with other systems in the platform. Liberty will also ease a future integration with IMS, since 3GPP has standardized several applications of the use of Liberty and SAML.

11.4.2 Context and Usage Feed: Publishing Information from Mobile and Fixed Devices

The Context information, handled in the User Information Management by the Context Enabler and Reasoner (CER), is provided by sources spread in the real word. These sources feed the system with “any" kind of raw context information in order to be analyzed and enriched. This internal process transforms the rough data into the high level context information. These sources of information are called *Context Source.*

There are two ways of gathering user-related information that can be distinguished as:

- Explicit: Based on direct feedback or input, no matter the user is aware of it (e.g. could be some application monitoring in the background, versus a web-based portal asking for user preferences (i.e. food, favourite movie types)
- Implicit: Based on some post-processing (reasoning/learning) that deduces some information out of other explicit information.

Another orthogonal dimension to be considered for gathering is related to the technological approach to be used to retrieving concretely this information:

- device-based, i.e. through some software in the end-user device (PC, mobile) that captures information related to the user (e.g. application usage, GPS location, Bluetooth information, etc).
- network-based, some system in the network that holds some information about the user or can generate some, e.g. presence info or reasoning/learning engine.

Typically, both approaches are valid and complementary, since some information resides on the end-user device only, other is only available in the network. Of course sources onboard devices need to optimize traffic and battery usage to only send meaningful changes. The CER typically supports both approaches allowing devices to send information, and network elements owning some user information to provide it as well.

Context Sources have been designed for both mobile and “desktop” devices. According to the ambient where they are running, Context Sources can collect different information. The implementation can use various kinds of technologies, in terms of frameworks, but it should have a rich access to the underlying operating system where it is executed. Indeed, most of the relevant context information needs to be extracted from internal or peripheral sensors such as connections parameters, device information etc. The approach of using a dedicated client application approach drastically limits the wide spread of such model in the real-world, unless standardization initiatives allow to embed similar functionalities in the devices themselves, being more and more sensor-equipped and with high processing and bandwidth capabilities.

The OPUCE project as developed 2 context sources: a light Context Source for desktop using the Yahoo! Widget framework, and one on Windows Mobile using the .Net Compact Framework. Both clients are able to collect a wide set of device information, where some of them are listed below:

- Device information (MAC address, CPU type, screen size, amount of memory, operating system and version, Java support, Flash support)
- Device status (Up and running, bandwidth usage, available bandwidth, battery level, free memory)
- Position (latitude, longitude, accuracy, localization mode)
- Device settings (language in use)
- Calendar (appointments in the user calendar)
- GSM/UMTS information (cell-id, connection type, Operator nameąŋ)
- Nearby Bluetooth devices
- Nearby WI-FI networks

It is important to note that in the User Generated Services area, context information can be heavily used for personalization, either managed by platform systems, or by end-users themselves when creating their own services. As such, the type of information to be considered has to be meaningful for the sake of service personalization, and be understandable by end-users themselves to some extent (the rest being dealt with by software).

Once the information has been collected, the Context Sources periodically send them to the Context Enabler and Reasoning (CER) using a REST-like interface, as described in the further paragraph.

11.4.3 Context Enabler

The Context Enabler and Reasoning (CER) is the subsystem in charge of gathering and providing context information (both as raw and "reasoned" information), or information related to matching contexts (e.g. users in the same context). Following the architecture defined in User Information Management, the CER provides its interfaces through the User Information Broker element, besides providing a suite of User Information Providers in the back-end.

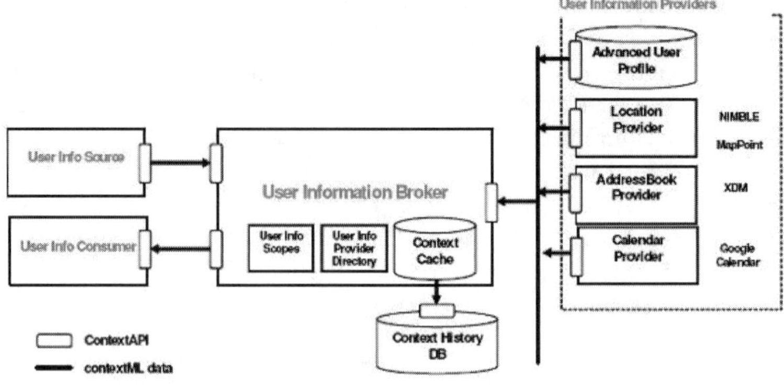

Figure 11.2: The Brokering Layer

According to this paradigm, the CER receives information through the different applications that implement the User Information Source paradigm. The CER can virtually connect to any platform to either extract or provide information, such as Wireless Sensor Networks or IMS/OMA enablers. For example, the CER can easily interconnect with IMS/OMA SIMPLE Presence system to import some presence information and export other presence information.

As illustrated in Figure, the Brokering layer is made of different entities, following some kind of SOA approach:

- Source: an entity producing/publishing some information. Typically a source sends out raw information spontaneously. Sources are typically in the end-user devices (context, usage information) or other devices (e.g. sensors) and send information to the broker. A source is asynchronous and pushes user information to the "Broker", but it does not expose any interface unlike the Provider.

- Provider: an entity that provides information upon request or subscription according to the specified language and interface. A Provider is a software entity that produces new user information and that can be collocated with a source. It registers its availability and type of user information ("scopes") by sending appropriate information to the Broker. A Provider exposes interfaces to provide user information to the broker and to consumers, that are both synchronous or asynchronous to be notified about information changes. Typical examples are presence provider, location provider, or other providers that rely on reasoning (such as situation provider for example).

- Broker: an entity that can potentially aggregate information from multiple providers or help find the right provider, also managing information lifecycle and dependencies. It acts as intermediary between sources, providers and consumers. The broker is the interface to the overall user information that is used by the service creation and service adaptation modules and/or building blocks.

- Consumer: an entity (e.g., a context-aware application) that requests user information for consumption. Following the previous example, a context-aware service adaptation or notification module would act as a consumer. The Context Base Service, or context-aware Base Services can act as consumers as well. Another example is the reasoning engine that acts as consumer to retrieve basic information and acts as provider/source to provide high-level information. The consumer can interact directly with a single provider or through the broker for aggregated information.

11.4.4 Context Web Service Interfaces

User information is exchanged within the Context Enabler and Reasoner (CER) and between the CER and Context Sources (like the CUF applications) or Consumers following a REST-like paradigm based on XML over HTTP, which eases the exchange of information with any type of software entity (also mobile phones) and simplifies the addition of (external) entities to the system. Requests are typically HTTP GET or POST requests that return XML-based content (the user information) according to the ContextML language.

Differently from the SIP protocol that was analysed initially as a possibility for carrying context information as mentioned above, the REST-Like protocol is stateless, and it follows the nowadays trends of the Web2.0 paradigm open to a variety of mash-ups. On the server side, both back-ends and front-ends can be easily implemented and the compatibility is widely guaranteed through HTTP. On the other hand the client/device side implementation is simplified as all the required libraries are natively supported by of the operating systems. Client then simply need to create or parse valid XML documents, following a single ContextML schema, independently from the data to address, in order to exchange context information with the platform.

Contrarily to the Presence data format, ContextML has the advantage of being open in terms of extension of the new context data discoverable. Context information is provided inside the ContextML using the scopes and parameters, which are simple strings in attributes in term of schema description. Actually a Scope is a simple well-defined aggregation of data with a semantic coherence grouped together and identified by a name, which is used in ContextML. Each single data within a scope is a parameter characterized with a nema, which can easily be mapped to ontology concepts in ontologies. The interfaces allow different interactions between the Context Source and the Broker/Providers:

- Publish Context : for the context source to provide context
- Retrieve Context : for a context consumer to ask for context
- Subscribe Context: to trigger a context notification.

Publish context is the interface used to push all the collected context data coming from end-user devices or other systems, namely sent by the CUF (Context Usage Feed) in OPUCE terminology. A HTTP POST message containing context information formatted according to the ContextML schema is posted to a specific URL of the Broker.

Retrieve context is instead a way to get context information from the Broker, which provides a URL where a Context Consumer can ask for specific data, using the scopes variables, and belonging to a specific entity (e.g. a specific user). The Broker will care about performing the appropriate queries on the available context providers, and the outcome is a ContextML message with the retrieved context.

Subscribe Context is the opportunity that the Broker provides in order to being notified whenever certain conditions are verified. For example, suitable situations can be based on specific Context Values, or when a threshold is reached. In such cases the subscribed applications receive a notification from the broker, with the information requested at the time of subscription (which may differ from the ones used as criteria).During the subscription the application should provide the call-back interface where the notification will be dispatched, the different condition in order to define the matching criteria, and the information to be retrieved whenever the condition are met. Different rules are supported by the broker, like more than, equal to, between x and y values. It is worth noting that subscriptions and notifications use the same ContextML schema.

11.5 Adapting User-Generated Services

The OPUCE platform embraces two different approaches to allow service adaptation to consumer's context: explicit adaptation and implicit adaptation. Explicit adaptation mechanisms are those in which the service creator specifies the adaptation actions dealing with content/variables or workflow. Conditions and actions are directly built by the creator. Implicit adaptation mechanisms on the contrary are driven automatically by the platform, based on limited input by the service creator who only has to specify adaptation points. At this adaptation points, the necessary information retrieval, condition definition and subsequent actions are carried out autonomously by the platform based on the expert knowledge it packs.

Explicit adaptation mechanisms in OPUCE are three: user sphere, Context-Aware Base Services and Context-Aware service composition, and are detailed in section 11.5.2. Implicit workflow adaptation and Dynamic Profile Retrieval (DPR) are the OPUCE implicit adaptation mechanisms, and are detailed in Section 11.5.3.

11.5.1 Faceted Service Specification

The OPUCE service creation and execution approach relies on a generic service specification. The OPUCE service specification (2) is an abstract meta-model that can be particularized for each service and type of service in a unique service description. The service specification conceptual model created in OPUCE provide sufficient information about all aspects of a service which could be needed while maintaining compatibility with standard approaches, but also being flexible enough to be able to respond to changes that occur in the near future. OPUCE leverages on a faceted service description: each service is specified as a set of separate facets, each of them focused on a specific separate aspect of the service. Facets have been defined to describe the service logic, the deployment activities required, the service level agreement, and also the context information to consider for each service in order to modify the service execution. A facet is formally described as a projection over one or more service properties that provide a partial description of a service. All of them are described by using an independent facet specification. The language used is flexible and can be easily adapted to the specific facet to be described. This model is suitable to describe both base and composed services. The only different is that some facets are mandatory for base services but not for composed services and vice versa.

Figure 11.3 shows the information contained in the XML service specification document and the relationships among the different elements of a service specification. The left of the figure is the master document that describes a service. It includes general information and links to the rest of facets. The structure of the facet specification is represented in the right side. It includes the identifier of the service that is described, the type of facet, the language used in this facet and the specific description of the facet in that language.

This document is generated at creation time. The portal compiles information from the user and from the editor and generates the master document and the correspondent facets. The master document will point to the rest of the facets, contained in separated XML documents, one for each facet.

11.5.2 Explicit Adaptation Mechanisms

The "User Sphere" concept and the DPR functionality, presented in Section 11.5.2 and Section 11.5.2 respectively, are the two sides of the parameter adaptation mechanism of the OPUCE platform. Context-Aware Base Services and Context-Aware service composition, presented in Section 11.5.2 and Sec-

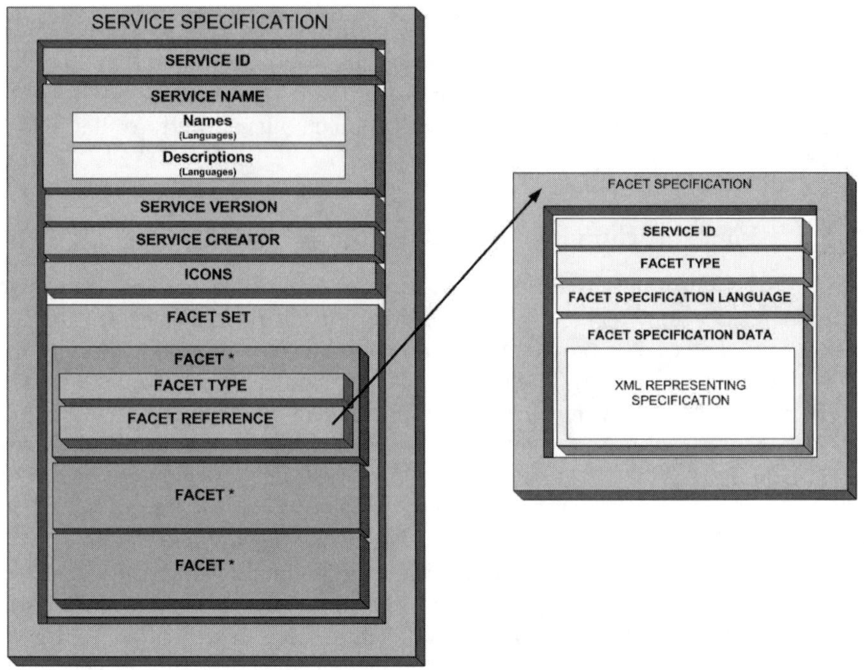

Figure 11.3: Faceted service specification structure

tion 11.5.2, on the other hand deal with workflow adaptation. Together, these four features form the set of explicit adaptation mechanisms of the OPUCE platform.

User Sphere

In OPUCE there are two distinct roles that an end-user can play: the service creator and the service consumer. In order to allow that a service composed by a creator could be used by several different consumers, it is necessary to choose among one of the following three approaches for parameterization: either there is no room for parameter personalization, so the service runs exactly with the same parameters for all users; or a human user manually personalizes those parameters; or the platform automatically retrieves the required parameters from the user profile of each of the consumers.

Consider for instance a service which monitors an email account and forwards the title of each received message as an SMS to a given phone number. If no personalization is allowed, the email address and the target phone number will be hardcoded in the service logic; they will always be the same regardless the consumer of the service and this renders the service almost useless. If manual personalization is allowed, the consumer of the service will manually introduce those parameters in a configuration stage. But if the service platform stores information about each end-user's email and mobile phone number, then the personalization could be carried out automatically.

The OPUCE platform supports both manual and automatic personalization of parameters. For manual personalization, at creation time the service creator is able to set a parameter as "open", meaning that the value will be introduced by the service consumer at execution time. And for automatic parameter personalization, OPUCE takes advantage of the "user sphere" concept, a construct called `$me` used to represent information about the consumer of a service at creation time.

If a service creator wants to build a service which sends an SMS to the consumer of the service, the parameter "Target_Mobile_Number" could be left in an "open" status, so it could be manually specified by the consumer at the configuration stage. But the creator can also use the $me construct to refer to the "end-user sphere". In this case the value of the "Target_Mobile_Number" parameter will be set to $me.Mobile_phone. This value will be resolved at execution time thanks to the Dynamic Profile Retrieval (DPR) functionality that will be presented in section 11.5.2.

The "user sphere", implemented in the $me construct, includes static information from the user present in the User Profile Repository, and dynamic Context information retrieved from the Context Enabler and Reasoning (CER) module, which have been introduced in the previous section..

Dynamic Profile Retrieval Functionality

The OPUCE Service Execution Environment (SEE) (6) is the platform subsystem responsible for the execution of the OPUCE Services that have been designed, specified and composed by means of the OPUCE Service Creation Environment (SCE).

Figure 13.3 depicts the high level architecture of the OPUCE SEE which is composed by the Orchestration System and a set of Base Services (BS). As already mentioned in Section 11.2, an OPUCE Service is specified through a set of blocks (i.e., the BSs) connected by means of arrows, where the arrows have the following semantics: "when event A occurs in the BS from which an arrow originates, action X is triggered in the BS at which the arrow points".

As showed in Figure 1.4, in the OPUCE execution platform we can identify two different software levels:

- *Base Services execution level*: each BS is implemented as a Tomcat Axis2 service that can be deployed in one or more nodes. Each BS is identified by its endpoint which contains the IP address/port of the hosting machine and the name of the BS (e.g. http://192.168.1.4:8080/axis2/services/SendSMS). BSs are independent in the sense that they are supposed to provide just one specific functionality and are not aware that they can be combined with other BSs.

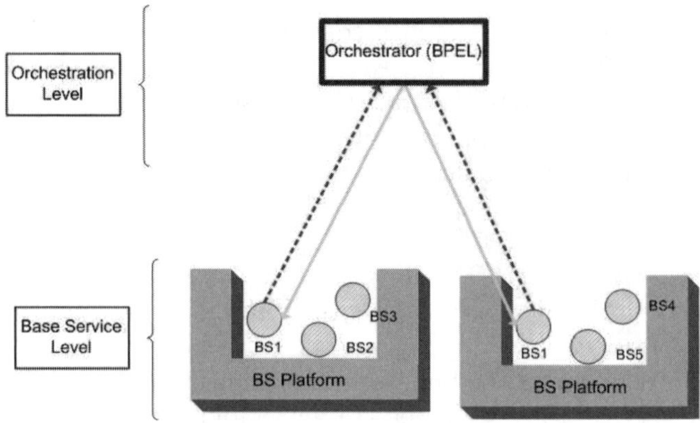

Figure 11.4: High level SEE architecture: continuous arrows refer to invokeAction calls and dashed arrows refer to notifyEvent calls.

Each BS is only supposed to expose the invokeAction Web Service by means of which they can provide their basics functionalities.

- *Orchestrator execution level*: this software component takes care of combining Base Services to execute the User Generated Services. It invokes BSs by using the invokeAction and it exposes the notifyEvent Web Service. The BPEL (Business Process Execution Language) technology is the widely used standard for the orchestration of web services and we chose it to implement our Orchestrator (in particular we used the Active BPEL Open Source implementation).

Therefore the invokeAction and the notifyEvent Web Services are the two Web Services that are continuously invoked at execution time.

In the SEE, the "SCE arrow" is implemented through the following pair of method calls:

- notifyEvent: Because of the fact that the Base Services (which live in different containers, even replicated in multiple instances) are not aware of the OPUCE Service they belong to, they only notify events to the Orchestration System which controls the execution of the OPUCE Service base on its knowledge of the service logic. Each notifyEvent call contains the output properties of the BS that is firing the event;

- invokeAction: when receiving a notifyEvent call, the Orchestration System invokes the execution of the appropriate action(s) in the appropriate base service(s) based on the service logic by means of the invokeAction

At service creation time, some of the OPUCE Service properties may have been filled with user profile information (e.g., first name, last name, telephone number, etc.) or with user context information (e.g., GPS data). All these user related data are referenced by the Service Creator (in the Service Creation Environment) by means of the `$me` keyword (e.g., $me.firstName, $me.CurrentLocation.GPS.Latitude, etc.) which need to be resolved with the actual corresponding value at service execution time: this functionality is called Dynamic Profile Retrieval (DPR). It is important to remark that it is not sufficient to retrieve this information from the UPR module and CER module at the start time of the OPUCE Service because both the user profile and the context information may change while the OPUCE Service is running. During the execution of the OPUCE Service the SEE interacts with these two external modules to update the $me variables appropriately.

The DPR functionality is performed at run time by the Orchestration System which takes care of communication between the Orchestration System and the UPR and between the Orchestration System and the CER in order to update the $me properties of the OPUCE Services in execution. The DPR functionality was implemented according to different policies as described next:

1. Each time an event reaches the Orchestration System, both the UPR module and the CER module are queried to update the $me properties. This invocation takes place even if no $me pròperties need to be updated.

2. The UPR and the CER modules are queried if and only if one or more $me properties need to be update.

3. DPR takes advantage of the "publish and subscribe" mechanism provided by the XDM Server described. This is the most efficient solution because the SEE and the UPR exchange data only if the user profile is modified (the UPR notifies the SEE about the user profile changes by sending a SIP notification message).

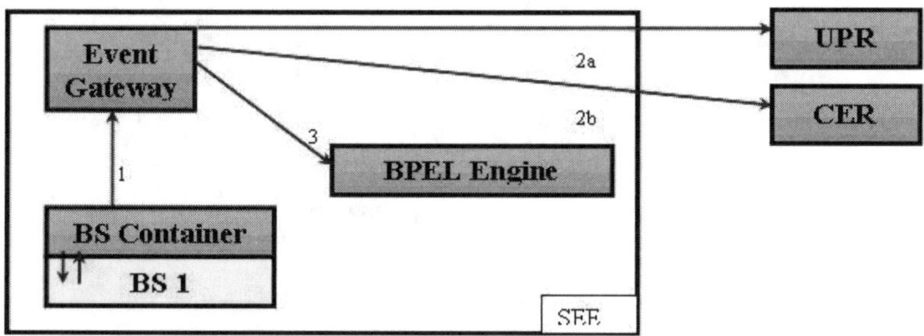

Figure 11.5: $me properties update in SEE1.0

Figure 11.5 summarizes the DPR mechanism in the OPUCE SEE which follows the first of the three aforementioned approches. The main steps are:

1. At a certain time during the execution of a OPUCE Service, one Base Service notifies a new event to the EG;

2. the EG receives the new event and retrieves the user profile information from the UPR module using the XCAP protocol (2a). It also retrieves the user context information from the CER using a Web Service REST call (2b). The UPR and the CER invocations are performed to fill two disjoint $me properties sets. The invocation of these two components of the OPUCE platform is done in parallel in order to minimize the latency related to communication. The time taken by a single DPR update is

   `t_DPR = max(t_access_UPR, t_access_CER)`

   instead of

   `t_DPR = t_access_UPR + t_access_CER.`

3. Once all the $me properties are retrieved, the EG passes the updated set of values to the BPEL process that will update its internal status accordingly to the new information and invokes the next base service(s).

It is worth noting that we also investigated two alternative DPR implementations to improve the performance of the system by applying the $me resolution only when strictly needed or by taking advantage of the SIP publish and subscribe mechanism. But for space limitations, these two mechanisms are not discussed in this chapter.

Here follows an example of how the OPUCE SEE executes a concrete service at run time.

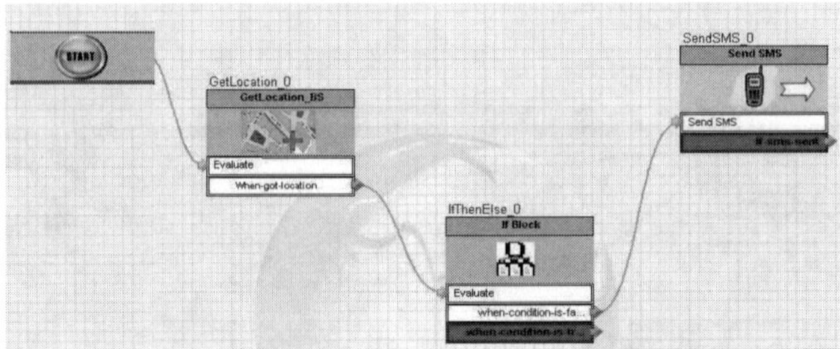

Figure 11.6: An example of an OPUCE service.

This User Generated Service allows parents to check the location of their son at a specific time of the day and it works as follow:

- It periodically starts its execution. It is possible to do so because the OPUCE platform allows users to schedule the execution of a service, for instance "every day at 2.00 p.m.". The service is used by the son so the username is equals to the son's name.

- The GetLocation_0 BS checks the location of the son because the *GetLocation_0.userName* is set to $me.userName.

- The *IfThenElse_0 BS* checks if the son's actual ZIP code (retrieved by the GetLocation_0 BS) equals to his home ZIP code ($me.postalCode, stored in the profile of the son - UPR call). If so, the service stops its execution (the son is near his home) otherwise the *SendSMS_0 BS* is invoked.

- The SendSMS_0 BS has the following input parameters:

  - SendSMS_0.destinationNumber = $me.mumMobileNumber (stored in the profile of the son - UPR call)
  - SendSMS_0.content = "Hi mum, I am currently in " + $me.currentLocation.city (CER - call)

  Because of this configuration, if the son is not near home, his mum will receive a SMS containing the name of the city where her son is in that moment.

Next figure shows how this service is executed by SEE 1.0.

1. The scheduler periodically issues the start command into the Orchestrator (BPEL engine)

2. The Orchestrator starts the execution of the composite service by invoking the GetLocation_0 BS invokeAction Web Service (actionName = "evaluate").

3. When the GetLocation_0 BS finishes its execution, it invokes the notifyEvent Web Service into the Orchestrator (eventName = "when-got-location").

4. The Orchestrator has to invoke next BS (IfThenElse BS) but the resolution of the $me.postalCode property is needed so the BPEL engine retrieves the user profile info from the UPR.

5. The Orchestrator invokes the IfThenElse_0 BS invokeAction Web Service (actionName = "evaluate").

6. The IfThenElse_0 BS notifies to the Orchestrator the result of the comparison between the user's ZIP code and the current ZIP code retrieved by the GetLocation_0 BS.

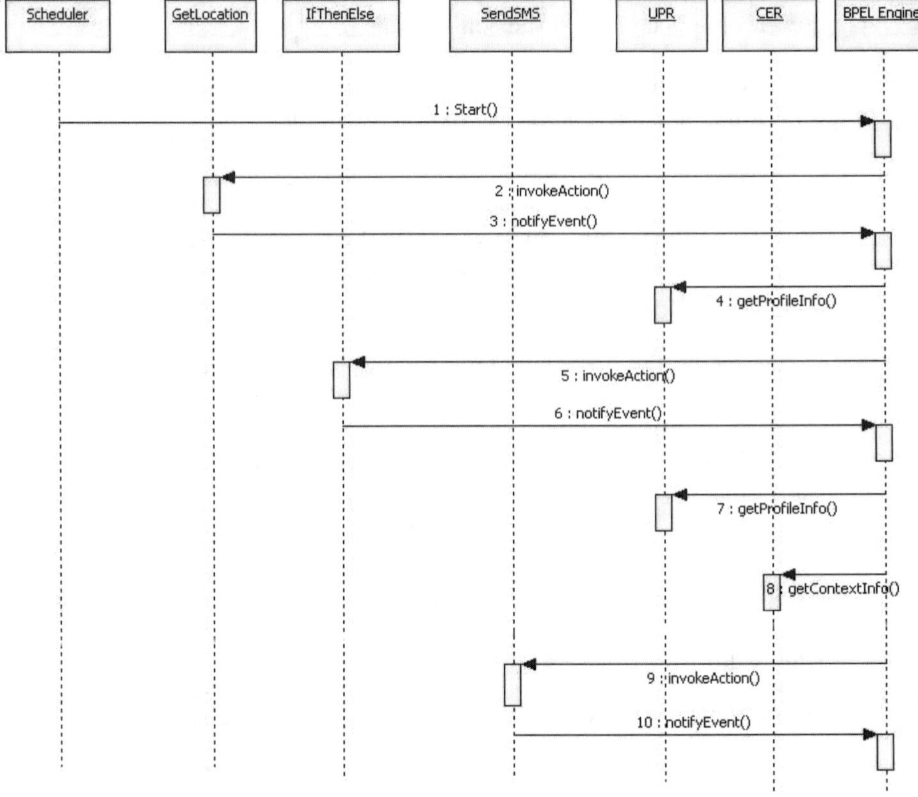

Figure 11.7: The service execution scenario of the former example.

7. If the result of the comparison that took place at step 6 is "true" (i.e. the two ZIP codes are the same) the service execution ends. Otherwise, as depicted in the UML communication diagram, the Orchestrator has to invoke next BS (SendSMS BS) but the resolution of the \$me.mumMobileNumber property is needed so the BPEL engine retrieves the user profile info from the UPR.

8. Since the \$me.currentLocation.city property is also needed, the Orchestrator also invokes the CER to retrieve the context information of the user.

9. The Orchestrator invokes the SendSMS_0 BS invokeAction Web Service (actionName = "sendSMS").

10. The SendSMS BS notifies to the Orchestrator that the SMS has been successfully delivered so the execution of the composite service ends.

Context-Aware Base Services

While the `User Sphere` concept and the DPR functionality deal with parameter adaptation, they cannot influence the actions to be carried out, an operation that is called workflow adaptation. One of the two ways to perform explicit workflow adaptation in OPUCE is the usage of Context-Aware Base Services.

These are Base Services which are directly hardcoded by their developers with context-aware functionalities. Therefore, the actions they carry out directly depend on the service consumer context. For instance, a "Context-Aware Message Sender" Base Service could be implemented to access the Context data of the service consumer upon invocation, and then send some data as an Instant Message if the presence status of the service consumer is "online", an email if it is "busy" or an SMS if it is "disconnected". This adaptation logic will be hardcoded at a low level inside the Base Service logic, and transparent to the end-users of the platform.

In order to compose a Context-Aware service, the service creator simply has to introduce this Base Service in the composition, instead of using the "Instant Message Sender", "Email Sender" and "SMS sender" Base Services separately. The adaptation is carried out automatically, so it is extremely simple. However, this approach is limited, because the adaptation logic is fixed and has to be implemented at a low level by the Base Service Developer.

Context-Aware Service Composition

In order to overcome the limitations of the Context-Aware Base Services, OPUCE also supports explicit Context-Aware Service Composition by means of If_Then and get_Context Base Services. The get_Context Base Service accesses the UPR and CER to retrieve the profile and context parameters of a given users, and makes them available for the service creator. Then, the output of the get_Context Base Service could be used as an input to a conditional Base Service (e.g.: If-Then) to drive the flow of the execution.

For instance, a service logic could involve a get_Context Base Service to retrieve the context of the service creator, linked to a "getDistance" Base Service to determine the distance between the "location" property of that user and $me.location (the location of the service consumer), and then an If-Then block will trigger the sending of an SMS if the distance is less than 200 meters informing of the proximity.

Although this approach is more flexible than the usage of hardcoded Context-Aware Base Services, it is more demanding on the creation side, because the service creator has to explicitly specify the Context-driven adaptation logic inside the service.

11.5.3 Implicit Adaptation

As already introduced, the implicit adaptation refers to the automatic adaptation of the service execution to the changing context and environment. Automatic adaptation here means that the user doesn't need to define the behavior of the service depending on the different context conditions.

The implicit adaptation can be achieved introducing an additional "Context Awareness" module in the OPUCE architecture:

1. The SEE redirects the Base Service invocations that potentially could be adapted to the Context Awareness subsystem, that is aware of the user context and platform status;

2. The Context Awareness subsystem infers the adaptation action to be performed (if any action is needed);

3. The Context Awareness subsystem discovers adaptation capabilities/actuators that could serve for performing the adaptations needed (actions from a different Base Service could even be invoked, as the Context Awareness is aware of every action offered by every BS);

4. Once performed the inferred adaptation (acting over the SEE and the adaptation capabilities), the Context Awareness notifies the SEE.

From a user point of view, the basis of this solution for implementing the implicit adaptation is based on the differentiation of two "views" of the Base Services at creation time:

- Normal view: the actions/events shown are the ones specified in the interface facet and the actions selected by the user will not be modified at execution time.

- Context Aware view: when selected, the actions/events shown are extracted from a Context Aware facet which provides a certain level of abstraction respect to the "normal" actions described in the interface facet. For example, there will be generic actions like "SendNotification" instead of actions like "sendIM" or "sendMail". The concrete action to be invoked will be inferred at execution time

The architecture of the Context Awareness subsystem is presented in Figure 13.4, also showing the interfaces with other OPUCE modules. The main blocks under the Context Awareness subsystem are:

- Information Manager: in charge of retrieving, storing temporarily and managing all the information about the user, the services and the platform itself.

- Adaptation Manager: provides the orchestration to the adaptation process. After receiving the adaptation request from the SEE, communicates with the Adaptation Inference Engine and invokes the needed actions.

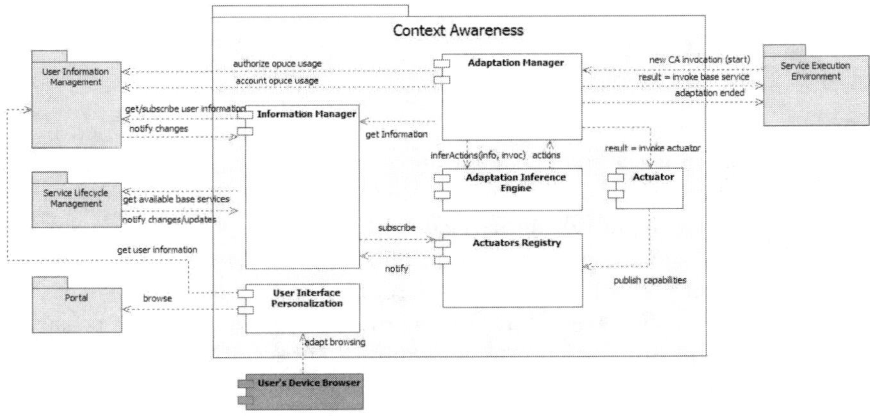

Figure 11.8: Context Awareness subsystem's architecture

- Adaptation Inference Engine: provides the intelligence of the adaptation process. It's able to infer the most appropriate actions (like invoking Base Services actions, actuators, etc.) analyzing the information related to the user, the available Base Services/actuators, etc.

- Actuators Registry: where the Actuators publish their descriptors, which are retrieved by the Information Manager.

- Actuators: an actuator is a block that is able to perform some actions to adapt the execution. There are different categories:

  - Media: change the media. E.g: from text to audio, from video to audio, etc.

  - Format: change the format. E.g: video codec, from plain text to HTML, etc.

  - Transport: change the transport. E.g.: from text to SMS.

- User Interface Personalization: responsible of adapting the user interface to the user's terminal, depending on the device inner (screen, keyboard, resolution, etc.) and/or environmental characteristics.

11.6 Conclusion

This chapter introduced a set of issues and first solutions proposed by the OPUCE project for the management and use of context information for personalizing and adapting services in the new field of User Generated Services. Our goal has been to make the user context, preferences and other dynamic variables such as presence, localization and device characteristics, influence the user experience in a pervasive way across the overall service lifecycle, and not only during service execution. We believe our design offers the possibility for end users to personalize interactions such as subscriptions, advertisements (notifications) and execution. The accomplishment of this goal required a deep interaction with other areas such as service description, service deployment and remote service management.

The user information management system introduced here provides the foundation for gathering, providing and managing user information, including user identity, profile and context, each of them treated separately for a more flexible and pervasive handling, but also to leverage existing operators' systems and standards. The system implemented includes user identity management (through the IdP/AAA module relying upon Liberty and Diameter), user profile management (OMA XDM-based UPR) and context management (CER and CUF). We have further enhanced the UIM for service and platform authorization rules and for gathering service "usage" information, for accounting purposes or potentially for user behavior learning, in order to improve implicit adaptation and personalization. The introduction of the user sphere approach happened to be necessary to address user-centric references to personal information as well as information portability across-users when editing or executing someone else's service. However, progress still has to be made on the sensing and gathering of contextual information from the device to limit client-server interactions and device-specific on-board clients, which require heavy battery consumption. On service execution side, the Dynamic Profile Retrieval function has triggered numerous challenges when addressing long-lived services, where freshness of information versus traffic optimization is a non-ending debate to best satisfy this kind of logic.

Other open issues have been identified to overcome current limitations and pave the way for future advances in platforms targeting the field. In particular, management of privacy information through flexible, yet simple means, is a crucial challenge that remains open for handling user-generated services that are natively portable and cross-editable. Advanced techniques of learning, reasoning and recommendation would also be welcome to optimize implicit adaptation, personalization and filtering amongst the thousands of Base Services or service compositions targeted.

Finally, such highly dynamic environment of combined short-lived and long-lived services created and managed by end-users would need to enhance the concept of context from user and service level down to the platform level: system overload or lack of resources would then be prevented by a context-aware adaptation framework that could dynamically reroute traffic or redeploy software components on an on-demand basis.

Acknowledgements The work presented in this chapter is executed as part of the OPUCE project and partly funded by the European Union under contract IST-034101. OPUCE is an Integrated Project of the 6th Framework Programme, Priority IST.

References

[1] OPUCE: Open Platform for User-Centric Service Creation and Execution. http://www.opuce.eu/.

[2] C. Baladron, J. Aguiar, B. Carro, J. Sienel, R. Trapero, J. Yelmo, J. del Alamo, J. Yu, and P. Falcarin. Service discovery suite for user-centric service creation. In *Proceedings of SOC@Inside'07, Service Oriented Computing: a Look at the Inside, Vienna, Austria*, 2007.

[3] C. Bolchini, C. A. Curino, E. Quintarelli, F. A. Schreiber, and L. Tanca. A data-oriented survey of context models. *SIGMOD Rec.*, 36(4):19–26, 2007.

[4] L. Capra, W. Emmerich, and C. Mascolo. Reflective middleware solutions for context-aware applications. In *Metalevel Architectures and Separation of Crosscutting Concerns - Proceedings of Reflection 2001, Kyoto, Japan, September 25-28, 2001. Lecture Notes in Computer Science (Vol. 2192). Springer Verlag*, 2001.

[5] CCPP. Composite capabilities / preferences profile (cc/pp). http://www.w3.org/Mobile/CCPP.

[6] D. Cipolla and J. Sienel et al. Web service based asynchronous service execution environment. In *Workshop on Telecom Service Oriented Architectures, Vienna, Sept 2007, Lecture Notes In Computer Science; Vol. 4907, Service-Oriented Computing - ICSOC 2007 Workshops*, 2009.

[7] L. Lamorte, C. A. Licciardi, M. Marengo, A. Salmeri, P. Mohr, G. Raffa, L. Roffia, M. Pettinari, and T. S. Cinotti. A platform for enabling context aware telecommunication services. In *Context Awareness for Proactive Systems(CAPS) Workshop, Guildford, UK.*, 2007.

[8] P. OTZTURK and A. AAMODT. Towards a model of context for case-based diagnostic problem solving. In *Proceedings of the interdisciplinary conference on modeling and using context*, 1997.

[9] A. Sanchez, C. Baladron, J. M. Aguiar, B. Carro, L. W. Goix, J. Sienel, I. Ordas, R. Trapero, and A. Bascunana. *At your service: An overview of results of projects in the field of service engineering of the IST Programme*, chapter User-centric service Creation and Execution. MIT Press Series on Information Systems, 2009.

[10] B.N. Schilit, N. Adams, and R. Want. Context-aware computing applications. In *Proceedings of the 1st International Workshop on Mobile Computing Systems and Applications*, pages 85–90, 1994.

[11] J. Sienel, A. Leon, C. Baladron, L. W. Goix, A. Martinez, and B. Carro. A telco driven service mash-ups approach. *Bell Labs Technical Journal*, 14-1, 2009.

[12] T. Strang and C. L. Popien. A context modeling survey. In *ACM Workshop on Advanced Context Modelling, Reasoning and Management, UbiComp 2004 - The Sixth International Conference on Ubiquitous Computing, Nottingham/England*, 2004.

[13] X. H. Wang, D. Q. Zhang, T. Gu, and H. K. Pung. Ontology based context modeling and reasoning using owl. In *Proceedings of the Second IEEE Annual Conference on Pervasive Computing and Communications Workshops (PERCOMWŠ04)*, 2004.

[14] WAPFORUM. User agent profile (uaprof). http://www.wapforum.org.

[15] S. S. Yau and J. Liu. Hierarchical situation modeling and reasoning for pervasive computing. In *Proceedings of 3rd Workshop on Software Technologies for Future Embedded & Ubiquitous Systems (SEUS 2006)*, 2006.

[16] J. Yu, P. Falcarin, J. M. del Alamo, J. Sienel, Q. Z. Sheng, and J. F. Mejia. A user-centric mobile service creation approach converging telco and it services. In *Eighth International Conference on Mobile Business(ICMB 2009), Dalian, China*, 2009.

Part III

Technology

Chapter 12

Context Coupling Techniques for Context-aware Web Service Systems: An Overview

Hong-Linh Truong and Schahram Dustdar

Abstract Distributed services in a context-aware Web service system or in a service composition need to couple context information from different sources in order to be able to interpret context information. Such interpretation is a prerequisite before services can adapt themselves to be context-aware. However, context coupling in context-aware Web service systems is a complex issue. It is related not only to how we represent and transfer context information among Web services, but also to how we support the context storage, and how we ensure security and privacy issues of context information. Context coupling techniques for Web service systems will be different from that of tightly coupled systems because, with Web service systems, context information is typically shared across the boundaries of organizations that host Web services. This chapter aims at presenting an overview of context coupling techniques, their related issues, and their implication for a context-aware Web service system. Our approach is to study existing techniques implemented in current systems, to present open context coupling issues in current and future context-aware Web service systems, and to suggest further research directions.

12.1 Introduction

At the time of writing, utilizing Web services to develop large-scale systems and distributed applications operating beyond the boundary of a single organization is the norm. Many Internet-based systems include several Web services, which are loosely coupled and owned by different providers. The concept of the Internet of Things and the Internet of Services (44), although still being shaped, puts this loosely coupled applications model further: future Internet-based applications will include various software services, things, and people interacting via standard protocols and models, which are highly dependent on SOA (Service-oriented Architecture) technologies. To date, we have observed the popularity of Web services, and user participation and customization in the Web. Certain types of these Web service based systems and applications require Web services to be aware of context associated with their operations. Such context could be associated with, for example, time, location, profile, and runtime status of services, things, and people. Obviously, such context could be individually handled by a single service and collectively processed by different services in multi-organizational environments.

From our previous study in (41), we have observed great challenging issues for supporting context-aware Web services. In our work, a context-aware Web service is a smart Web service which, defined by Manes, *"can understand situational context and can share that context with other services"* (31). Being smart or context-aware is important for Web services because they could not only effectively match user's needs to their capabilities, but also be able to adapt themselves with situation changes to improve their availability and reliability. As we discussed in (41), when systems and applications are built from different Web services provided by and hosted in multiple organizations, it is challenging to make the systems and applications context-aware, as this requires services to be aware of each other and aware of the context of customers and applications. This challenge is due to the distributed, large-scale, and diverse nature of Web service-based environments. Unlike past context-aware systems in which components are tightly coupled and in a closed environment, such as (47), as indicated by (31, 33, 42) and others, solutions for enabling context sharing in Web services-based environments must be open and interoperable to ensure that they can be applicable in Web service-based systems.

To support the concept of context-aware Web services, distributed Web services in a context-aware Web service system or in a service composition need to couple context information from different sources in order to be able to interpret context information. Such interpretation is a prerequisite before services can adapt themselves, a condition to be context-aware. However, context coupling in context-aware Web service systems is a complex issue. It is related not only to how we represent and transfer context information among Web services but also to how we provide context information to different interesting services and how we ensure security and privacy issues of context information. Context coupling techniques for Web service systems will be different from that of tightly coupled systems because, with Web service systems, context information is typically shared across the boundaries of organizations that host Web services.

To examine this issue, this chapter aims at presenting an overview of context coupling techniques, their related issues, and their implication to the success of a context-aware Web service system. Our approach is to study existing techniques implemented in current systems, to present open context coupling issues in current and future context-aware Web service systems, and to suggest further research directions. To this end, in Section 12.2 we will discuss fundamental assumptions, present scenarios, and explain what context coupling means. From well-established coupling techniques, we concentrate our study on four different models, named structure, data, message and common couplings, which in turn strongly impact the design of existing context-aware Web services. In Section 12.3, we study how existing context-aware Web service systems support context coupling techniques and analyze strength and weakness of existing techniques. To further analyze how context coupling techniques are implemented in a real-world scenario, we present a case study based on the EU FP6 inContext project. Based on our analysis, we discuss some open issues and suggest few recommendations for future research in Section 12.5. Related work of this study is also given in Section 12.6 and we conclude the chapter with an outline of some of our future steps in Section 12.7.

12.2 Fundamental Concepts

12.2.1 Context-aware Web Services

What context information and context-aware systems are, has been defined and discussed in various papers (13, 19, 23, 17). Context information is dependent on individual systems, as a type of information might be considered as context information in one system but not in another one. Context-aware Web services are a subtype of context-aware systems defined in these papers. As given in (41), we consider context information as any additional information that can be used to improve the behavior of a service in a situation. Without such additional information, the service should be operable as normal but with context information, it is arguable that the service can operate better or more appropriately.

A Web service might be context-aware as it can adapt its operations according to the context of its clients and environment. Naturally, context-aware Web services sharing context information will operate in a distributed environment. However, it might never need to share context information, e.g., to other services it depends on. This case is not the focus of the environment we assume in this study. Instead, we consider the case in which different services will share context information. We also consider whether these services belong to the same (virtual) organization or not. By (virtual) organization, we mean that these services will follow certain policies established and enforced by the same (virtual) organization, such as security, privacy, and data governance. For example, when services are provided by the same company, they might not worry about how sensitive the context information transferred among the company's services is, thus the services might relax some privacy conditions which must be implemented when the services do not belong to the same organization.

12.2.2 What is Context Coupling?

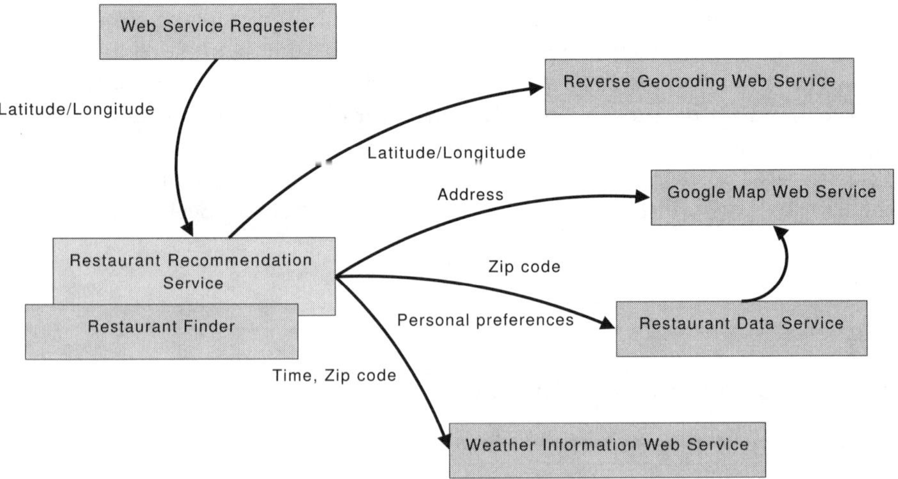

Figure 12.1: Example of context information sharing.

Before we discuss context coupling, let us consider the following scenarios. The first scenario is described in Figure 12.1. Assume that a user would like to use a PDA (Personal Digital Assistant) which is equipped with a GPS (Global Positioning System) to find relevant restaurants with/without open gardens when the user is in Vienna[1]. To date, all information relevant to this search could be provided by Web services and this user's need could be

[1]Although this scenario is imaginary, restaurant recommendation is a popular scenario that can be found in many papers.

fulfilled by different means. For example, a `Restaurant Recommendation Service` could offer recommendations to restaurants based on several criteria. The `Restaurant Recommendation Service` can also utilize various other services, such as the `Reverse Geocoding` for mapping GPS information to addresses, the `Google Map` for finding businesses close to an address, the `Restaurant Data Service` for searching restaurants based on user preferences, and the `Weather Information Service` for obtaining weather information. Many types of context information could be exchanged in this scenario. First, the user's PDA has GPS, thus user's location (latitude/longitude) information can be captured and utilized. In a simple form, in order to find a restaurant, a `Restaurant Finder` in the PDA would, for example, be implemented as a Web service requester that directly invokes other services, e.g., by utilizing mashup techniques to aggregating content from different services. The `Restaurant Finder` can automatically obtain GPS information and send this information to the `Reverse Geocoding Web service` to get the address associated with the user's location. Then it uses the address to call the `Google Map Web service` to find restaurants close to the address. Furthermore, it uses the zip information to get the weather information and then recommends restaurants with gardens if the weather is nice. In this case the `Restaurant Finder` has to model and couple all context information used for finding restaurants. In another possibility, a more complex form, the user just uses the `Restaurant Recommendation Service` which is provided by a service provider because the provider can get more benefits (e.g., acquiring more user information or providing the service as an value-added offer). This service requires the user to provide only one parameter that is the restaurant search command, but it manages several user-related context information. When the user uses this service, the user's `Web service requester` may automatically pass the user's GPS information to the service via SOAP headers. This service then utilizes the user's location to determine the corresponding address, thus being able to locate relevant restaurants based on the address. Furthermore, it can pass user preferences/behaviors, which it has in its database, to the `Restaurant Data Service` in order to obtain a better match of restaurants to user's profile.

In this scenario, possible types of context information are, for example, location (latitude/longitude), time (when the request is issued), weather status, and personal preferences. Depending on capabilities of services, specific configurations, and software development processes, not all of these types of information might be used. For example, when the weather status is unknown, it is probably hard to recommend an outdoor restaurant. These types of context information are shared between different clients and services spanning different organizations. Sharing methods are different, depending on available services and compositions. Therefore, some fundamental questions about sharing mechanisms or how context is coupled, arise. For example, how latitude/longitude information is transferred to the `Restaurant Recommendation Service` (using parameter invocation or embedded information in a SOAP header)? Does `Restaurant Recommendation Service` manage an ontology and instance information of context information about location, time, weather, etc.? Will the `Restaurant Recommendation Service` pass a structure of personal preferences to the `Restaurant Data Service` or just give a link to a person's preferences? If a link is passed, how can we be sure that sensitive information, such as user identity, is not accessed by the `Restaurant Data Service`? Generally speaking, as a Web service may manage, process, and transfer different types of context information, it has to couple different types of context. How can it deal with this? Which techniques should be used? There are many questions and this chapter aims to answer some of them by studying current state of the art systems. Besides many possible types of context information might be shared, this sharing is also conducted across different services that are not designed to work together on purpose and may be deployed in different geographic locations. This scenario reflects the case of Web services belonging to different organizations that share context information. Context coupling techniques have to work with the assumption that the policies are governed by different laws, countries, and distributed environments.

The second scenario is about a smart home, shown in Figure 12.2, and is based on the EU FP7 SM4All project (10). In Clara's home, there is a system that can operate lights and doors automatically based on the simple presence information so she has bought this system and installed it in her home. Unfortunately, she got an accident and she is now in a wheelchair but she is doing some rehabilitation every day. Thus, she needs more help in order to deal with other difficulties during her rehabilitation. First, she is equipped with a health monitoring system that will connect to a distributed health-care system managed by her caretakers. Second, she installs a wearable activity recognition system that can detect her gesture. This activity recognition system can interact not only with her light and door controllers but also with the distributed health-care system. Furthermore, to support her rehabilitation, a home multimedia system is installed. This system can be controlled by her activity recognition system and interact with external multimedia services provided by, for example, Yahoo!, YouTube, and a weather service. Since intruder breaks are increasing, she also install sensors to detect intruders and to control the alarm process, including calling the police, turning on/off lights, and opening/closing doors. This human recognition system also allows her to control the access to her house (automatic block or access). Furthermore, her other properties, such as her car, are also monitored and monitoring information will influence her activities.

This illustrative scenario presents many possibilities in exchanging and consuming context information. For example, to ensure that she selects suitable multimedia contents according to her health status, health status information can be used to filter the content. To ensure that external multimedia services offer best selections for a patient, her profile and status can be used to optimize the content delivery. Furthermore, when she has a severe health con-

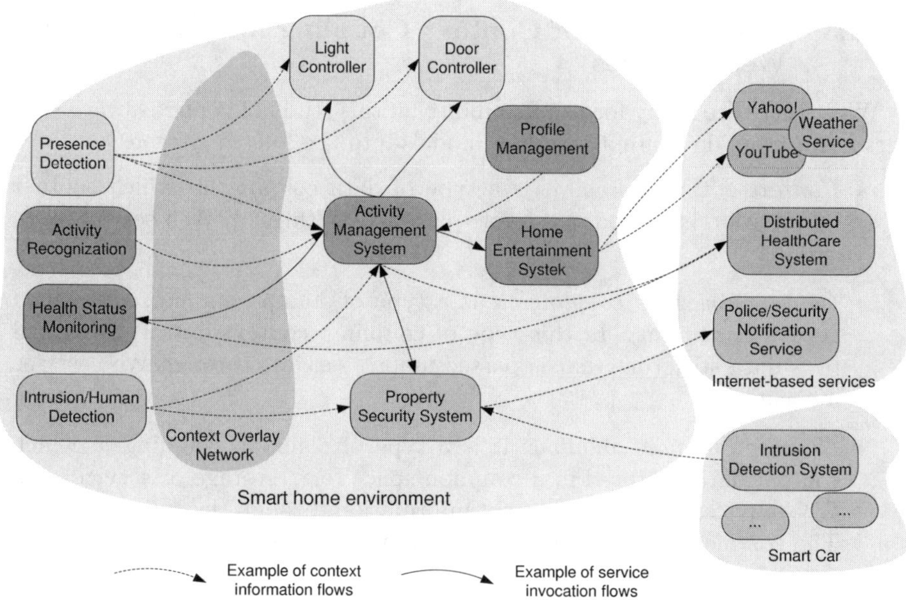

Figure 12.2: Context providers and consumers in Clara's home and other spaces.

dition or based on her gesture, health caretakers will be informed and doors are automatically opened when they enter her house based on pre-defined profiles. This scenario shows that different types of context information are coupled over the time. For example, at the beginning, in Clara's home, we may have only a presence sensor, a service to turn lights (built based on a light controller), and a service to open/close doors. Then, as an activity recognition system is installed, we have wearable sensors, an activity management system, and a user profile management system. This activity management system will have to interact with light/door controllers and planned multimedia entertainment system. With this scenario, we see that (1) a context model covering many context sources cannot be pre-defined due to the diversity and opportunistic usage of context sources, (2) context middleware cannot assume all context sources use the same protocol (e.g., push or pull) to provide context information. Thus, they raise the questions of how context information could be encapsulated and propagated over such a network of services.

In this paper, context coupling refers to the degree how a Web service/client relies on another Web service/client with respect to context information. In our view, context coupling is a special topic of *coupling* (2, 1). Understanding context coupling techniques is not only useful for selecting suitable techniques to transfer context but also for ensuring concerns associated with context information, such as privacy, to be guaranteed.

12.2.3 Basic Models of Context Coupling in Context-aware Web Services

While there are many forms of coupling, in this study of context-aware Web services, we divide context coupling models into the following types:

- *Context data coupling*: it is a type of data coupling in which context information is passed, as simple parameters, through Web service invocations.
- *Context structure coupling*: it is a type of stamp coupling, also called type use coupling. In this type of coupling, context data is described by a data structure that is passed among services through Web service invocations.
- *Context common coupling*: it is a type of common coupling in which context data is stored in a common space (e.g., storage or service) and Web services access the context information through the common space. The common space could be also extended to a distributed space (e.g., a distributed system of repositories).
- *Context message coupling*: it is a type of message coupling in which context-aware Web services use messages to transfer context information. It can be built based on publish-subscribe and P2P (Peer-to-Peer) systems.

The main reason for considering only four types of context coupling is because we observed those as the main types used in existing systems. Other coupling models are either not used in or not applicable for context-aware Web service systems. Note that these types of coupling are not orthogonal, because, as we discuss them in Section 12.3, one type of coupling might use another one.

Context coupling techniques are related to several other techniques in a context-aware system, such as context representation, context dissemination, and context storage. For example, context representation describes how context data is structured. Thus, it has a strong impact on context structure coupling. Context dissemination protocols describe which protocols are used to conduct context message coupling. Context storage can be part of coupling techniques, for example, context message coupling can be performed via centralized context storage. A Web service entity stores context in the storage, which is then accessed by another. In addition, context coupling techniques for context-aware Web services deal with the transfer of context information of which many types are sensitive, such as user information and location. Therefore, security and privacy issues in context sharing are of paramount importance (14, 34). Thus, when studying context coupling techniques, we will also discuss possible privacy and security issues. However, we will not discuss how Web services will utilize context information to become "smart" (refer to (41) for context adaptation techniques, for example).

12.3 Context Coupling Techniques in Current Context-Aware Web Service Systems

12.3.1 Structure and Data Coupling

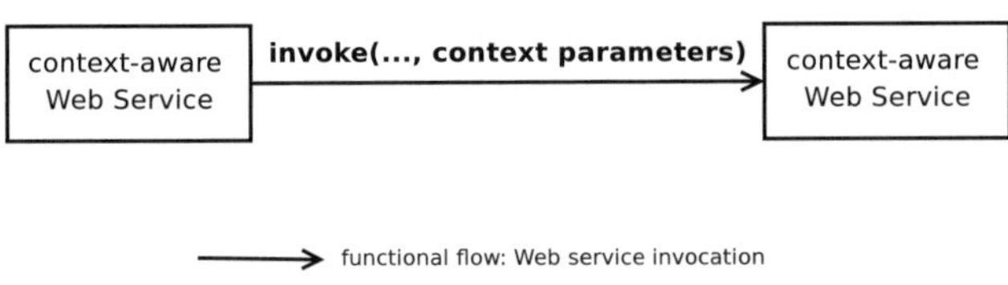

Figure 12.3: Data coupling model.

In structure and data coupling techniques, context information is described in structured and/or simple parameters. A context-aware Web service will explicitly pass context information to another context-aware Web services by invoking appropriate service invocations. Figure 12.3 describes the interaction between two context-aware Web services supporting structure and data coupling techniques in which `context parameters` can be simple or structured input variables. Since context information is shared through service invocation, a context-aware Web service has to model and express its input context parameters explicitly in its service operations, although it is possible that values of these parameters might never be set in service invocations.

One technique to support such couplings is to develop context-aware Web services based on the Model-driven Engineering (MDE) approach. The MDE approach is increasingly used for structure and data coupling of context information in various context-aware Web services systems. The basic idea is to model the context parameters and their structures explicitly and to associate them with Web services descriptions. Examples of systems utilizing this way are CoWSAMI (16), ContextUML (38), CSOA (45), and their applications and works built atop them. Another technique of structure and data coupling is to develop/utilize agreed context models to pass context information. In this case, mostly a context model is developed based on ontologies (22, 42) or XML schemas (43), and context-aware Web services just provide service operations that accept context information following the model. Typically, this case of structure and data coupling is utilized together with common coupling techniques (see Section 12.3.3).

Using structure and data coupling techniques, a developer of a callee context-aware Web service will focus on describing what his/her context-aware Web services need and let his/her caller deal with the integration issue. Thus, it is more flexible with respect to the development process, in particular, when context-aware Web service systems are built from existing ones. For example, it is relatively straightforward to compose several context-aware Web services because their operations are known. However, when context parameters are changed, the composition has to be reworked. When a context-aware Web service needs to be improved, e.g., able to handle a new type of context information, its developer freely performs his/her work, but the developer of the composition has to rebuild the composition. Thus, it is also not suitable for large-scale context-aware systems, e.g., envisaged in the Internet of Services. In such a Internet, diverse types of context sources exist and modeling all of them is very difficult.

With respect to privacy and security issues, these types of couplings offers many possibilities to strongly enforce privacy and security controls. The caller has a total control over what kind of context information it could share. Security enforcement is the same as what is for other functional invocations of the service, for example WS-Security standards (11) can be used for SOAP-based context-aware Web services while HTTP Authentication (4) and HTTPS (5) can be used for REST-based context-aware Web services. Thus, security enforcement does not impose a large cost for security implementation for context information. Although, many systems studied do not deal with privacy issues or do not describe how they address privacy issues, data and structure coupling techniques could utilize MDE techniques and pre-/post-conditions assertion techniques to enhance the assurance of privacy issues (e.g., eliminating user identity from context parameters). Some basic work (28, 46) have been done in this direction. In the privacy-aware context profile (28), context is associated with owner and authorization for accessing context information can be associated with group or user. Thus, this model can be used to describe some privacy-related information. While privacy-related information can be described, technically, privacy compliance can only be performed at the caller side because there is no way to ensure that the callee will follow the same policy. In fact, the systems studied in this section do not describe how they ensure privacy compliance. Some work, such as (30), have presented the use of information about privacy concerns and policies to enforce the privacy compliance within Web services. However, the connection between context modeling (at design time) and privacy compliance (at runtime) in an end-to-end software engineering process is not well established in existing work.

12.3.2 Message Coupling

In context message coupling techniques, context information is exchanged via messages. A context-aware Web service passes context information to another Web service by sending a message consisting of context information or links to the information. The message can be relayed through a message middleware.

Using SOAP headers to transfer context information is one technique in

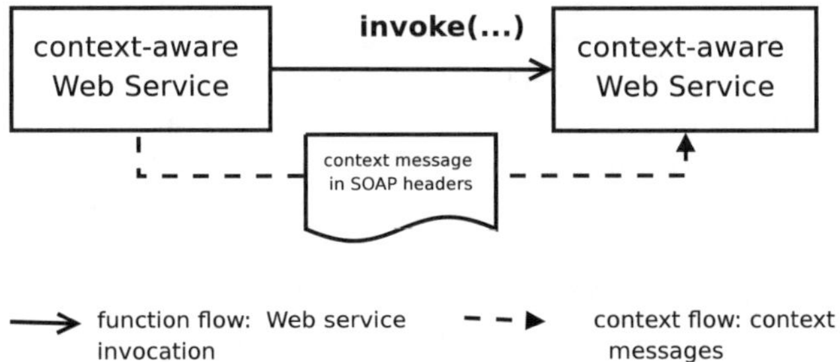

Figure 12.4: Model of context message coupling using SOAP headers.

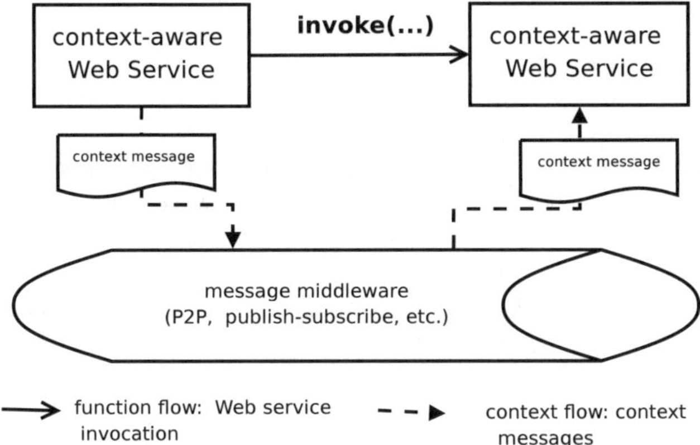

Figure 12.5: Model of context message coupling through message middleware.

this type. In this technique, context information or links to the information are encapsulated in SOAP header messages that will be transferred from a caller Web service to its callee Web service. As shown in Figure 12.4, context message is separated from messages of functional calls. Context messages can be directly sent from one service to another one (see Figure 12.4) or indirectly relayed through message middleware (see Figure 12.5). This technique has been proposed and implemented in several systems. Keild and Kemper present a generic framework to support the development of context-aware adaptable Web services (29). This framework separates clients/Web services from the context framework that supports clients and services. The transfer of context information is performed through the SOAP message header. Context information can be explicitly and directly processed by clients or Web services or be automatically handled by the context framework. Similarly, in the CASD (Context-aware Service Directory) system (21) SOAP headers are used to transfer context information that influences the operation of CASDs. In the inContext system (40), URIs (Uniform Resource Identifiers) specifying the location of context information are transferred and based on that context information can be obtained. The above-mentioned systems can actually be considered as particular cases that WS-Context (12) aims to support. WS-Context specifies mechanisms to transfer context messages among Web services, supporting transferring context messages as well as references to context messages. In the first case, context information is wrapped into a message and transferred among Web services using SOAP headers. In the second case, only the references are transferred. The real context information will be obtained from an external service (context manager). WS-Context, therefore, supports different message coupling techniques. Similar to the above-mentioned tools, when transferring messages, WS-context implementations will face similar issues. When transferring references, corresponding context information could be retrieved from dedicated repositories or services. There are some implementations utilizing WS-Context concepts, such as (35).

While, this technique does not require change of service interfaces, thus it will not require changes of service composition, it requires a mutual understanding between the caller and the callee. Context-aware Web services sharing similar context information have to know schema of the message in advance. Thus, it is suitable for service providers who have a strong agreement (e.g., are in the same organization). One advantage of this technique is that it supports very loosely coupling by means of messages. The use of references instead of message content allows context-aware Web services to access context information from many context providers. However, this technique works only with SOAP-based Web services.

Another message coupling technique is to support the exchange context information through overlay networks (shown in Figure 12.5). There are many possibilities in this model. First, context-aware services may know the structure of the context information and just send different query or subscription requests over a network of services in order to obtain the context information. Second, context information may be published through an overlay network, possibly, of context management services. For example, in the ERMHAN system (36) distributed context management services relay context information to a central context management service, which then propagates selected context information fulfilling specific conditions to appropriate applications. Using messages to couple context information through overlay networks could also be useful for context-aware Web services which have a low temporal coupling degree, such as mobile Web services, because it does not require that both services have to be present at the same time.

With respect to privacy and security issues, context providers have the flexibility to enforce which information is to be shared and protected. Even though encryption can be applied to SOAP headers, most existing systems do not apply security methods to context information in SOAP headers. In message middleware for context coupling, security can be established based on WS-Security, HTTP Authentication, and HTTPS. However, security enforcement could be complicated in cases of transferring references because the context provider might not be the same as the caller (thus this requires some forms of security configuration between the caller, the provider, and the callee). The systems in our study do not show how they solve this problem. In this respect, some protocols could be utilized for context-aware Web services. For example, the OAuth protocol (6) can be used for authorizing resource access in Web services. With the OAuth protocol, a user can grant service consumers to access user's private resources hosted in another service. Therefore, it can be used to grant access to privacy context information among context-aware Web services. The OpenID protocol (7) is another one that can be used for managing identities associated with the caller, the provider, and the callee when transferring context references is employed.

12.3.3 Context Common Coupling

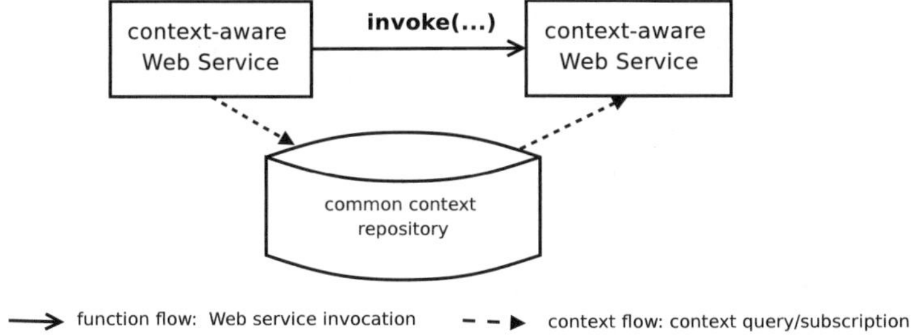

Figure 12.6: Common coupling model.

The common coupling model for context information is widely used in practice, especially when Web service providers agree on context information to be shared, or when services are managed by a single organization. This type of coupling is typically achieved by using a common repository to store, manage, and provide context information. The repository can be implemented as a Web service, such as in (26, 15, 35, 27, 32, 8, 37, 3), a tuple space, such as in the Ubiquitous Semantic Space (USS) (39), and agents, such as in (18).

This model could serve as a fast solution for context sharing in Web services. The common repository can provide well-defined interfaces for storing, querying, and manipulating context information. In many cases, context common coupling techniques are implemented based on a centralized model. However, this type of coupling is not limited to a centralized model. For example, in the ESCAPE framework (43), context information is hosted by different context management services. In order to retrieve context information, various different languages and service operations can be utilized, depending on specific implementations and context data representations.

This model is also usually used together with other models in two main ways. In the first case, it can be used with message coupling techniques. Context-aware services exchange only URIs indicating the context information and the information is retrieved from the common repository. One example of this case is the inContext system (40). In the second case, it can be used with context structure/data coupling. For example, in the CATIS system (37), the user's location is stored into a context manager. An application server serving user's request will retrieve the context information from the manager and passes the context information to relevant services.

With this type of coupling, in principle, Web services can exchange context information regardless of how the repository is implemented. However, when the repository is not based on Web services, such as in (18), from the integration point of view, there is a problem to support context-aware Web service systems comprising different services from different providers. One of the main issues for this type of coupling is the performance and scalability of the common repository, in particular, when ontology-based repository is used. To overcome this issue, distributed repositories are proposed. Another issue is that all services have to agree on the context structure. Furthermore, querying context information will bring overheads and more interactions are needed in order to retrieve context information. On the other hand, with this model, context information is easily coupled and strong reasoning capabilities can be built. Moreover, historical context information might be stored for other purposes.

With respect to privacy and security issues, this type of coupling allows a strong control of accessing and sharing context data at a central point. In addition to transport security enforcement that can be based on common techniques, as discussed in other types of couplings, such as WS-Security, HTTP Authentication and HTTPS, context access control can be based on rules. Such rules can also be strongly related to privacy preferences. For example, the Google Latitude (3) supports different privacy policies for a particular instance or type of context information to a particular people/application. In (46), privacy preferences, privacy rules and context ownership information are used in the context manager to ensure the privacy-aware context sharing. In (26), access policies can be established based on user's time, location, and event participation. Overall, this common repository model allows utilizing rich semantic representations for context information and well-integration of access control rules.

12.3.4 Summary

Table 12.1 presents the summary of how systems studied utilize context coupling techniques. Overall, structure, data and common coupling techniques are widely employed in current context-aware Web services. It is understandable because it is fast and easy to perform the integration with these techniques. Furthermore, they allow the developer to be flexible in developing their services without worrying much on common agreement in sharing context information. Until now only WS-Context is proposed as a standard for exchanging context in Web services, but it is not widely adopted yet.

Systems	**Types of Coupling**				**Privacy**	**Multi-organization**
	Structure	Data	Common	Message		
MDE approach	+	+				
inContext (40)	+			+		+
WS-Context (12)				+		+
USS (39)		+				+
CATIS (37)		+	+			+
CASD (21)				+		
ESCAPE (43)	+		+			
CA-SOA (20)	+		+		+	
Akogrimo (8)			+		+	+
ConServ (26)			+		+	+
ERMHAN (36)				+		
Keild and Kemper (29)				+		

Table 12.1: Summary of context coupling techniques supported in existing context-aware Web services. We use + for a feature when a system explicitly supports the feature. When it is unknown or unclear (some systems mention a feature but do not show how they implement the feature), the feature was left blank.

In general, security and privacy issues are not well addressed in studied context-aware systems. Some of systems studied provide authentication and authorization access controls, but most of them neglect privacy issues. This is probably due to the fact that many types of context information in these systems are (i) not classified as private and sensitive information, and (ii) actually used only within a single organization. However, these issues have to be addressed when context coupling is performed via open, Internet-wide context-aware services.

Table 12.2 summarizes our perspectives on the utilization of context coupling techniques. Overall, structure/data and common couplings offer less complexity and effort for the implementation but the interoperability degree is low, while message coupling is complex but offers a better interoperability degree.

Category	Types of Coupling			
	Structure	Data	Common	Message
Development flexibility	high	high	medium	low
Integration complexity	low	low	medium	high
Security enforcement effort	low	low	medium	high
Privacy enforcement effort	low	low	medium	high
Interoperability degree	low	low	high	high

Table 12.2: Perspectives on the utilization of context coupling techniques

---

12.4 A Case Study: Context Coupling in the inContext Project

The EU FP6 inContext project (9) aims at supporting highly dynamic forms of human collaboration such as Nimble (short-lived collaboration to solve emerging problems), Virtual (spanning different geographical place and having diverse professionals), and Mobile (collaboration with mobility capabilities) teams. Therefore, it deals with many types of context information. Furthermore, the inContext project proposes its solutions based on the SOA model. Context-aware services enabling team interactions are built as Web services (42). Thus, in the inContext project, various types of context coupling techniques are employed, namely context structure, context data, message, and common coupling techniques, and context-aware Web services are strongly supported.

In this section, we analyze techniques that are used by the inContext to design and implement context coupling. As described in (42, 40), the inContext project proposes four core capabilities dealing with context. Figure 12.7 describes these capabilities:

- Manage Structure: within this capability, the inContext project describes and associates different types of context, such as team activities, user profile, and resource information. Therefore, according to our classification, this capability is actually the context coupling based on the context structure technique.
- Create Correlation and Extract Correlation: the inContext project uses SOAP headers to transfer URIs indicating context information among Web services. These capabilities provide libraries for handling context URIs extraction, propagation, and correlation. Therefore, in our analysis, these two capabilities are dealing with context message techniques.
- Manage Context: the inContext project provides tools for storing, managing, and querying context information. Thus, this capability supports context common coupling.

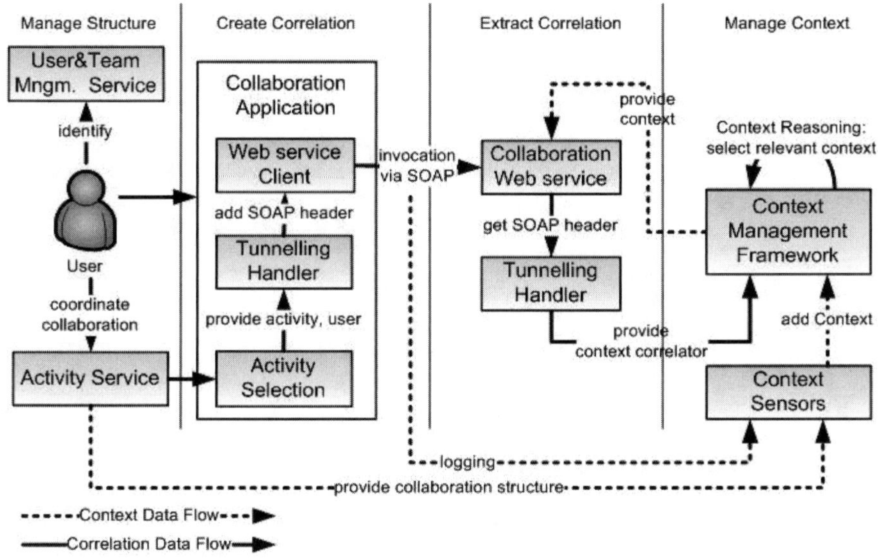

Figure 12.7: Four capabilities for supporting context in the inContext project.

12.4.1 Context Structure and Data Coupling

The approach to support context structure and data coupling techniques in the inContext project is to design an ontology describing all possible types of context information used in teamwork. To this end, the inContext project has reused many existing ontologies and developed new ones and linked them together in the so-called inContext context model. Figure 12.8 shows the inContext ontology that is able to seamlessly unify individual, team and activity context (42). Individual context includes not only most of the traditional context types, such as location, available devices, and communication online status, but also more collaborative work related information, such as team membership, activities (within the different teams), available resources, skills, and team members (from different teams). Team context includes information about interactions, projects, organizations, and locations that are associated with members of a team. Activity context describes tasks and their associated information in different levels of detail, for example, work breakdown structures of a project or user current activities.

The inContext context model is described and implemented using the RDF (Resource Description Framework) and OWL (Ontology Web Language). The main advantage of the inContext approach is that it allows for flexibility and extensibility of the context model, for instance, by inclusion of domain-specific data or reuse of Web data already available in common RDF formats. Furthermore, the ontology-based model provides common coupling with reasoning capabilities that will be discussed in Section 12.4.3.

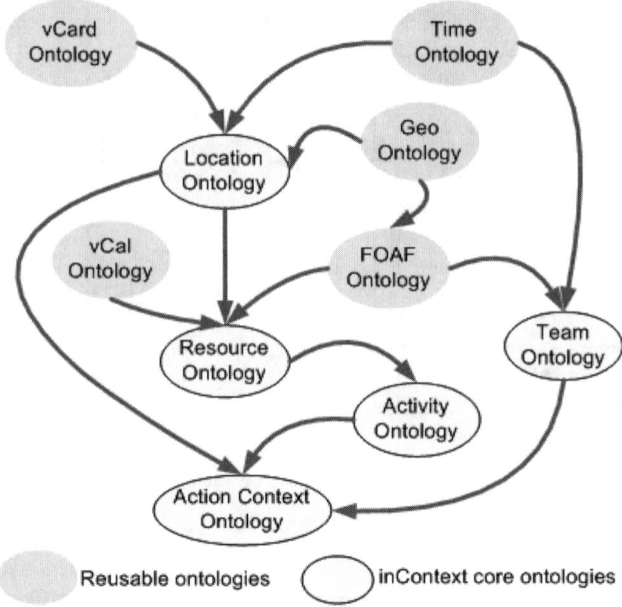

Figure 12.8: The inContext context model used to describe individual, team, and activity context.

12.4.2 Message Context Coupling

Message context coupling techniques in the inContext project are developed based on the propagation of URIs indicating context information via SOAP headers. Unlike (29), in the inContext project, only URIs specifying the location of context information are transferred. This assumes that by using the URIs, a context-aware Web service can retrieve context information from appropriate context providers. Figure 12.9 depicts the message context coupling implemented in the inContext project (40). URIs specifying context information, such as `ActivityURI` for activity information, and `UserURI` for user identifiers, are embedded into SOAP headers in SOAP messages exchanged between context-aware Web services. A service will use the URI to access the context information which is stored in a separate service (e.g., a `Context Store`, see Section 12.4.3) in the inContext's `Context Management Framework`. Listing 12.1 presents a simplified example in which context information related to activity `act1` and user `Rossi` is transferred. The `ActivityURI` and `UserURI` are `http://www.in-context.eu/pcsa#act1` and `http://www.in-context.eu/pcsa#Rossi.E54`, respectively. The context information itself is stored in the `Context Store`.

The approach of sending URI only facilitates the access to context information from any services. Thus, it is possible to support runtime binding:

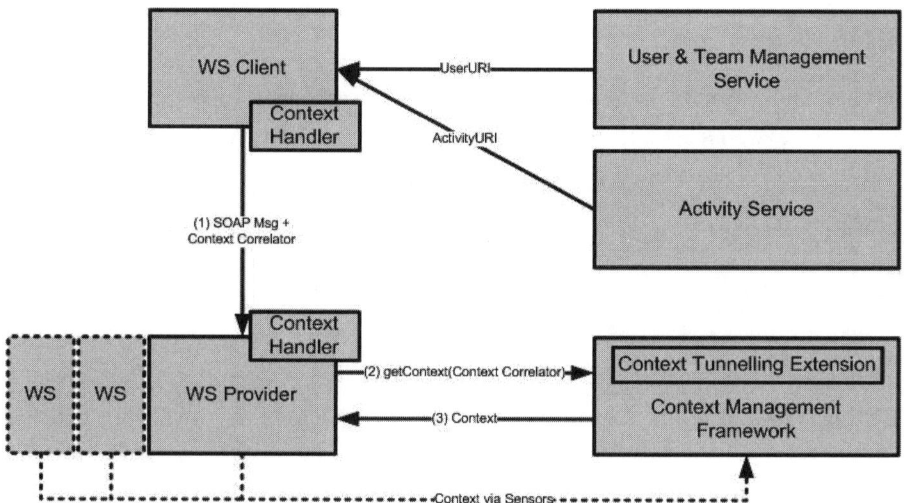

Figure 12.9: Conceptual model of the inContext's context message coupling.

each time a context can be accessed from a different service. Currently, the inContext project has not supported any privacy protection for this sharing.

12.4.3 Context Common Coupling

Context common coupling is achieved in the inContext project by means of a centralized `Context Store` implemented atop an RDF store with added OWL and RDFS inference capabilities. With a strong reasoning capability, the inContext approach offers better retrieval mechanisms for context-aware Web services. However, as noted in (40), it requires Web services being able to process RDF content, which is not common in Web services yet, and is hard to achieve in mobile applications. To master this problem, the inContext project also provides a transformation solution that returns the results in XML format. While offering a strong reasoning capability, the inContext's `Context Store` does not provide any privacy policies at the time of writing.

12.4.4 Summary

The inContext case study shows that for large scale and complex systems dealing with various types of context information, a single technique might not be enough. On the one hand, the inContext project employs structure and common coupling techniques together to support a context storage that can be used by any context-aware Web services. On the other hand, it utilizes message coupling to exchange context information. Using multiple types of couplings together could help deal with the diversity of context-aware Web

```
<?xml version="1.0" encoding="UTF-8"?>
<soapenv:Envelope
 <soapenv:Header>
   <ns1:ctxtunnelling soapenv:actor="http://schemas.xmlsoap.org/soap/
        actor/next"
      soapenv:mustUnderstand="0" xmlns:ns1="www.in-context.eu">
   <ns1:Activity>
     http://www.in-context.eu/pcsa#act1
   </ns1:Activity>
   <ns1:User>
     http://www.in-context.eu/pcsa#Rossi.E54
   </ns1:User>
  </ns1:ctxtunnelling>
 </soapenv:Header>
 <soapenv:Body>
 </soapenv:Body>
</soapenv:Envelope>
```

Listing 12.1: Simplified example of SOAP header message including context coupling information in the inContext project.

services in different situations and configurations. For example, the common context storage could be utilized by context-aware Web services which are deployed in strong platforms and which require complex types of context information, possibly, inferred from other types of information, because the context storage has strong capabilities to support advanced reasoning. On the other hand, message coupling could be used to share context information to mobile Web services deployed in mobile devices or to those requiring simple context information and minimum numbers of interactions. For example, a notification service just needs to be aware if the user is available in an instant messaging service or not in order to use instant messages to notify the user with some information. This does not need to acquire complex context information using reasoning techniques. Combining structure, common and message couplings could be, for example, suitable for the scenario of smart homes in which there are a multitude of varying services.

12.5 Open Issues and Recommendations

From our study, we draw some conclusions. First, software engineering techniques for building context-aware Web services should take into account privacy and security issues in the whole development and deployment of these services. On the one hand, privacy and security features should be provided to the user when context information being shared is sensitive to the user. This is particularly important for common coupling techniques. On the other hand, in case of using structure and data coupling techniques, constraints could be added into the design of context-aware Web services, e.g., based on the MDE approach, to automatically check and ensure privacy issues when context information is passed through service invocation.

Second, it is hard to achieve a common context model but the use of rich semantic representations, agreed concepts, and extensible message specifications to describe context information would facilitate the context coupling in the Internet of Services. It could enhance the runtime context structure coupling by means of reasoning and enrichment. Using reasoning techniques one could correlate many different types of context information from a common repository and couple these types into a message specification.

Third, currently there is a lack of techniques to couple existing context information described in different specifications and provided by different sources during runtime. For example, in structure and common coupling, context information is mostly assumed to follow the same specification, forcing, for example, a new context-aware Web service to use the same specification for its sharing context information, even if its context information is described in a different form. Consider the scenario of smart homes in which each service (e.g., light control, entertainment, and activity recognition systems) has its own context model. Then, a newly-built context-aware Web service retrieves context information from existing services would need to solve many interoperability issues among context specifications.

Fourth, related to the third issue is the need to have open Internet-scale context exchange protocols and models. The WS-Context could be a good starting point but its assumption of transferring context information using SOAP would be a limitation as certain context-aware Web services will not be based on SOAP. We suggest to agree, like WS-Context proposed, that operations of obtaining context are well defined. However, context messages could be transferred by different means based on SOAP and REST (Representational State Transfer). Furthermore, these protocols and models should also address the interoperability of different context-aware Web service systems. A similar approach to a recent work on bridging context management systems (24) could be investigated for context-aware Web services.

Fifth, existing protocols for authorizing resources access in Web services and for supporting identity management and single sign-on, such as the OAuth and OpenID protocols, could be useful for context coupling in a large-scale network of context-aware Web services. We suggest to integrate and extend these protocols for context coupling among context-aware Web services, especially those provided by different vendors.

12.6 Related Work and Further Reading

In our previous work (41), we have analyzed different techniques employed in context-aware Web services. This paper is a further step to detail how context information could be coupled. However, as (41) gives a broad view of existing techniques, it also covers some systems studied in this chapter. Furthermore, in (41), we mainly study existing techniques based on context supporting components, such as context presentation, context sensing, context storage, context distribution and adaptation. In this chapter, we examine only context coupling techniques that are of course related to different context supporting components. Thus, some results found in (41) are also given in this chapter.

Coupling techniques are well known (2, 1), thus they have presented in several papers. When study coupling techniques in Web services, in general, and in context-aware Web service systems, in particular, we are able to find only four common types of coupling techniques namely structure, data, message, and common couplings. Context coupling techniques are a subset of these techniques, thus they have some common properties. However, to our best knowledge, there exists no study of context coupling techniques in context-aware Web services.

The interaction models between context-aware Web services that are employed in context coupling techniques are also related to SOA and enterprise integration patterns (25). Thus generic concerns associated with these models with respect to integration issues could also be learned from these patterns. However, these patterns are generic, while using them for coupling context information should be driven by the properties of and compliance rules applied to context information.

12.7 Conclusion

Enabling context-aware Web services, or "smart Web services", is important as this will substantially improve how services could adapt to complex, situational behaviors of humans, things, and services on the Internet. Coupling context information across multiple Web services is a nontrivial problem because it requires well-agreed context models and protocols. In this chapter, we have studied a particular topic "context coupling" that is a must for enabling context-aware Web services. We have analyzed existing systems and presented a case study. Overall, many open issues have not been addressed yet in current literature. This is understandable because context-aware Web services research is relatively new. We need to invest more efforts on establishing open protocols for exchanging context information in a large-scale system. Furthermore, privacy and security issues for context coupling should be treated as first entities.

Our future work is to focus on message specifications that can be used to transfer different types of context information and open protocols for context coupling over overlay networks. In particular, we will target our work in multiple spaces connecting smart homes to distributed health-care systems and Internet-based Web services.

Acknowledgments This work has been partly funded by the European Commission through the EU FP6 inContext (Interaction and Context-based Technologies for Collaborative Teams) (9) and the EU FP7 SM4All (Smart Homes for All) (10) projects. We thank all inContext members for their contribution on the discussion and the implementation of the context coupling in inContext, which serves as the case study in this chapter. We thank Fei Li for fruitful discussion on open issues and the scenario of the smart home, which is built based on the SM4All research. inContext's coupling techniques are partially published in (40). We thank Christoph Dorn, Giovanni Casella, Axel Polleres, and Stephan Reiff-Marganiec for their contribution in (40) which is extracted for the case study in this chapter.

References

[1] Coupling. http://www.site.uottawa.ca:4321/oose/coupling.html. Last access: 27 May 2009.

[2] Coupling (computer science). http://en.wikipedia.org/wiki/Coupling_(computer_science). Last access: 27 May 2009.

[3] Google Latitude. http://www.google.com/latitude/intro.html.

[4] HTTP Authentication: Basic and Digest Access Authentication. http://tools.ietf.org/html/rfc2617. Last access: 24 July 2009.

[5] HTTP Over TLS. http://tools.ietf.org/html/rfc2818. Last access: 24 July 2009.

[6] OAuth. http://oauth.net. Last access: 24 July 2009.

[7] OpenID. http://openid.net. Last access: 24 July 2009.

[8] The Akigrimo project. http://www.mobilegrids.org/.

[9] The inContext project,http://www.in-context.eu.

[10] The SM4All project: Smart Homes for All, http://www.sm4all-project.eu.

[11] WS-Security. http://www.oasis-open.org/committees/tc_home.php?wg_abbrev=wss. Last access: 24 July 2009.

[12] Web Services Context Specification (WS-Context) Version 1.0. http://docs.oasis-open.org/ws-caf/ws-context/v1.0/wsctx.html, April 2007. Last access: 27 May 2009.

[13] Gregory D. Abowd, Anind K. Dey, Peter J. Brown, Nigel Davies, Mark Smith, and Pete Steggles. Towards a Better Understanding of Context and Context-Awareness. In *HUC*, pages 304–307, 1999.

[14] Mark Ackerman, Trevor Darrell, and Daniel Weitzner. Privacy in Context. *Hum.-Comput. Interact.*, 16(2):167–176, 2001.

[15] Chimay J. Anumba and Zeeshan Aziz. Case Studies of Intelligent Context-Aware Services Delivery in AEC/FM. In *EG-ICE*, pages 23–31, 2006.

[16] Dionysis Athanasopoulos, Apostolos V. Zarras, Valerie Issarny, Evaggelia Pitoura, and Panos Vassiliadis. CoWSAMI: Interface-aware context gathering in ambient intelligence environments. *Pervasive Mob. Comput.*, 4(3):360–389, 2008.

[17] Matthias Baldauf, Schahram Dustdar, and Florian Rosenberg. A Survey on Context-aware Systems. *International Journal of Ad-Hoc and Ubiquitous Computing*, Jan 2006.

[18] Harry Chen Baltimore and Harry Chen. Semantic Web in a Pervasive Context-Aware Architecture. In *In Artificial Intelligence in Mobile System 2003 (AIMS 2003), in conjuction with Ubicomp*, pages 33–40, 2003.

[19] Guanling Chen and David Kotz. A Survey of Context-Aware Mobile Computing Research. Technical report, Hanover, NH, USA, 2000.

[20] Irene Y. L. Chen, Stephen J. H. Yang, and Jia Zhang. Ubiquitous Provision of Context Aware Web Services. In *SCC '06: Proceedings of the IEEE International Conference on Services Computing*, pages 60–68, Washington, DC, USA, 2006. IEEE Computer Society.

[21] Christos Doulkeridis, Vassilis Zafeiris, Kjetil Norvag, Michalis Vazirgiannis, and Emmanouel A. Giakoumakis. Context-based Caching and Routing for P2P Web Service Discovery. *Distrib. Parallel Databases*, 21(1):59–84, 2007.

[22] Tao Gu, Hung Keng Pung, and Da Qing Zhang. A Service-oriented Middleware for Building Context-aware Services. *J. Netw. Comput. Appl.*, 28(1):1–18, 2005.

[23] Karen Henricksen, Jadwiga Indulska, Ted McFadden, and Sasitharan Balasubramaniam. Middleware for distributed context-aware systems. In *OTM Conferences (1)*, pages 846–863, 2005.

[24] Cristian Hesselman, Hartmut Benz, Pravin Pawar, Fei Liu, Maarten Wegdam, Martin Wibbels, Tom Broens, and Jacco Brok. Bridging context management systems for different types of pervasive computing environments. In *MOBILWARE '08: Proceedings of the 1st international conference on MOBILe Wireless MiddleWARE, Operating Systems, and Applications*, pages 1–8, ICST, Brussels, Belgium, Belgium, 2007. ICST (Institute for Computer Sciences, Social-Informatics and Telecommunications Engineering).

[25] Gregor Hohpe and Bobby Woolf. *Enterprise Integration Patterns: Designing, Building, and Deploying Messaging Solutions.* Addison-Wesley Longman Publishing Co., Inc., Boston, MA, USA, 2003.

[26] Gearoid Hynes, Vinny Reynolds, and Manfred Hauswirth. Enabling Mobility Between Context-Aware Smart Spaces. In *The 5th International Symposium on Web and Mobile Information Services, IEEE Workshop at IEEE international conference on advanced information networking and applications*, Bradford, England, 2009.

[27] Carlos R. E. Arruda Jr., Renato Bulcão Neto, and Maria da Graça Campos Pimentel. Open Context-aware Storage as a Web Service. In *Middleware Workshops*, pages 81–87, 2003.

[28] Georgia M. Kapitsaki and Iakovos S. Venieris. PCP: privacy-aware context profile towards context-aware application development. In *iiWAS '08: Proceedings of the 10th International Conference on Information Integration and Web-based Applications & Services*, pages 104–110, New York, NY, USA, 2008. ACM.

[29] Markus Keidl and Alfons Kemper. A Framework for Context-Aware Adaptable Web Services. In *EDBT*, pages 826–829, 2004.

[30] Zakaria Maamar, Qusay H. Mahmoud, Nabil Sahli, and Khouloud Boukadi. Privacy-Aware Web Services in Smart Homes. In *ICOST*, pages 174–181, 2009.

[31] Anne Thomas Manes. Enabling Open, Interoperable, and Smart Web Services - The Need for Shared Context, April 2001. `http://www.w3.org/2001/03/WSWS-popa/paper29`.

[32] Renato Bulcão Neto, Carlos Jardim, José Camacho-Guerrero, and Maria da Graça Pimentel. A Web Service Approach for Providing Context Information to CSCW Applications. *Web Congress, Joint Conference Brazilian Symposium on Multimedia and the Web & Latin America*, 0:46–53, 2004.

[33] Katsumi Nihei. Context Sharing Platform. *NEC Journal of Advanced Technology*, pages 200–204, 2004.

[34] Ciaran O'Driscoll. Privacy in context: Privacy issues in ubiquitous computing applications. In Pit Pichappan and Ajith Abraham, editors, *ICDIM*, pages 827–837. IEEE, 2008.

[35] Sangyoon Oh and Geoffrey Fox. Optimizing Web Service Messaging Performance in Mobile Computing. *Future Generation Comp. Syst.*, 23(4):623–632, 2007.

[36] Federica Paganelli, Emilio Spinicci, and Dino Giuli. ERMHAN: A context-aware service platform to support continuous care networks for home-based assistance. *Int. J. Telemedicine Appl.*, 2008(5):1–13, 2008.

[37] Ariel Pashtan, Remy Blattler, Andi, Andi Heusser, and Peter Scheuermann. CATIS: A Context-Aware Tourist Information System, 2003. `http://www.ece.northwestern.edu/~peters/IMC.CATIS.pdf`.

[38] Quan Z. Sheng and Boualem Benatallah. ContextUML: A UML-Based Modeling Language for Model-Driven Development of Context-Aware Web Services Development. In *ICMB '05: Proceedings of the International Conference on Mobile Business*, pages 206–212, Sydney, Australia, 2005. IEEE Computer Society.

[39] R. Sudha, M. R. Rajagopalan, M. Selvanayaki, and S. Thamarai Selvi. Ubiquitous Semantic Space: A context-aware and coordination middleware for Ubiquitous Computing. In *COMSWARE*, 2007.

[40] Hong-Linh Truong, Christoph Dorn, Giovanni Casella, Axel Polleres, Stephan Reiff-Marganiec, and Schahram Dustdar. inContext: On Coupling and Sharing Context for Collaborative Teams. In *Proceedings of the 14th International Conference of Concurrent Enterprising (ICE 2008)*, pages 225–232, 2008.

[41] Hong-Linh Truong and Schahram Dustdar. A Survey on Context-aware Web Service Systems. *International Journal of Web Information Systems*, 5(1):5–31, 2009.

[42] Hong-Linh Truong, Schahram Dustdar, Dino Baggio, Stephane Corlosquet, Christoph Dorn, Giovanni Giuliani, Robert Gombotz, Yi Hong, Pete Kendal, Christian Melchiorre, Sarit Moretzky, Sebastien Peray, Axel Polleres, Stephan Reiff-Marganiec, Daniel Schall, Simona Stringa, Marcel Tilly, and HongQing Yu. inContext: A Pervasive and Collaborative Working Environment for Emerging Team Forms. In *SAINT*, pages 118–125, 2008.

[43] Hong-Linh Truong, Lukasz Juszczyk, Atif Manzoor, and Schahram Dustdar. ESCAPE - An Adaptive Framework for Managing and Providing Context Information in Emergency Situations. In *EuroSSC*, pages 207–222, 2007.

[44] Georgios Tselentis, John Domingue, Alex Galis, Anastasius Gavras, David Hausheer, Srdjan Krco, Volkmar Lotz, and Theodore Zahariadis, editors. *Towards the Future Internet - A European Research Perspective.* IOS Press, May 2009. ISBN 978-1-60750-007-0.

[45] Samyr Vale and Slimane Hammoudi. Towards Context Independence in Distributed Context-aware Applications by The Model Driven Approach. In *SIPE '08: Proceedings of the 3rd international workshop on Services integration in pervasive environments*, pages 31–36, New York, NY, USA, 2008. ACM.

[46] Ryan Wishart, Karen Henricksen, and Jadwiga Indulska. Context Privacy and Obfuscation Supported by Dynamic Context Source Discovery and Processing in a Context Management System. In *UIC*, pages 929–940, 2007.

[47] Hao Yan and Ted Selker. Context-aware office assistant. In *IUI '00: Proceedings of the 5th International Conference on Intelligent User Interfaces*, pages 276–279, New York, NY, USA, 2000. ACM.

Chapter 13

Context-Aware Semantic Web Service Discovery through Metric-Based Situation Representations

Stefan Dietze, Michael Mrissa, John Domingue, and Alessio Gugliotta

Abstract Semantic Web Services (SWS) enable the automatic discovery of distributed Web services based on comprehensive semantic representations. However, although SWS technology supports the automatic allocation of Web services for a given well-defined task, it does not entail their discovery according to a given situational context. Whereas tasks are highly dependent on the situational context in which they occur, SWS technology does not explicitly encourage the representation of domain situations. Moreover, describing the complex notion of a specific situation in all its facets is a costly task and may never reach sufficient semantic expressiveness. Particularly, following the symbolic SWS approach leads to ambiguity issues and does not entail semantic meaningfulness. Apart from that, not any real-world situation completely equals another, but has to be matched to a finite set of semantically defined parameter descriptions to enable context-adaptability. To overcome these issues, we propose Conceptual Situation Spaces (CSS) that are aligned to established SWS standards. CSS enable the description of situation characteristics as members in geometrical vector spaces following the idea of Conceptual Spaces. Semantic similarity between situations is calculated in terms of their Euclidean distance within a CSS. Extending merely symbolic SWS descriptions with context information through CSS enables similarity-based matchmaking between real-world situation characteristics and predefined resource representations as part of SWS descriptions. To prove its feasibility, we apply our approach to the E-Learning and E-Business domains and provide a proof-of-concept prototype.

13.1 Introduction

Context-aware discovery and invocation of *Web services* is highly desired across a wide variety of application domains and subject to intensive research throughout the last decade (9, 36, 20). According to Dey's well-known definition (7), we define *context* as the entire set of surrounding characteristics that characterize an entity relevant to the interaction between a user and an application (including the user and the application themselves). Each individual *situation* represents a specific state of the world, and more precisely, a particular state of actual context. A *situation description* defines the context in a particular situation, and is described by a combination of *situation parameters*, each representing a particular situation characteristic. Following this definition, context-adaptation can be defined as the ability to adapt to distinct possible situations.

Semantic Web Services (SWS) technology (16) supports the automatic discovery of distributed Web services for a given task based on comprehensive semantic descriptions. Concretely, a SWS is a Web service with a description that contains explicit semantic information, and SWS technology consists of tools and technologies that enable getting the benefits from these semantically-explicit service descriptions. First results of SWS research are available, in terms of reference ontologies, e.g., OWL-S (28) and WSMO (1), as well as tools and comprehensive frameworks (e.g. DIP project[1] results).

However, whereas SWS technology supports the allocation of appropriate resources based on semantic representations, it does not entail the discovery of appropriate SWS representations for a given situation, i.e., the actual context. Even though tasks, as semantically described through SWS representations, are highly dependent on the situation in which they occur, current SWS technology does not explicitly encourage the representation of domain situations related to task representations. Furthermore, describing the complex notion of a specific situational context in all its facets is a costly task and may never reach sufficient semantic expressiveness. The symbolic approach, i.e., describing symbols by using other symbols without a grounding in the real world, of established SWS and Semantic Web representation standards in general, such as RDF[2], OWL[3], OWL-S, or WSMO leads to ambiguity issues and does not

[1]DIP Project: `http://dip.semanticweb.org`

[2]`http://www.w3.org/RDF/`

[3]`http://www.w3.org/TR/owl2-primer/`

entail semantic meaningfulness, since meaning requires both the definition of a terminology in terms of a logical structure (using symbols) and grounding of symbols to a cognitive or perceptual level (5, 24). Moreover, whereas not any situation or situation parameter completely equals another, the number of predefined semantic representations of situations and situation parameters within a SWS description is finite. Consequently, to enable context-adaptive resource discovery, a potential infinite set of (real-world) situation characteristics has to be matched to a finite set of semantically defined situation parameter descriptions. Therefore, rather fuzzy classification and matchmaking techniques are required to classify a real-world situation based on a limited set of predefined parameter descriptions to support the discovery of the most appropriate SWS representation within a given situation context.

Conceptual Spaces (CS), introduced by Gärdenfors (19, 18), follow a theory for describing entities in terms of their natural characteristics similar to natural human cognition in order to avoid the symbol grounding issue. CS enable representation of objects as vector spaces within a geometrical space which is defined through a set of quality dimensions. For instance, a particular color may be defined as point described by vectors measuring the quality dimensions hue, saturation, and brightness. Describing instances as vector spaces where each vector follows a specific metric enables the automatic calculation of their semantic similarity, in terms of their Euclidean distance, in contrast to the costly representation of such knowledge through symbolic SW representations. Even though several criticisms have to be taken into account when utilizing CS (Section 15.7) they are considered to be a viable option for knowledge representation.

In this paper, we propose *Conceptual Situation Spaces (CSS)* which utilize CS to represent situational contexts. CSS are mapped to standardized SWS representations to enable, first, context-aware discovery of appropriate SWS descriptions, and finally, automatic discovery and invocation of appropriate Web services to achieve a given task within a particular situation. Extending merely symbolic SWS descriptions with context information on a conceptual level through CSS enables fuzzy and similarity-based matchmaking between real-world situation characteristics and predefined SWS representations. Whereas similarity between situation parameters, as described within a CSS, is indicated by the Euclidean distance between them, real-world situation parameters are classified along predefined prototypical parameters that are implicit elements of a SWS description. Whereas current SWS technology addresses the issue of allocating resources for a given task, our approach supports the discovery of SWS representations within a given situational context. Consequently, the expressiveness of current SWS standards is extended through CSS in order to enable fuzzy matchmaking mechanisms when allocating resources for a given situation.

To prove the feasibility of our approach two proof-of-concept prototypes are provided. The first prototype relates to the domain of E-Learning and uses CSS to describe learning styles, following the Felder-Silverman Learning

Style theory (14), as a particular learning situation parameter. The second prototype illustrates a CSS application to the E-Business domain, and uses CSS to describe business actors' requirements according to current situational contexts.

The paper is organized as follows. The following Section 13.2 provides background information on SWS and the discovery problem, and gives an overview on related works in the field. Section 13.3 introduces our approach of Conceptual Situation Spaces that are aligned to current SWS representations. Section 13.4 illustrates the application of CSS to the E-Learning domain and introduces a Conceptual Learning Situation Space, particularly, a CSS subspace representing learning styles. Utilizing CSS, we introduce our approach to similarity-based classification of a given situation based on distance calculation at runtime in Section 13.5. Section 13.6 shows an application of CSS in the E-Business examples, and details how CSS helps discovering SWS depending on business actors' requirements at runtime, before giving some insight on the use of CSS for semantic mediation of data in business processes. Finally, we conclude our work in Section 15.7 and provide an outlook to future research.

13.2 Background and Motivation

In this section, we provide some background information and motivate our approach by reporting on the current state of the art in Semantic Web Services discovery.

13.2.1 Semantic Web Services (SWS) and Context-Dependent SWS Mediation

SWS technology aims at the automatic discovery, orchestration, and invocation of distributed services for a given user goal on the basis of comprehensive semantic descriptions. SWS are supported through representation standards such as WSMO (1) and OWL-S (28). In this paper, we particularly refer to the Web Service Modelling Ontology (WSMO), a well-established SWS reference ontology and framework. WSMO is currently supported through several software tools and runtime environments, such as the Internet Reasoning Service IRS-III (2) and WSMX (23). The conceptual model of WSMO defines the following four main entities:

- **Domain ontologies** not only support Web service-related knowledge representation but semantic knowledge representation in general. They provide the foundation for describing domains semantically, and they are used by the three other WSMO entities.

- **Goals** define the tasks that a service requester expects a Web service to fulfill. In this sense they express the requester's intention.
- **Web service descriptions** represent the functional behavior of an existing deployed Web service. They also outline how Web services communicate (choreography) and how they are composed (orchestration). In this paper, a SWS represents the semantic description of a particular Web service and is synonymous with the term SWS description.
- **Mediators** handle data and process interoperability issues that arise when handling heterogeneous systems.

A SWS description (either the description of the Web service or the description of the service request) is formally represented within a particular ontology that complies with a certain SWS reference model such as OWL-S (28) or WSMO (1). By adopting a common formalisation of an ontology (11, 12), we define a populated *service ontology* O – as utilized by a particular SWS representation – as a tuple:

$$O = \{C, I, P, R, A\} \subset SWS$$

With C being a set of n *concepts* where each concept C_i is described through $l(i)$ *concept properties* pc, i.e.:

$$PC_i = \left\{ (pc_{i1}, pc_{i2}, \ldots, pc_{l(i)}) \left| pc_{ix} \in C_i \right. \right\}.$$

I represents all m *instances* where each instance I_{ij} represents a particular instance of a concept C_j and consists of *l(i) instantiated properties pi* instantiating the concept properties of C_j:

$$PI_{ij} = \left\{ (pi_{ij1}, pi_{ij2}, \ldots, pi_{l(i)}) \left| pi_{ijx} \in I_{ij} \right. \right\}.$$

Hence, the properties P of an ontology O represent the union of all concept properties PC and instantiated properties PI of O:

$$P = \{(PC_1, PC_2, \ldots, PC_n) \cup (PI_1, PI_2, \ldots, PI_m)\}$$

Given these definitions, we would like to point out that properties here exclusively refer to so-called data type properties. Hence, we define properties as being distinctive to relations R. The latter describe relations between concepts and instances. In addition, A represents a set of *axioms* that define constraints on the other introduced notions. Since certain parts of a SWS ontology describe certain aspects of the Web service (request), such as its capability *Cap*, interface *If* or non-functional properties *Nfp* (4), a SWS ontology can be perceived as a conjunction of ontological subsets:

$$Cap \cup If \cup Nfp = O \subset SWS$$

The semantic capability description, as central element of a SWS description, consists of further subsets, describing the assumptions *As*, effects *Ef*, preconditions *Pre* and postconditions *Post*. However, given the lack of a clear distinction between assumption/effect and pre-/postcondition, we prefer the exclusive usage of assumptions/effects:

$$As \cup Ef = Cap \subset O \subset SWS$$

Given that a SWS ontology by its very nature always captures the semantics of a service from a specific perspective, it represents a specific context in which the annotated service is meant to be used. Hence, even when explicitly representing information about the Web service context, the nature of ontologies - being symbolic representations of conceptualizations from a specific viewpoint - leads to highly heterogeneous SWS descriptions.

SWS mediation aims at addressing heterogeneities among distinct SWS to support all stages that occur at SWS runtime, namely *discovery, orchestration* , and *invocation.* In contrast to (4, 33), we classify the mediation problem into (i) *semantic level* and (ii) *data level mediation* (Figure 13.1).

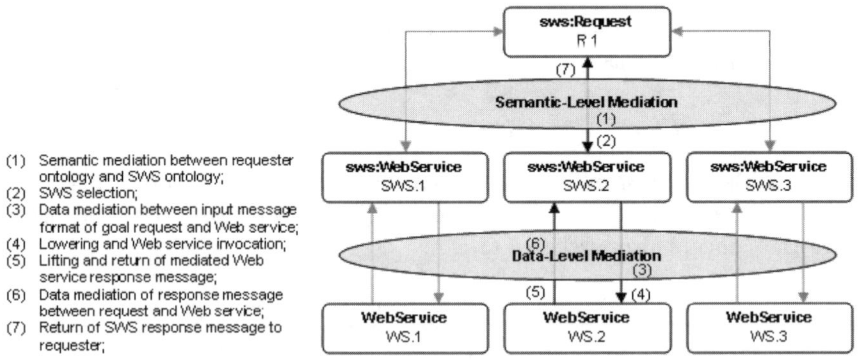

Figure 13.1: Semantic level and data level mediation as part of SWS discovery, orchestration and invocation

Whereas (i) refers to the resolution of heterogeneities between concurrent semantic representations of services and service contexts – the actual SWS representations – (ii) refers to the mediation between mismatches related to the Web service implementations themselves, i.e., related to the structure, value, or format of I/O messages. Hence, semantic level mediation primarily supports the discovery stage, whereas data level mediation occurs during orchestration and invocation. As shown in Figure 13.1, semantic level mediation occurs before a particular SWS is selected and aims at aligning distinct semantic vocabularies, for instance, the ones used by a SWS requester and a

SWS provider. Please note that, for the sake of simplification, Figure 13.1 just depicts mediation between a SWS request and multiple SWS, while leaving aside mediation between different SWS or between different requests.

13.2.2 Semantic Web Services Discovery – An Ontology Mapping Problem

In this chapter, we exclusively address semantic level mediation between distinct context-representations, which is perceived to be a fundamental requirement for context-adaptive SWS discovery, and hence, to further exploit SWS approaches on a Web scale. In order to better understand the needs of semantic level mediation, it is necessary to understand the requirements of the SWS discovery task to which semantic level mediation is supposed to contribute. In order to identify whether a particular SWS S_1 is potentially relevant for a given request S_2, also representing a particular context, a SWS broker has to compare the capabilities of S_1 and S_2, i.e., it has to identify whether the following holds true:

$$As_2 \subset As_1 \cup Ef_2 \subset Ef_1$$

However, in order to compare distinct contextual annotations of available SWS that each utilize a distinct vocabulary, these vocabularies have to be aligned. For instance, to compare whether an assumption expression of one particular S_1 is the same as the one of another service S_2, where I_i represents a particular instance, matchmaking engines have to perform two steps: (a) identification of relationships between concepts/instances involved in distinct SWS representations; (b) evaluation whether the semantics of the logical expressions used by each SWS match each other, i.e., represent the same fact (i.e., capability).

Whereas current SWS execution environments exclusively focus on (b), semantic level mediation also requires mediation between different SWS context ontologies, as in (a), and can be perceived as a particular instantiation of the *ontology mapping problem* (40). With respect to (3), we define ontology mapping as the creation of structure-preserving relations between multiple ontologies. I.e., the goal is, to establish formal relations between a set of knowledge entities E_1 from an ontology O_1 - used to represent a particular SWS S_1 - and entities E_2 that represent the same or a similar semantic meaning in a distinct ontology O_2 (11, 12), which is used to represent an additional S_2. The term set of entities here refers to the union of all concepts C, instances I, relations R, and axioms A defined in a particular SWS ontology. In that, semantic mediation strongly relies on identifying *semantic similarities* between entities across different SWS ontologies. Hence, the identification of similarities is a necessary requirement to solve the heterogeneity problem for multiple SWS representations (34, 40). However, in this respect, the following issues have to be taken into account:

1. Symbolic SWS and context representations lack grounding to the conceptual level: similarity-detection across distinct SWS descriptions requires semantic meaningfulness which inherently describes semantic similarity between represented entities. However, the symbolic approach, i.e., describing symbols by using other symbols, without a grounding in the real world, of established SWS representation standards, leads to ambiguity issues and does not fully entail semantic meaningfulness, since meaning requires both the definition of a terminology in terms of a logical structure (using symbols) and grounding of symbols to a conceptual level (5).

2. Lack of automated similarity-detection methodologies: Describing the complex notion of specific SWS contexts in all their facets is a costly task and may never reach sufficient semantic expressiveness due to the above. While contextual representations across distinct SWS representations, even those representing the same real-world entities, hardly equal each other, semantic similarity is not an implicit notion within SWS representations. But manually or semi-automatically defining similarity relationships is costly. Moreover, such relationships are hard to maintain in the longer term.

Given the lack of inherent similarity representation, current approaches to *ontology mapping* could be applied to facilitate context-aware SWS discovery. These approaches aim at semi-automatic similarity detection across ontologies mostly based on identifying linguistic commonalities and/or structural similarities between entities of distinct ontologies (3, 32). Work following a combination of such approaches in the field of ontology mapping is reported in (32, 13, 21, 29). However, it can be stated, that such approaches require manual intervention, are costly and error-prone, and hence, similarity-computation remains a central challenge. In our vision, instead of semi-automatically formalizing individual mappings, we rely on methodologies to automatically compute or implicitly represent similarities across distinct SWS representations, which are better suited to facilitate SWS mediation.

13.2.3 Spatial Approaches to Knowledge Representation

Distinct streams of research approach the automated computation of similarities through spatially oriented knowledge representations. *Conceptual Spaces (CS)* follow a theory of describing entities at the conceptual level in terms of their quality characteristics similar to natural human cognition in order to bridge between the neural and the symbolic world. (19) proposes the representation of concepts as multidimensional geometrical *Vector Spaces*, which are defined through sets of quality dimensions. Instances are supposed to be represented as vectors, i.e., particular points in a space CS. For instance, a particular color may be defined as point described by vectors measuring the quality dimensions hue, saturation, and brightness. Describing instances as points within vector spaces where each vector follows a specific metric enables the automatic calculation of their semantic similarity by means of distance metrics such as the Euclidean, Taxicab, or Manhattan distance (27) or the Minkowsky Metric (38). Hence, in contrast to the costly formalization of such knowledge through symbolic representations, semantic similarity is implicit information carried within a CS representation what is perceived as the major contribution of the CS theory. *Soft Ontologies (SO)* (26) follow a similar approach by representing a knowledge domain D through a multi-dimensional *ontospace* A, which is described by its so-called *ontodimensions*. An item I, i.e. an instance, is represented by scaling each dimension to express its impact, presence or probability in the case of I. In that, a SO can be perceived as a CS where dimensions are measured exclusively on a ratio-scale.

However, although CS and SO aim at solving SW(S)-related issues, several issues still have to be taken into account. For instance, similarity computation within CS requires the description of concepts through quantifiable metrics even in case of rather qualitative characteristics. Moreover, CS as well as SO do not provide any notion to represent any arbitrary relations (37), such as *part-of* relations which usually are represented within symbolic knowledge models such as SWS representations. In this regard, it is even more obstructive that the scope of a dimension is not definable, i.e., a dimension always applies to the entire CS/SO (37).

13.3 Conceptual Situation Spaces for Semantic Web Services

In this section, we describe the formalisms developed to backup our approach to context-aware SWS discovery, which is based on describing situational contexts as members within a domain-specific *Conceptual Situation Space (CSS)*, which are incorporated into SWS descriptions.

13.3.1 Approach: Grounding SWS Contexts in Conceptual Situation Spaces

CSS enable the description of a particular context within a particular situation as a member of a dedicated CS that are used to describe SWS capabilities. CSS enable the implicit representation of semantic similarities across heterogeneous SWS context representations provided by distinct agents. Hence, refining heterogeneous SWS context descriptions into a set of shared CSS supports similarity-based mediation at the semantic level and consequently facilitates context-aware SWS discovery. Whereas CSS allow the

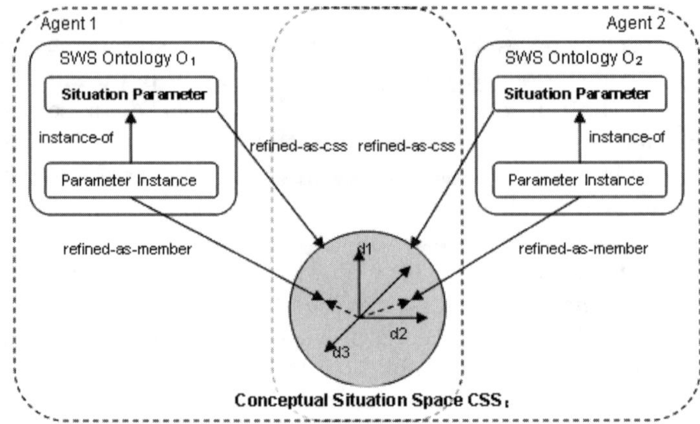

Figure 13.2: Representing distinct situation parameters, being part of heterogeneous SWS representations, through shared CSS.

representation of semantic similarity as an implicit notion, it can be argued, that representing an entire SWS context through a coherent CSS might not be feasible, particularly when attempting to maintain the meaningfulness of the spatial distance as a similarity measure.

Therefore, we claim that CSS are a particularly promising model when being applied to individual situation parameter concepts, as part of SWS descriptions, instead of representing an entire SWS ontology in a single CSS. In that, we would like to highlight that we consider the representation of a set of n situation parameters S of a SWS ontology O through a set of n CSS 13.2. Hence, instances of parameters are represented as members in the respective CSS. While still taking advantage from implicit similarity information within a CSS, our hybrid approach – combining SWS descriptions with multiple CSS – allows to overcome CS-related issues by maintaining the advantages of ontology-based SWS context representations. Please note that our approach relies on the agreement on a common set of CSS for a given set of distinct SWS ontologies O_1 and O_2, instead of a common agreement on set of shared ontologies. Hence, while in the latter case two agents have to agree on a common ontology at the concept and instance level, our approach only requires agreement at the concept level, since instance similarity becomes an implicit notion. Moreover, we assume that the agreement on ontologies at the concept level (Figure 13.2) becomes an increasingly widespread case, due to, on the one hand, increasing use of upper-level ontologies such as DOLCE (17), SUMO[4] or OpenCyc[5], which support a certain degree of

[4]`http://www.ontologyportal.org/`

[5]`http://www.opencyc.org/`

commonality between distinct ontologies. On the other hand, SWS ontologies often are provided within closed environments, for instance, virtual organizations, where a common agreement to a certain extent is ensured. In such cases, the derivation of a set of common CSS is particularly applicable and straightforward.

13.3.2 A Formalization for Conceptual Situation Spaces

Our approach formalizes the notion of CSS via a dedicated *Conceptual Situation Space Ontology (CSSO)*, based on OCML (30) and aligned to SWS in order to enable SWS description. Since both metamodels, WSMO as well as CSS, are represented based on the OCML representation language, the alignment was accomplished by defining relations between concepts of both ontologies. The design, formalization, and alignment processes have been detailed in previous work. To make this paper self-contained and for a good understanding, we provide the necessary results obtained from this previous work in the following. For additional details, we refer the reader to (10). Hence, a CSS is defined as a vector space with weighted dimensions:

$$C^n = \{(p_1c_1, p_2c_2, \ldots, p_nc_n)|c_i \in C, p_i \in P\}$$

where c_i being the quality dimensions of C and P is the set of real numbers. The prominence value p is attached to each dimension in order to reflect the impact of a specific quality dimension on the entire CSS. We enable dimensions to be detailed further in terms of subspaces. Hence, a dimension within one space may be defined through another conceptual space by using further dimensions (35, 10).

Semantic similarity between two members of a space is perceived as a function of the Euclidean distance between the points representing each of the members. Hence, given a CSS definition C and two members represented by two vector sets V and U, defined by vectors $v_0, v_1, \ldots, v_n$ and $u_1, u_2, \ldots, u_n$ within C, the distance between V and U can be calculated as:

$$|d(u,v)|^2 = \sum_{i=1}^{n}(z(u_i) - z(v_i))^2$$

where $z(u_i)$ is the so-called Z-transformation or standardization (6) from u_i. Z-transformation facilitates the standardization of distinct measurement scales that are utilized by different quality dimensions in order to enable the calculation of distances in a multidimensional and multimetric space. The z-score of a particular observation u_i in a dataset is calculated as follows:

$$z(u_i) = \left(\frac{u_i - \bar{u}}{s_u}\right)$$

where $\overline{u}$ is the mean of a dataset U and s_u is the standard deviation from U. Considering prominence values p_i for each quality dimension i, the Euclidean distance $d(u,v)$ indicating the semantic similarity between two members described by vector sets V and U can be calculated as follows:

$$d(u,v) = \sqrt{\sum_{i=1}^{n} p_i \left(\left(\frac{u_i - \bar{u}}{s_u} \right) - \left(\frac{v_i - \bar{v}}{s_v} \right) \right)^2}$$

13.4 A Conceptual Learning Situation Space

In (8) we introduce a general-purpose procedure for refining arbitrary ontologies, such as SWS, through CS-based representations such as a CSS. However, in this section we would like to illustrate CSS through a particular example from the e-Learning domain, a *Conceptual Learning Situation Space*, which also demonstrates the applicability of our metamodel even to rather qualitative parameters. As described in (10) a learning situation is defined by parameters such as the technical environment used by a learner, his/her competency profile or the current learning objective. This section focuses exemplarily on the representation of one parameter through a CSS subspace, which is of particular interest within the E-Learning domain: the learning style of a learner. A learning style is defined as an individual set of skills and preferences on how a person perceives, gathers, and processes learning materials (25). Whereas each individual has his/her distinct learning style, it affects the learning process (14) and consequently has to be perceived as an important parameter describing a learning situation. To describe a learning style, we refer to the Felder-Silverman Learning Style Theory (FSLST) (14) approach to describe learning styles within computer-aided educational environments (15), where a learning style is described with four quality dimensions (14) defined here with 4 quality dimensions l_i that hold metric scale, datatype, value range, and prominence values in a CSS L, as presented in Table 13.1:

	Quality Dimension	Metric Scale	Data-Type	Range	Prominence
l_1	Active-Reflective	Interval	Integer	-11..+11	1.5
l_2	Sensing-Intuitive	Interval	Integer	-11..+11	1
l_3	Visual-Verbal	Interval	Integer	-11..+11	1.5
l_4	Global-Sequential	Interval	Integer	-11..+11	1

Table 13.1: Quality dimensions l1 - l4 describing learning styles following FSLST

As depicted in Table 13.1, each quality dimension is ranked on an interval scale with a value range being integers between -11 and +11. This particular measurement scale was defined with respect to an established assessment method, the Index of Learning Styles (ILS) questionnaire defined by Felder and Spurlin (15), aimed at identifying and rating the particular learning style of an individual. The authors would like to highlight, that prominence values have been assigned which rank the first (l_1) and the third dimension (l_3) higher than the other two, since these have a higher impact with respect to the purpose of the learning situation, which is focused on the aim to deliver appropriate learning material to the learner.

In order to classify an individual learning style, we define prototypical members in the FSLST-based vector space L. To identify appropriate prototypes, we utilized existing knowledge about typical correlations between the FSLST dimensions, as identified throughout research studies such as (22, 39). Details on the construction methodology of these prototypes are given in (10). This resulted in the following 5 prototypical members and their characteristic vectors shown in Table 13.2.

Prototype	Act/Ref	Sen/Int	Vis/Ver	Seq/Glo
P1: Active-Visual	-11	-11	-11	+11
P2: Reflective	+11	-11	-11	0
P3: Sensing-Seq.	-11	-11	-11	-11
P4: Intuitive-Glob.	-11	+11	-11	+11
P5: Verbal	-11	+11	+11	+11

Table 13.2: Prototypical learning styles defined as prototypical members in the CSS ontology

13.5 Fuzzy SWS Goal Discovery and Achievement at Runtime

To prove the feasibility of our approach, a proof-of-concept prototype application[6] was provided, which utilizes the CSS metamodel and ontology framework introduced in Sections 13.3 and 13.4 to implement a use case from the E-Learning domain.

13.5.1 Runtime Reasoning Support for CSS and SWS

In order to describe situations within the domain of E-Learning, a CSS specific for the domain of E-Learning was provided that is able to represent domain-specific situations described by concepts defined within a particular WSMO domain ontology. Linking each situation parameter, defined within

[6]The application is utilized within the EU FP6 project LUISA (`http://www.luisa-project.eu/www/.`)

a WSMO SWS description, to a particular CSS, and defining prototypical instances within each CSS enables the automatic classification of situation parameters in terms of their similarity with a set of prototypical parameters. Figure 13.3 depicts the architecture used to support reasoning on CSS and SWS in distinct domain settings through a Semantic Execution Environment (SEE), which is in our case implemented through IRS-III (Section 13.5).

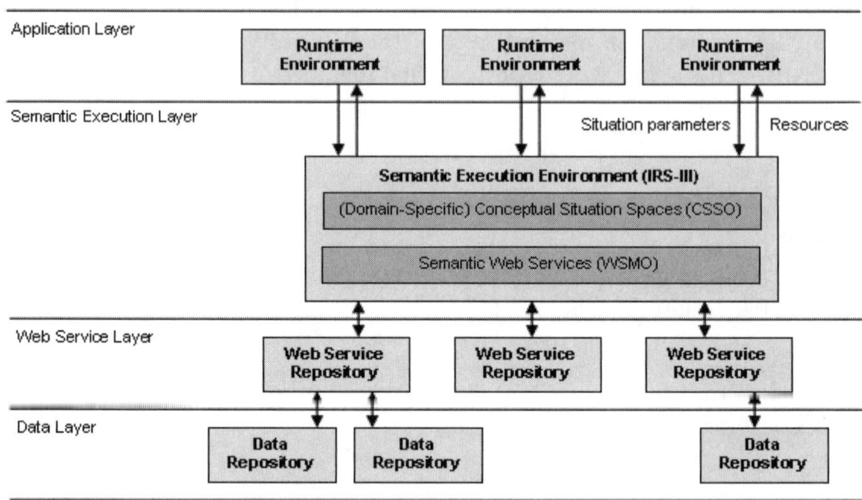

Figure 13.3: Architecture to support runtime reasoning on CSS and SWS models.

SEE utilizes a semantic representation of the CSS metamodel (CSSO), which is derived for specific domains, and of the SWS metamodel based on WSMO. Both are represented utilizing the OCML representation language (30). IRS-III dynamically classifies a given situation based on the CSSO and provides resources, represented based on WSMO, which suit a specific runtime situation. Distinct runtime environments can serve as user interfaces to enable users to interact with SEE and to provide knowledge about the current real-world situation. Given a set of real-world situation parameters, their semantic distance to predefined prototypical situation parameters, defined within a domain-specific CSS, is calculated to enable classification of a set of real-world situation parameters. The SEE finally discovers and orchestrates appropriate Web services that show the capabilities to suit the given situation.

13.5.2 SWS Goal Discovery Based on Context Classification

In order to reach situation awareness, the application automatically detects semantic similarity of specific situation parameters with a set of predefined prototypical parameters to enable the allocation of context-appropriate resources through the SEE. Referring to a CSS subspace L described previously, given a particular member U in L, its semantic similarity with each of the prototypical members is indicated by their Euclidean distance. Since we utilize a CSS described by dimensions that each use the same metric scale (ordinal scale), the distance between two members U and V is calculated disregarding a Z-transformation (Section 13.3) for each vector.

The calculation of Euclidean distances is accomplished by a standard Web service that is exposed as SWS and is invoked through IRS-III at runtime. Given a particular CSS description, a member (representing a specific parameter instance) as well as a set of prototypical member descriptions (representing prototypical parameter instances), the Web service calculates similarities at runtime in order to classify a given situation parameter. For instance, a particular situation description includes a learner profile indicating a learning style parameter, which is defined by a member U in the specific CSS subspace to describe learning styles following FSLST with the following vectors:

$$U = \{(u_1 = -5, u_2 = -5, u_3 = -9, u_4 = 3)\,|u_i \in L\}$$

Learning styles such as the one above, could be assigned to individual learners by utilizing the ILS Questionnaire as assessment method. Calculating the distances between U and each of the prototypes described in Table 13.2 of Section 13.4 led to the following results:

Prototype	**Euclidean Distance**
P1: Active-Visual	12.649110640673518
P2: Reflective	20.85665361461421
P3: Sensing-Sequential	17.08800749063506
P4: Intuitive-Global	19.493588689617926
P5: Verbal	31.20897306865447

Table 13.3: Euclidean distances between U and prototypical learning styles

As depicted in Table 13.3, the lowest Euclidean distance between U and the prototypical learning styles applies to $P1$, indicating a rather active and visual learning style described as in Table 13.2 of Section 13.4.

Classified contexts are utilized to discover the most appropriate SWS goal representation for a given context, by utilizing the alignment of CSS and SWS (Section 13.3). Given a specific situation description, IRS-III first identifies SWS goal representations (*wsmo:Goal*), which suit the given situation and finally selects and orchestrates SWS that are appropriate to suit the given runtime situation. For instance, in the proposed use case, distinct SWS goal representations are available, each retrieving content that addresses a distinct learning style (Figure 13.4).

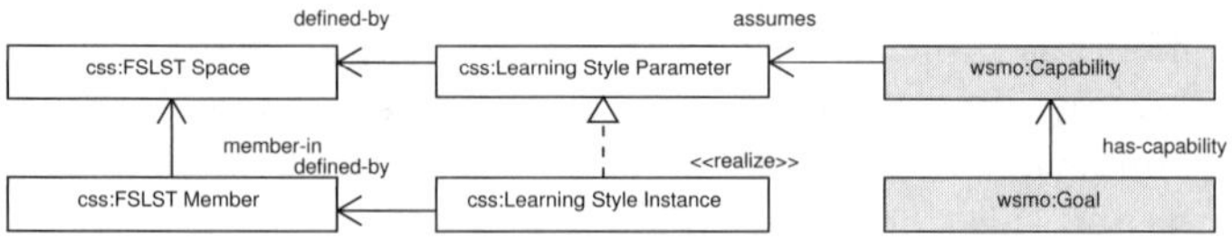

Figure 13.4: SWS Goals assuming learning styles described as members in a CSS.

Given the similarity-based classification of a set of real-world parameters, e.g. learning styles, a SWS goal representation that assumes matching prototypical parameter instances is selected and achieved through IRS-III (Figure 13.3, Section 13.4). Finally, IRS-III utilizes the SWS goal capability to identify a SWS that suits the given goal. For instance, given a classified learning style together with classifications of all further situation parameters, a SWS goal representation that assumes matching prototypical situation parameter instances is selected and achieved at runtime. Consequently, following the alignment of CSS with established SWS frameworks, context-aware SWS applications are enabled, which automatically discover not only Web services for a given task but also SWS goal descriptions for a given situation.

13.6 Applying CSS to the E-Business Domain

As demonstrated in the previous section, refining SWS through CSS helps describing and reasoning about SWS capabilities, which in turn enhances context-aware selection/matchmaking of SWS. In this section, we demonstrate the relevance of CSS with respect to the E-Business application domain. We detail how CSS dimensions are designed and utilized, according to the requirements of involved business actors. In addition, we discuss in the end of this section how CSS could be used in conjunction with a context model presented in (31), in order to enhance semantic mediation of data between Web services. Hence, this section demonstrates the independence of CSS with respect to applications domains and underlying semantic models, and offers some insights on foreseeable use of CSS for other purposes such as semantic mediation of data.

13.6.1 The Ordering Example

In contrast to the previous example from the E-Learning domain, in this section we develop another use case of CSS application, illustrated with a typical E-Business scenario. Let us assume a UK-based goods reseller who aims at ordering manufactured goods from a Japanese producer via a broker that selects the best producer WS according to its client's requirements. For the purpose of this example, we identify the requirements of the reseller with the following set of criteria:

- shipping delay, measured in days (comprised between 0 and ∞);
- type of packaging, set to fragile, normal or secured and identified with the values 1, 2 and 3 in the CSS dimension;
- quantity of items per delivery (comprised between 0 and ∞);
- payment security level, set to low (no encryption), medium (weak encryption method) or high (strong encryption such as SSL) and identified with 1, 2 or 3;
- authentication security level, set to 0 (no authentication) or 1 (password-based authentication) or 2 (certificate-based authentication).

For the purpose of SWS discovery, we evaluate the quality offered by each producer SWS with respect to current reseller's requirements represented as a CSS member. Indeed, the selected producer SWS is "only" the closest match to the reseller's requirements and may not completely satisfy them. In order to give priority to some dimensions over others, weighted prominences were setup on quality dimensions. Table 13.4 summarizes the different dimensions composing the CSS and proposes prominences that are useful to the selection task.

	Quality Dimension	Metric Scale	Data-type	Range	Prominence
l_1	shipping delay	Interval	Integer	$+1..+\infty$	1.5
l_2	type of packaging	Interval	Integer	$+1..+3$	1
l_3	number of items/delivery	Interval	Integer	$+1..+\infty$	0.5
l_4	payment security level	Interval	Integer	$+1..+3$	2
l_5	authentication security level	Interval	Integer	$+0..+2$	1.5

Table 13.4: Quality dimensions l_1 - l_5 describing transaction options

The values of these requirements may often change as they depend on external (environmental) conditions related to business/economical issues. Therefore, in order to select the best producer, the realization of SWS discovery is relevant at runtime via a broker. The SWS discovery and ordering process involves simple interactions between business actors. These interactions are modeled as a business process, as described in Figure 13.5.

The reseller sends a SWS request by means of a WSMO goal to the broker (step 1). This request includes a set of parameters that describe the aforementioned requirements. Then, the broker searches for SWS that provide the functionality (step 2). It sends back either (step 3a) a positive answer containing access information to the best WS found, or (step 3b) a negative answer (no SWS found). In the latter case, the business process either goes back to step 1 or ends with a failure notification (step 4b). If the answer was positive, the actual order is sent to the broker, that acts as a proxy (step 4a) and the transaction is performed (step 5), before ending the business process with a successful notification (step 6). Interactions between Web services are typically specified in a WS-BPEL[7] file, as a sequence of WSMO goal invocations which are resolved by the broker (i.e. IRS-III) [8]. For the sake of brevity we do not detail the technical aspects related to the business process that is sketched in Figure 13.5.

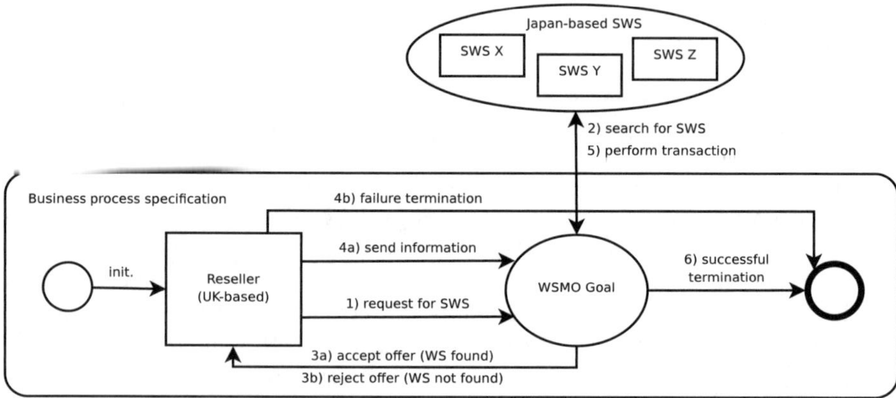

Figure 13.5: Overview of the ordering business process.

13.6.2 Overview of SWS Discovery with CSS

In the following, we provide details about the SWS discovery process to be achieved in step 2) of Figure 13.5, according to the aforementioned reseller's requirements, which are described as CSS members (each requirement is expressed as a vector valueing one of the dimensions). Each producer SWS is also represented as a CSS member (a set positions on vectors) and matched with the requirements of the reseller. The discovery process takes place in a

[7] www.oasis-open.org/committees/wsbpel/

[8] See the SUPER project for additional details `http://www.ip-super.org/`.

similar way to the one described in Section 13.4. In order to illustrate the selection process, we provide a sample of the evaluation process on the basis of our developed example. The CSS is described as illustrated in Table 13.4. The reseller's requirements are described with the following CSS member:

$$V_{Optimal} = \{(l_1 = 1, l_2 = 3, l_3 = 20, l_4 = 2, l_5 = 2)\,|l_i \in L\}$$

where L is the CSS subspace defined with Table 13.4. For the purpose of the demonstration, we described the providers X, Y, and Z shown in Figure 13.5. They each present specific characteristics that are described in CSS as follows:

$$V_{Prov_X} = \{(l_1 = 3, l_2 = 1, l_3 = 20, l_4 = 1, l_5 = 0)\,|l_i \in L\}$$
$$V_{Prov_Y} = \{(l_1 = 2, l_2 = 2, l_3 = 15, l_4 = 2, l_5 = 1)\,|l_i \in L\}$$
$$V_{Prov_Z} = \{(l_1 = 3, l_2 = 3, l_3 = 5, l_4 = 2, l_5 = 2)\,|l_i \in L\}$$

With this information, it is possible for the broker to calculate the distance between the CSS representation of the reseller's requirements and each SWS. In our example, we obtain the following result:

Provider	**Euclidean Distance**
Provider X	4.242640687119285
Provider Y	4.06201920231798
Provider Z	10.88577052853862

Table 13.5: Euclidean distances between requirements and discovered SWS

Table 13.6.2 shows the respective distance from each SWS to the given requirements. In our case, *providerY* is the best SWS for these requirements and is then selected by the broker.

Concretely, the broker makes use of a Web service that calculates the Euclidean distance between members of a CSS according to a given base member (here $V_{optimal}$). The response from the Web service returns the respective distances of the members and lastly, a SmallestDistance tag indicates the member that happens be the most appropriate for the requirements of the request.

13.6.3 An Insight on Context-Based Data Mediation with CSS

In (31), a context model is proposed to enable mediation of data exchanged between Web services. According to Section 13.2.1, where semantic-level and data-level mediation are introduced, we provide in this section some insight on how CSS can be used with this context model to offer novel possibilities for data-level mediation.

At the step 4a and 5 of our E-Business example, certain data (i.e., input/output data) is exchanged between the reseller and the best producer selected by the broker. Indeed, the data exchanged is interpreted according to a certain context[9], which is related to the environment of the business actors. In step 4a, the reseller sends a price proposal as part of the order. At this point, data-level mediation is needed at runtime in order to preserve the semantic consistency of the data exchanged between services. The "price" information has to be converted from a specific British context (in our example, GBP as a currency, included VAT and a scalefactor of 1) to a specific Japanese context (in our example, JPY as a currency, excluded VAT, and a scale factor of 1000) to reach semantic correctness in the business process.

In the following we give some insight on how to use CSS as a support to enhance mediation of the data exchanged in a business process, on the basis of the context model developed in (31). We first remind the reader with the basics of the context model, before showing how its integration with CSS could be possible and to which extent this integration could ease the mediation task.

A Reminder on the Context Model

In (31), a context model dedicated to semantic, data-level mediation has been developed. This model helps make explicit the underlying semantic assumptions related to Web service input and output data. Its objective is to provide mediators with the required semantic information to reason about and mediate between alternative representations (i.e., following different contexts). This model builds around the notion of the semantic object, which consists of a concept c from a domain ontology, an XML type t that grounds the data to a particular physical representation, a value v that represents the value of the data and is instantiated at runtime, and a set C of semantic attributes called context, that is made of other semantic objects. Context is organized as a tree structure that details the semantic information required for a correct data interpretation.

This model has been tested with the representation of monetary values (i.e., prices). The representation of prices depends on a tree of contextual attributes (i.e., date, date format, country, VAT rate, etc.) that play a role in the interpretation of a price instance and form together its "context". As an example, a simple "price" semantic object p of value 15 expressed in Euro with VAT included is described as follows:

```
p = {ns:price, xsd:float, 15, {
    currency = Euro,
    VATincluded = true {VATRate = 19,6%},
    scaleFactor = 1}}
```

We show in the following how a semantic object in a specific context can also be described as a specific CSS member.

[9]Here, context refers to any piece of information required in order to correctly interpret the data.

Additional Features of CSS

Context is built-up as a tree in order to refine the semantic description of context attributes, which complies with the definition of CSS that allows a dimension to be further refined via sets of conceptual subspaces. However, the use of CSS implies a couple of enhancements to the context model. The following features of CSS are relevant to mediation, and added as extensions to the original context model:

- the notion of **quality dimension**: involves describing context attributes as points on a scale, which is particularly useful for mediation purposes, as detailed hereafter.

- the notion of **prominence**: enhances mediation by allowing to setup specific prominence levels that play a role during the mediation process. For example, Web services may agree on a minimum prominence level, under which a quality dimension is considered as negligible for the mediation purpose.

An Insight on CSS Mediation with the Ordering Example

As demonstrated previously, in order to preserve the semantic consistency of our illustrative business process, we need to make explicit the different contextual aspects related to prices.

In our example, data-level mediation consists in applying various tranformation operations to the price value transmitted from the reseller to the producer, in order to converting context dimensions from their original values to the targeted values. Price currencies correspondences can be described by associating GBP, USD, EUR, or JPY to values on the "currency" scale, corresponding to the actual relative values of these currencies with respect to each other. For instance, when USD is worth 1,2 Euro, then the USD is placed on 1 and Euro on 1,2 on the currency scale. Accordingly, scalefactors are placed on 1 and 1000. The VAT boolean is set to 1 if a VAT applies to the price, and the VAT rate is expressed as a float. In our example, we show that a price of value 15 expressed in GBP with a scale factor of 1 and VAT included. This context information is represented as a CSS member over the dimensions expressed previously:

$$Context_1 = (currency = GBP, sf = 1, VATincluded = true, (VATrate = UKrate))$$

The target context can also be expressed as a CSS member as follows:

$$Context_2 = (currency = JPY, sf = 1000, VATincluded = false, (VATrate = JPYrate))$$

Therefore, the data-level mediation task can be reduced to converting the "price" value according to the operations to be applied to the CSS dimensions for $Context_1$ to become equal to $Context_2$. In our example, in order to change the sf value from 1 to 1000 we need to multiply the price by 1000. In order to change the currency we need to convert the price from GBP to JPY according to their values on the CSS dimensions, and similarly for the VAT rates.

We have shown in this section that CSS could be utilized to simplify data-level mediation. It replaces complex conversion rules stored in a knowledge repository involved in the original context-based mediation work (31) with simple value conversions, according to the positions of context elements over the CSS dimensions. Also, the use of CSS adds the notion of prominence that is missing in the existing context model, and further refines the mediation strategy by explicitly stating the relative importance of context elements with respect to the mediation purpose.

13.7 Conclusions

In this paper, we proposed an approach to support fuzzy, similarity-based matchmaking between real-world context characteristics and predefined SWS capability descriptions by incorporating semantic context information on a conceptual level into symbolic SWS representations utilizing a metamodel for Conceptual Situation Spaces (CSS). By utilizing the CSS and its alignment to SWS technology, the most appropriate resources, whether data or services, for a given situation are identified based on the semantic similarity, calculated in terms of the Euclidean distance, between a given real-world situation and predefined resource descriptions as part of SWS capability representations. Consequently, by aligning CSS to established SWS frameworks, the expressiveness of symbolic SWS standards is extended with vector-based context information to enable fuzzy context-aware discovery of services and resources at runtime. Whereas current SWS frameworks such as WSMO and OWL-S address the allocation of distributed services for a given (semantically) well-described task, the CSS approach particularly addresses the similarity-based discovery of the most appropriate SWS task representation for a given context. To prove the feasibility of our approach, two proof-of-concept prototype applications were presented. Whereas the first one applies the CSS metamodel to enable context-adaptive resource discovery in the domain of E-Learning, the second one applies CSS to an E-Business scenario.

However, although our approach aims at solving Semantic Web (Services)-related issues such as the symbol grounding problem, several criticisms still have to be taken into account when applying CSS. While defining situations, respectively instances within a given CSS appears to be a straightforward process of assigning specific values to each quality dimension of a CSS, the definition of the CS itself is not trivial and strongly dependent on individual perspectives and subjective appraisals. Whereas the semantics of an object are grounded to metrics in geometrical vector spaces within a CS, the quality dimensions itself are subject to one's perspective and interpretation, which may lead to ambiguity issues. With regard to this, the approach of CSS does not appear to fully solve the symbol grounding issue but to shift it from the process of describing context instances to the definition of a CSS. Indeed, distinct semantic interpretations and vector-based groundings of each dimension may be applied by different individuals. Apart from that, whereas the size and resolution of a CS is indefinite, defining a reasonable CSS for a specific purpose or domain may become a challenging task. Nevertheless, distance calculation as major contribution of CSS, relies on the fact, that entities are described in the same geometrical space.

Consequently, CS-based approaches such as CSS may be perceived as step forward but do not fully solve the issues related to symbolic Semantic Web (Services)-based knowledge representations. Hence, future work has to deal with the aforementioned issues. For instance, we foresee to enable adjustment of prominence values to quality dimensions of a specific CSS to be accomplished by a user him/herself, in order to most appropriately suit his/her specific priorities and preferences regarding the resource allocation process, since the prioritization of dimensions is a highly individual and subjective process. In addition, it is intended to apply different distance metrics in order to evaluate the most appropriate similarity measure within CSS. Nevertheless, further research will be concerned with the application of our approach to further domain-specific situation settings.

References

[1] Sinuhé Arroyo and Michael Stollberg. WSMO Primer. WSMO Deliverable D3.1, DERI Working Draft. Technical report, WSMO, 2004. http://www.wsmo.org/2004/d3/d3.1/.

[2] Liliana Cabral, John B. Domingue, Stefania Galizia, Alessio Gugliotta, Barry Norton, Vlad Tanasescu, and Carlos Pedrinaci. Irs-iii: A broker for semantic web services based applications. In *Proceeding of the 5th International Semantic Web Conference (ISWC2006)*, 2006.

[3] Namyoun Choi, Il-Yeol Song, and Hyoil Han. A survey on ontology mapping. *SIGMOD Rec.*, 35(3):34–41, September 2006.

[4] E Cimpian, A Mocan, and M Stollberg. Mediation enabled semantic web services usage. In *1st Asian Semantic Web Conference (ASWC2006), September 2006*, page 2006, 2006.

[5] Anne Cregan. Symbol grounding for the semantic web. In *ESWC*, pages 429–442, 2007.

[6] J Devore and R Peck. Statistics: The exploration and analysis of data, 1999.

[7] Anind K. Dey. Understanding and using context. *Personal and Ubiquitous Computing*, 5:4–7, 2001.

[8] S. Dietze and J. Domingue. Exploiting conceptual spaces for ontology integration. In *Proceedings of Workshop on Data Integration through Semantic Technology (DIST2008) @ 3rd Asian Semantic Web Conference 2008*, Bangkok, Thailand, 2007.

[9] S. Dietze, A. Gugliotta, and J. Domingue. A semantic web services-based infrastructure for context-adaptive process support. In *Proceedings of IEEE 2007 International Conference on Web Services (ICWS)*, 2007.

[10] Stefan Dietze, Alessio Gugliotta, and John Domingue. Towards context-aware semantic web service discovery through conceptual situation spaces. In *CSSSIA '08: Proceedings of the 2008 International Workshop on Context enabled Source and Service Selection, Integration and Adaptation*, pages 1–8, New York, NY, USA, 2008. ACM.

[11] M Ehrig and S Staab. Qom - quick ontology mapping. pages 683–697. Springer, 2004.

[12] M Ehrig and Y Sure. Ontology mapping – an integrated approach. pages 76–91. Springer Verlag, 2004.

[13] J Euzenat, P GuÃĺgan, and Valtchev. P.: Ola in the oaei 2005 alignment contest. In *Proceedings of the K-CAP Workshop on Integrating Ontologies*, pages 61–71, 2005.

[14] R. M. Felder and L. K. Silverman. Learning and teaching styles in engineering education, 1988.

[15] R.M. Felder and J. Spurlin. Index of learning styles questionnaire, 1997.

[16] D. Fensel, H. Lausen, A. Polleres, J. de Bruijn, M. Stollberg, D. Roman, and J. Domingue. Enabling semantic web services - the web service modelling ontology, 2006.

[17] A Gangemi, N Guarino, C Masolo, A Oltramari, and L Schneider. Sweetening ontologies with dolce. pages 166–181. Springer, 2002.

[18] P. Gärdenfors. *How to Make the Semantic Web More Semantic*, pages 17–34.

[19] Peter Gärdenfors. *Conceptual Spaces: The Geometry of Thought.* The MIT Press, March 2000.

[20] Hans W. Gellersen, Albrecht Schmidt, and Michael Beigl. Multi-sensor context-awareness in mobile devices and smart artifacts. *Mob. Netw. Appl*, 7:341–351, 2002.

[21] F Giunchiglia, P Shvaiko, and M Yatskevich. S-match: an algorithm and an implementation of semantic matching. In *In Proceedings of ESWS*, pages 61–75, 2004.

[22] Sabine Graf, Silvia Rita Viola, Kinshuk, and Tommaso Leo. Representative characteristics of felder-silverman learning styles: An empirical model. In *Proceedings of the IADIS International Conference on Cognition and Exploratory Learning in Digital Age, Barcelona, Spain, 2006.* IADIS Press, 2006.

[23] Armin Haller, Emilia Cimpian, Adrian Mocan, Eyal Oren, and Christoph Bussler. Wsmx - a semantic service-oriented architecture. In *ICWS*, pages 321–328, 2005.

[24] Stevan Harnad. The symbol grounding problem. *CoRR*, cs.AI/9906002, 1999.

[25] C. Johnson and C. Orwig. What is learning style?, 1998.

[26] Mauri Kaipainen, Peeter Normak, Katrin Niglas, Jaagup Kippar, and Mart Laanpere. Soft ontologies, spatial representations and multi-perspective explorability. *Expert Systems*, 25(5):474–483, November 2008.

[27] E.F. Krause. *Taxicab Geometry.* Dover, 1987.

[28] David L. Martin, Massimo Paolucci, Sheila A. McIlraith, Mark H. Burstein, Drew V. McDermott, Deborah L. McGuinness, Bijan Parsia, Terry R. Payne, Marta Sabou, Monika Solanki, Naveen Srini-

vasan, and Katia P. Sycara. Bringing Semantics to Web Services: The OWL-S Approach. In *SWSWPC*, pages 26–42, 2004.

[29] P Mitra, N F Noy, and A R Jaiswal. Ontology mapping discovery with uncertainty. In *Fourth International Conference on the Semantic Web (ISWC-2005*, pages 537–547, 2005.

[30] E. Motta. An overview of the ocml modelling language. In *8th Workshop on Methods and Languages*, 1998.

[31] Michael Mrissa, Chirine Ghedira, Djamal Benslimane, and Zakaria Maamar. A context model for semantic mediation in web services composition. In *ER*, pages 12–25, 2006.

[32] N F Noy and M A Musen. The prompt suite: Interactive tools for ontology merging and mapping. *International Journal of Human-Computer Studies*, (59), 2003.

[33] M Paolucci, N Srinivasan, and K Sycara. Expressing wsmo mediators in owls. In *Proceedings of the workshop on Semantic Web Services: Preparing to Meet the World of Business Applications held at the 3rd International Semantic Web Conference (ISWC 2004*, 2004.

[34] Y Qu, W Hu, and G Cheng. Constructing virtual documents for ontology matching. In *Proceedings of the 15th International World Wide Web Conference*, pages 23–31. ACM Press, 2006.

[35] Martin Raubal. Formalizing conceptual spaces. In Varzi A. Vieu and L., editors, *Proc. of the Third International Conference on Formal Ontology in Information Systems*, Frontiers in Artificial Intelligence and Applications, pages 153–164. IOS Press, Amsterdam, NL, 2004.

[36] A Schmidt and C Winterhalter. C.: User context aware delivery of e-learning material: Approach and architecture. *Journal of Universal Computer Science (JUCS*, (10):28–36, 2004.

[37] Angela Schwering. Hybrid model for semantic similarity measurement. In Robert Meersman, Zahir Tari, Mohand S. Hacid, John Mylopoulos, Barbara Pernici, özalp Babaoglu, Hans A. Jacobsen, Joseph P. Loyall, Michael Kifer, and Stefano Spaccapietra, editors, *OTM Conferences (2)*, volume 3761, pages 1449–1465. Springer, 2005.

[38] P. Suppes, D. H. Krantz, R.D. Luce, and A. Tversky. *Foundations of Measurement*, volume 2: Geometrical, threshold and probabilistic representations. Academic Press, New York, 1989.

[39] S. R. Viola and S. Graf. Investigating relationships within the index of learning styles: A data-driven approach. *International Journal of Interactive Technology and Smart Education*, 4(1):7–18, 2007.

[40] Z Wu, K Gomadam, A Ranabahu, A Sheth, and J Miller. Automatic composition of semantic web services using process mediation. In *Proceedings of the 9th Intl. Conf. on Enterprise Information Systems ICES 2007*, 2007.

Chapter 14

Privacy Protection in Context-Aware Web Services: Challenges and Solutions

Georgia M. Kapitsaki, Georgios V. Lioudakis, Dimitra I. Kaklamani, and Iakovos St. Venieris

Abstract Context-aware Web Services comprise a technological trend that has become the catalyst for the provision of innovative services, paving the way for the brave new world of the "Future Internet." On the other hand, they bring the issue of personal privacy to the spotlight; not only context awareness natively depends on the collection and processing of personal information, but also the Web Services architectural concepts increase the leakage-proneness of the data flows. This chapter aims at contributing to break the false dichotomy between the provision of advanced services and personal privacy, fostering the introduction of privacy awareness in this class of services. Having as a starting point the personal data protection legislation and fair practices, the chapter discusses the associated challenges and identifies the stemming technical requirements. It puts in place the notion of "privacy context," targeting to make the fundamental privacy principles integral to contextual information, and proposes an abstract architecture for its enforcement. This way, the chapter aims in serving as a guide for developers and providers, by setting the focus on the privacy challenges and solutions related with context-aware Web Services.

14.1 Introduction

"On the Internet, nobody knows you are a dog" according to the famous Pat Steiner cartoon in *The New Yorker* in 1993, which has been very frequently cited in order to emphasize the potential for anonymity and privacy that the Internet was supposed to offer. However, the reality seems to be rather different and, more than a century after the first essay identifying that privacy as a fundamental human right was endangered by technological advances (56), never before in history the citizens have been more concerned about their personal privacy and the threats by the emerging technologies (27). Among the several threats to personal privacy caused by the emerging Information and Communication Technologies (ICT), an outstanding position is held by context-aware Web Services (WS) that bring about severe implications.

Intuitively, since context awareness implies the collection and processing of contextual information regarding the service recipient and his/her environment and situation, issues concerning the disclosure, dissemination, and use of personal data related to the user are raised. Compared to other ICT domains, context-aware WS have characteristics that make them in particular interesting from a privacy point of view. On the one hand, the data that services mostly deal with fall in the category of "semi-active" data (34), in the sense that their collection occurs transparently for the user; this type of transparent, implicit collection tends to raise greater privacy concerns than those initiated by the user (16). On the other hand, the advent of WS has resulted in increased complexity of the service provisioning chain; a service can be a puzzle comprised of different WS components, possibly involving a number of different providers and sometimes even dynamic. That is, context-aware WS create complex and leakage-prone information flows where context data travel between different connection points. The enforcement of fair data practices is a challenging aspect and the establishment of the corresponding mechanisms is deemed necessary, whereas the fact that the users should maintain control over their own data is not to be neglected.

Taking all these into account, the issue of privacy protection in such environments constitutes the subject of this chapter. As the privacy domain is increasingly becoming a legislated area, the provision of context-aware WS is surrounded by related issues. Therefore, in order to identify the requirements posed to the services' provision chain, the chapter begins with a codification of the underlying legal and regulatory principles, followed by a discussion of the consequent technical challenges (Section 14.2). It then goes on to an overview of existing approaches that aim at protecting privacy, examining both frameworks that directly refer to context-aware systems and services, as well as technologies that can be leveraged to this end (Section 14.3). In Section 14.4, the chapter introduces the concept of "privacy context," considering two additional contextual sources, specifically devised for injecting privacy-awareness to the service provision chain; for the enforcement of the privacy context, an abstract architecture is outlined in Section 14.5. Before concluding, the chapter discusses in Section 14.6 issues related to the combination of the described framework with context adaptation schemes.

14.2 Privacy Regulations and Technical Requirements

The design of a context-aware system or service cannot be considered as a purely technical activity, as it clearly has a considerable societal impact in terms of its implications on personal data protection and the right to privacy. Moreover, public policies and regulatory requirements do affect to a significant extent the technical requirements of such a system. Therefore, the provision of an overview of the underlying principles is deemed necessary.

>From a philosophical perspective, the notion of privacy has broad historical roots; for instance, consider Aristotle's distinction between the public sphere of political activity and the private sphere associated with domestic life, or the Hippocratic Oath, the seminal document on the ethics of medical practice that explicitly included privacy among the medical morals. Nowadays, privacy is recognized as a fundamental human right by the Universal Declaration of Human Rights of the United Nations (51), as well as the Charter of Fundamental Rights of the European Union (23), and is protected by relevant legislation in all the democratic countries throughout the world. The first data protection act, adopted in 1970 by the West Germany state of Hesse, set in motion a trend towards adopting privacy legislation.

The first influential text was the U.S. Privacy Act (52), adopted by the Congress in 1974. A significant milestone in the privacy literature has been the codification of the fundamental privacy principles by the Organization for Economic Co-operation and Development (OECD) (41) in 1980, as this codification lays out the basis for the protection of privacy. The OECD principles are reflected in the European Directive 95/46/EC (21), "*on the protection of individuals with regard to the processing of personal data and on the free movement of such data.*" The Directive 95/46/EC enforces a high standard of data protection and constitutes the most influential piece of privacy legislation worldwide, affecting many countries outside Europe in enacting similar laws.

Under Article 2, the Directive 95/46/EC defines personal data as "*any information relating to an identified or identifiable natural person ('data subject'); an identifiable person is one who can be identified, directly or indirectly, in particular by reference to an identification number or to one or more factors specific to his physical, physiological, mental, economic, cultural or social identity.*" This definition stresses on the explicit reference to indirect identification data, implying any information that may lead to the identification of the data subject through association with other available information (thus indirectly), that may be held by any third party. The Directive 95/46/EC is further particularized and complemented with reference to the electronic communication sector by the Directive 2002/58/EC (22), which imposes explicit obligations and sets specific limits on the processing of users' personal data by network and service providers in order to protect the privacy of the users of communications services and networks. Especially with respect to contextual data, the Directive 2002/58/EC has granted a specific and high degree of protection to traffic and location data, imposing strict limits and requirements to their processing due to their peculiar nature. Indeed, traffic data allow knowing user's activities and behavior and defining the user's personality; traffic data allow user's localization and tracking. Combined traffic and location data enable to build user's profile enriched with geographical information, thus resulting in a significant encroaching into the individual's personal life and in invasive surveillance.

The impact of contextual data collection and processing has been further elaborated in subsequent documents from the European Article 29 Data Protection Working Party, which have focused on the use of location data with a view to providing value-added services (e.g., (7)), while having emphasized several problems with the introduction of RFIDs and other sensors and have stressed the importance of indicating what kind of data should be processed and under which conditions (e.g., (8)).

Since contextual information constitutes personal data and is subject to the data protection legislation, in the following, the fundamental legal and regulatory principles and requirements are summarized. The focus is on the legislation of the European Union, since it comprises the most representative, influential and mature approach worldwide, that seems to pull a general framework and has been characterized as an "engine of a global regime" (12). For some insights in the frameworks of the United States and other countries, the reader is referred to Solove's excellent essay (48). Having the legal and regulatory framework as a starting point, the section concludes with a description of the technical requirements stemming from the legislation that should be satisfied by the corresponding systems.

14.2.1 Legal and Regulatory Requirements for Context-Aware Systems

The legal and regulatory requirements that should be taken under consideration when developing context-aware systems or providing context-aware services can be summarized as follows:

- *Lawfulness of the data processing*: The systems/services should be able to examine whether the data processing complies with applicable laws and regulations.

- *Purposes for which data are processed*: The systems/services should provide the means for identifying the data processing purposes, which must be lawful and made explicit to the data subject. Moreover, they should be able to check these purposes to avoid that data processed for a purpose may be further processed for purposes that are incompatible with these for which data have been collected.

- *Necessity, adequacy, and proportionality of the data processed*: The systems/services should be able to guarantee that only the data that are functional, necessary, relevant, proportionate and not excessive with regard to the sought processing purpose are processed.

- *Quality of the data processed*: The systems/services should provide that the data processed are correct, exact, and updated. Inaccurate data must be deleted or rectified; outdated data must be deleted or updated.

- *Identifiable data*: The systems/services should provide the means for keeping the data processed in an identifiable form only for the time necessary to achieve the sought processing purpose.

- *Notification and other authorizations from competent Privacy Authority*: The systems/services should be able to monitor compliance with the notification requirement and with the provisions on the authorizations of competent Privacy Authority. Moreover, the systems/services should provide for means that allow communications between the systems/services and the competent Privacy Authority.

- *Information to the data subjects*: The systems/services should be able to provide for informing the data subject that the data are processed according to applicable data protection legislation.

- *Consent and withdrawal of consent*: The systems/services should guarantee that when requested by applicable data protection legislation, the data subject's consent to the data processing is required, and that the data processing is performed according to the preferences expressed by the data subject. Further, withdrawal of consent and an objection to data processing by the data subject should be handled appropriately.

- *Exercising rights of the data subject*: The systems/services should enable the data subject to exercise the rights acknowledged by applicable data protection legislation in relation to intervention in the data processing (for example the right to access data, to ask for data rectification, erasure, blocking, the right to object to the data processing, etc.).

- *Data security and confidentiality*: The systems/services should be secure in order to guarantee the confidentiality, integrity, and availability of the data processed. Moreover, the systems/services should provide that the listening, tapping, storage, or other kinds of interception or surveillance of communications and the related traffic data may be performed only with the data subject's consent or when allowed by applicable legislation for public interest purposes.

- *Traffic data and location data other than traffic data; special categories of data*: The systems/services should be able to guarantee that the processing of special categories of data (for example traffic or other location data, sensitive, and judicial data) is performed in compliance with the specific requirements that the applicable data protection legislation sets forth for said categories of data.

- *Access limitation*: The systems/services should provide for an authorization procedure that entails differentiated levels of access to the data and also for recording the accesses to the data.

- *Data storage*: The systems/services should be able to automatically delete (or make anonymous) the data when the pursued processing purpose is reached or in case of elapse of the data retention periods specified under applicable legislation.

- *Dissemination of data to third parties*: When components of the service logic are outsourced to third-party providers and, in this context, personal data are disseminated for being processed, the systems/services should be able to provide certain guarantees that the consequent processing of information complies with the underlying fair data practices and the contract with the data subject.

- *Transfer of data to third countries*: The data dissemination principle described above applies especially when data are transferred to third countries, possibly with essentially different legislation regarding personal data collection and processing. The systems/services should be able to provide for compliance with the specific provisions ruling on transfer of data. For instance, consider the Safe Harbour Principles regulating data transfer between the European Union and the United States (19).

- *Supervision and sanctions*: The competent Privacy Authority should be provided with the means for supervising and controlling all actions of personal data collection and processing.

- *Lawful interception*: The competent Privacy Authority should be provided with the means to perform interception only when this is allowed by applicable laws and regulations and according to the conditions therein set forth. The necessary "hooks" for the lawful interception should under no circumstance become available to other not authorized third parties.

14.2.2 Technical Challenges

"The machine is the problem: the solution is in the machine" according to Poullet (44) and, following this principle, context-aware systems and services should be built in order to meet certain technical challenges that stem from the legal and regulatory requirements as outlined above. Elaborating the fundamental principles, the following challenges are derived:

- *The need for access control*: Any privacy violation certainly includes illicit access to personal data and, intuitively, access control constitutes a fundamental aspect of privacy protection. Beyond the traditional access control models, such as the well-adopted Role-Based Access Control (RBAC) (24), the incorporation of features that are specific to privacy protection has resulted in the emergence of the research field referred to as Privacy-Aware Access Control. The next section will provide a brief survey of the area.

- *Privacy-aware information flows*: The trend of outsourcing components of services leads to the execution of a single business process by a number of collaborating organizations, creating virtual organizations and "globalized" data streams. In fact, Web Services particularly rely on and foster this concept. Therefore, the introduction of privacy awareness in these complex and leakage-prone information flows constitutes a challenge of paramount importance.

- *The importance of semantics*: Evidently, the particular type characterizing each data item constitutes a parameter of significant importance for the determination of the procedures to be applied on the data and, consequently, of the adaptation of a context-aware service behavior. The semantics of the service itself, the underlying collection and processing purposes, as well as the entities involved, are of equal importance and altogether are part of what we call the "privacy context."

- *Complementary actions*: In several cases, access to the data should be accompanied by certain behavioral norms of the system or service. These are often referred to in the literature as "privacy obligations" (14) and include the interaction with the data subjects and/or the Authorities when mandated by the legislation, as well as the enforcement of data retention provisions.

- *The role of the users*: The users are granted certain rights, including the right to be informed regarding the collection or processing of personal data, to be asked about their explicit consent, to access their data. Additionally, they should be able to specify their privacy preferences, which can be later taken into account in the service execution. This way the users affect the service provision procedure, with respect to privacy.

- *The role of the Authorities*: The legislation grants the Personal Data Protection Authorities with certain rights and competences. These include the notification of the Authority, the supervision of the procedures and the means for performing Lawful Interception of data. As a result, the Authority should be able to interact with the system.
- *Adequacy of security protection means*: The deployment of strong information and network security mechanisms always constitutes the bottom line for personal data protection. The associated technologies are "neutral," i.e., the systems and services comprising the focal point of this chapter are not characterized by any particular security requirements. Given this neutrality, as well as the technical maturity of the contemporary security means, this chapter will take their availability as granted.

14.3 Related Work

Several approaches have been recently documented that either directly target the protection of privacy during the provision of context-aware services or have a significant, yet side impact in the field. Section 14.3.1 briefly introduces the former, while Sections 14.3.2 and 14.3.3 focus, respectively, on proposed technologies for privacy-aware access control and privacy preferences specification, since such technologies can be essentially exploited for protecting privacy in WS.

14.3.1 Privacy and Web Services

An interesting case suitable for service provision in organizations is described in (4). The authors propose a framework towards the protection of the sensitive user data in the interaction with Web Services. The procedure exploits SemanticLIFE, an information management system that allows the acquisition and storage of data in a semantic form through the use of ontologies. The service requester may or may not access specific properties depending on privacy policies imposed by the organization that offers the services, whereas he/she has the right to define further policies for specific services related to personal information. A further approach that combines ontologies and user preferences in the WS technology is presented in (50), where user preferences are described in DAML-S (42) and agents are used in the negotiation between the user and the service regarding the amount of information to be revealed.

The work of (28) introduces e-Wallet in the provision of context-aware services. E-Wallet includes a number of privacy rules specified by the user regarding access control properties for different conditions and data abstraction.

The privacy preferences of the user form part of a wider preference set, where the users specify different properties based on different contextual ontologies. An editor that enables users to create, edit, and delete their privacy preferences expressed as rules in OWL format (55) is also proposed by the authors. The editor, which is suitable for advanced users, is linked to the ontologies contained in e-Wallet.

Other approaches to privacy preservation refer either in general to WS protection mechanisms (e.g., (45)) or to context-aware services or applications taking into account general or specific context properties, such as the group of Location Based Services (LBS), where pseudonyms can be applied (31). In fact, the protection of location privacy currently comprises a field of intense research and diverse associated mechanisms have been proposed. A comprehensive survey (6) has identified identity, position, and path privacy as the three categories of location privacy and has classified the protection techniques in anonymity-, obfuscation- and policy-based. The latter form the basis of a trend referred to as Location-Based Access Control (LBAC), which is discussed along with several other access control approaches in the next section.

14.3.2 Access Control Solutions

Location-Based Access Control (LBAC) systems integrate traditional access control mechanisms with access conditions based on the physical position of users and other attributes related to the users' location (5). For instance, in (17) several issues related to LBAC are presented, while dynamic context representations and their integration with approaches to access negotiation in mobile scenarios are discussed.

Beyond location, access control constitutes as aforementioned a fundamental aspect of privacy protection and, in that respect, there are several approaches for privacy-aware access control, comprising now a quite mature research area. A milestone in the area has been the concept of Hippocratic Databases (2). Nevertheless, since they target stored data and make use of database tables for their operation, this approach, as well as other database-oriented ones (11), are not suitable for context-aware systems.

The provision of the automation of the privacy policies enforcement has been the focus of several approaches, including the ones proposed by IBM (9, 10), OASIS (40), and Hewlett Packard (14, 15), the concept of Purpose Based Access Control (13) and the family of Privacy-aware Role Based Access Control Models (P-RBAC) (38). All these frameworks enhance traditional Role-Based Access Control models (24) with additional, privacy-related aspects, such as the purpose for data collection and the automation of retention periods' enforcement, in order to put in place privacy-aware access control. Recently, some similar approaches with the additional feature of being grounded on a semantic basis have been proposed (e.g., (26, 39, 36, 25)) identifying the importance of contextual semantics.

14.3.3 Privacy Preferences Formalization

To some extent, the basis for the privacy-aware access control models has been the Platform for Privacy Preferences (P3P) (53) and its user-side complement called A P3P Preference Exchange Language (APPEL) (54). On the other hand, P3P and APPEL have been exploited in other related contexts; in (18) users are given with more flexibility regarding the control over privacy issues through preferences and weighting mechanisms. Preferences are expressed through data practices extracted from P3P privacy policies, whereas weights are applied on the specified user rules in order to enable dynamic decision making on privacy preferences for different context situations. The amount of privacy threat expected in each situation is computed through a rule engine. Other documented approaches for enabling users to formally express their privacy preferences include (3) and (34), while in (47), the authors describe an approach for intelligent automatic form filling for context-aware services, which is based on user's preferences.

14.4 Privacy Context

In context-aware services, contextual information is evaluated at execution time in order for the service to be dynamically adapted to the environment and situation reflected by the context. In that respect, privacy-related regulatory provisions and user preferences can become part of the context and be used for injecting privacy-awareness to the service. Along this line, we introduce here the concept of "privacy context," comprised of two different sources: the privacy preferences that should be integrated to the user contextual profile and a source of privacy policies that represents fair information practices stemming from the legislation. Considering the importance of the underlying semantics, a semantic information model is firstly introduced, serving as the domain for defining privacy context.

14.4.1 Semantic Information Model

As discussed in Section 14.2, the specific type characterizing a data item constitutes a critical parameter for its treatment. Consequently, a semantically uniform formal way for describing the specific type of every possible piece of personal information should be considered. The consideration of the notion of "purpose" in order to reason upon an access decision is typical to the family of privacy-aware access control models. That is, similarly to the data types, a common vocabulary of the different purposes behind information collection and processing is deemed necessary.

The most convenient way to formally express the notions of data and purpose types is by means of an ontology. This gives high expressive power of formalisms used to describe the underlying concepts. Additionally, the use of an ontology enables the semantic interoperation or integration with existing models for context-aware systems. In fact, there are many approaches in the literature that have used ontologies as the means for describing context (e.g., (49, 57)). On the other hand, in context-aware WS, purpose is essentially represented by the semantic identity of the service; the advent of semantic Web Services (37) provides a good ground for the specification of a semantic graph of the services to become part of the model.

The semantic information model should be complemented by the representations of the different entities involved in a service provisioning chain. Essentially, this implies the integration of the roles' structures that constitute the basis of the role-based models; as has been mentioned in Section 14.3, modern access control models have started adopting semantic technologies.

Putting together these ideas, the privacy context should be grounded on an ontology comprised of three classes:

- *ContextualData*, containing the semantic representations of the different types of contextual data.
- *Services*, reflecting the purpose of data collection and processing.
- *Roles*, representing the different entities involved.

These semantic subgraphs comprising the ontology should be organized in hierarchies in order to enable the inheritance of rules and properties among the contained elements. For all three subgraphs, there should be both AND- and OR-tree hierarchies. In addition, the instances of the *ContextualData* class must incorporate a relationship that expresses the detail level of the same concept, e.g., location. For instance, when a user asks for a location-based weather forecast, giving his/her exact position (e.g., geographical coordinates) is rather redundant, since less accurate data would suffice. These relationships between the instances of the classes can be implemented by means of OWL object properties (55).

14.4.2 User Preferences Specification

Apart from the established data practices governing the execution of a service's logic, the users/data subjects should have an active role in determining the fate of their data. In that respect, a technical problem to be approached is how to enable the user to that direction, i.e., to control the disclosure, storage, and processing of personal information, when this information is traveling through the various system and service components. The privacy preferences should become part of the "privacy context," i.e., integral to the user's contextual profile. There exist several specifications for user profile descriptions, such as the Liberty Alliance Personal Profile (33) and the earlier 3GPP Generic User Profile (1), as well as several individual research works like the user profile used in the IST Simplicity project (46, 30); while they do not explicitly address the issue of privacy preferences description, they can be leveraged and extended with the ideas for the preferences' specification and distribution along the composite WS chain that are introduced here. Under this perspective, the user preferences incorporate information related with:

- Disclosure permission and level of disclosure for different user properties (no disclosure, conditional disclosure, unconditional, unlimited, etc.).
- Further restrictions on disclosure properties (such as spatial and temporal constraints).
- Data processing and dissemination control, including data retention, abstraction, and modification.

The main preferences schemes on data disclosure can be specified in the form of allow/block properties. When the user's desire is to allow the disclosure of specific contextual data to specific services or service providers, appropriate "allow" elements are defined, appointing the corresponding data type, service type, and recipients. In the contextual data contained in these elements the user can specify a high-level data category or more specific types. The use of ontologies is essential, since ontologies capture the nature and the meaning of the data provided and various reasoning operations can be performed. In all cases the categorization depends on the information space that needs to be captured. However, coping with the detail level of a model such as the one described above would result in high complexity for the user and, therefore, a more general categorization that could be managed by a simple user interface (e.g., the "Privacy Badge" (29)) should be exploited. Such a categorization can be represented by the first ontological layer, including – among others – the following elements:

- Identity data (e.g., name, date of birth, marital status, etc.)
- Data related to occupation and education
- Location data
- Communication data
- Health data
- Financial information (e.g., credit card details)
- Personal preferences (e.g., preferences regarding travel destinations, book genre, etc.)

In the description of the service end recipient, the user can similarly specify a service category or service name or service provider or all of them depicting which services or service providers he/she trusts for the disclosure of his/her context details. The same process is followed for the case where the user wants to block specific properties by indicating the specifics of the context data and the service elements.

As a supplement to the above it is also possible to specify the way in which each contextual data element will be used. To this end, the user can include information on data retention, which can vary among *No retention*, *Indefinite*, *Law defined*, and *User defined*. In the latter case, the user also specifies the specific retention period desired. In some cases it is useful to allow the context data to arrive to the service provider in an abstract form (e.g., the user may allow a service to know whether he/she is above 18 years old, but at the same time forbid the disclosure of the exact age). For this purpose, further properties may include whether data modification and abstraction is allowed or not, whereas different abstraction levels may be applied (e.g., full or as abstract as it gets, one level up, etc.). Nevertheless, if an abstraction level is given, it is presupposed that the WS has taken into account in its functionality the case, where the data may be provided in different forms. Otherwise, the service provision will fail due to this type of restriction defined by the user.

Further restrictions can be applied to each allow field in the form of separate condition elements, where the user can specify temporal or/and spatial restrictions that formulate a geospatial space about when/when not and where/where not the contextual data will be visible to the specified service ones (e.g., a user may desire the available services to gain the exact location information, only when the location lies somewhere in his/her birthplace and block location views, when he/she is on vacations).

Various data stemming from the above generic preferences and rules can be incorporated in an XML Schema or in ontological representations. A simple example case in XML format is depicted in Figure 14.1. In the example, two "allow" properties are used specifying that: 1) the user wishes to disclose location information, only when he/she is in Madrid and only during working hours, and 2) all age-related information will be abstracted to the top level when provided to a service without allowing any data retention. A further block property denies the view of any personal user preferences. No service data are specified demonstrating that the allow/block elements are applicable to all services.

Of course, the user should have the right to edit the corresponding preferences at any time, leading to a reevaluation procedure by the system operating as privacy enforcer. The privacy enforcer is responsible for receiving the user preferences and incorporating them in the WS provision.

```
<?xml version="1.0" encoding="UTF-8"?>
<upp:privacyPreferences xmlns:upp="http://www.privacy.org/UPP">
  <upp:allow>
    <upp:dataEnd dataCategory="Location"/>
    <upp:condition>
      <upp:timePeriodIn>
        <upp:startWeekDay>Monday</upp:startWeekDay>
        <upp:endWeekDay>Friday</upp:endWeekDay>
        <upp:time>
          <upp:startTime>08:00:00</upp:startTime>
          <upp:endTime>18:00:00</upp:endTime>
        </upp:time>
      </upp:timePeriodIn>
      <upp:spaceLimitIn>
        <upp:Country>Spain</upp:Country>
        <upp:City>Madrid</upp:City>
      </upp:spaceLimitIn>
    </upp:condition>
  </upp:allow>
  <upp:allow>
    <upp:dataEnd dataCategory="PersonalInfo" dataName="age"/>
    <upp:use>
      <upp:retention>NoRetention</upp:retention>
      <upp:abstraction>Full</upp:abstraction>
    </upp:use>
  </upp:allow>
  <upp:block>
    <upp:dataEnd dataCategory="PersonalPreferences"/>
  </upp:block>
</upp:privacyPreferences>
```

Figure 14.1: User privacy preferences example in XML format.

14.4.3 Privacy Policies Provider

The second privacy-related context source proposed by this chapter concerns a provider of privacy policies that have their roots at the legislation and reflect fair information practices. Indeed, as previous research has discussed (e.g., (35)), the personal data protection legislation can be modeled by means of fine-grained semantic access control rules with a high degree of granularity. In that respect, there can be a WS specifically devised for playing the role of being the contextual source of privacy-aware access control rules.

The privacy policies should be grounded on the ontology described above. Essentially, each rule must be an object that links together a {*data type*, *service type*, *role type*} triad and assigns to it access control properties. More specifically, each rule should define:

- Provisions regarding the "read" and "write" access rights of the *role type* over the *data type* in the context of the provision of the *service type* under consideration.

- The period for which the data should be retained.

- Potential complementary actions that should accompany the enforcement of the rule; these may include the information of the user or/and the Privacy Authority about an event, the initiation of an interactive procedure for asking the explicit consent of the user, some activity logging, etc.

- Meta-properties, regarding the inheritance of the rule to the descendants of the specified *data type*, *service type*, and *role type*.

- Whether the provisions of the rule override the corresponding privacy preferences of the user, if any.

The rules can be implemented as instances of an additional class of the afore-described ontology, with OWL object and annotation properties serving for the definition of the rule's body. For instance, Figure 14.2 illustrates an example of an access control rule. What this rule states is that "when the service under consideration is an adult service (*AdultServices*), and when the service provider (*ServiceProvider*) requests access to the personal data of *IsAdult* type (a binary data type, reflecting whether the user is an adult or not), the data should be given to the provider, while the data should not be further retained. The rule applies for the descendants of the *AdultServices* service type, while it does not apply for the descendants of the *IsAdult* data type and of the *ServiceProvider* role type."

Something that should be noted here is the side impact that the concept of privacy context and especially the provision of privacy policies in this manner would have to the WS providers. A significant problem that the latter confront with concerns compliance with legal regulations. The providers seem to be helpless in their effort to compile and put in practice unclear or incoherent data protection laws. Moreover, privacy laws are frequently complemented by explicit obligations for registering and retaining additional data beyond the ones needed for operational purposes, in order for other provisions to be enforced (e.g., counterterrorist laws), thus creating a "regulatory jungle" (43). The cost of compliance as well as the administrative burden imposed can be onerous. According to the Eurobarometer (20), the lack of knowledge of the legislation constitutes the main reason for explaining the disrespect of the data protection law by the European organizations, while another major reason is the fact that the adaptation to the new requirements of the data protection is very time consuming. The European organizations indicate that the best way to improve and simplify the implementation of the legal framework lies in the provision of further clarification on the practical application of its key definitions and concepts. It should be further noted that in this context, the

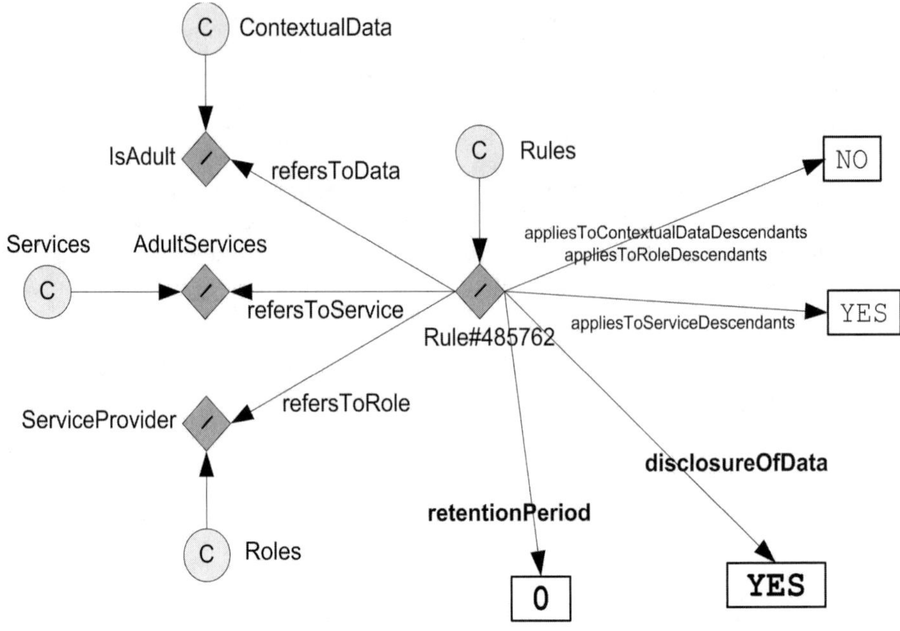

Figure 14.2: Example of access control rule.

small- and medium-sized organizations have a competitive disadvantage, since typically they do not have the resources to develop the technical means for implementing privacy protection standards, nor are able to conduct extensive data protection audits. The concept of privacy context described here would provide the WS providers with the desired degree of automating the enforcement of fair information practices.

14.5 Enforcement Framework

Taking into consideration the fundamental privacy principles and requirements, as well as leveraging the existing technologies, a framework for the enforcement of privacy protection in context-aware WS is presented. As made clear in Section 14.4, the approach lies on a twofold basis: 1) the access and use of information resources should comply with the well-established privacy rules and user preferences, and 2) the resulting service provision needs to be compatible with the respective privacy policies. The above can be seen as some sort of privacy configuration based on privacy context that accompanies the context-aware WS provision and execution. The service needs to comply with the principles indicated, in order to foster the service delivery in a privacy-aware fashion respecting the needs and potential restrictions imposed by the user and the international regulations.

The architecture of a generic framework for privacy enforcement is illustrated in Figure 14.3. It includes a number of actors participating in the interaction between the user or service requester and a number of WS providers. The component denoted as "privacy enforcer" is the main mediator between the service requester and the service providers, assigned with the task of injecting privacy-aware behavior in the service provision procedure, which may also include interactions between different service providers. It takes into account in its functionality the two privacy-related context sources as also indicated above: the privacy preferences of the end-user and the policies provider, whose separate roles were detailed in the preceding section. Based on the configuration parameters imposed by these privacy entities, the privacy enforcer takes the appropriate actions, in order to retrieve the necessary context information from the individual sources and assist the context awareness mechanism in the provision of a context-aware WS. In other words, instead of having a direct communication link between the context awareness mechanism, the user and the service providers, the privacy enforcer comes among them in order to verify that all ends respect the technical requirements introduced by the legislation and the preferences specified by the user.

It should be noted that the mechanism responsible for injecting context-awareness into the WS behavior is independent from the privacy-related entities. The aim of the framework is to provide the adequate level of abstraction and flexibility, in order to support the integration with various adaptation mechanisms as described in the next section. In that respect, in the generic framework architecture the context awareness mechanism acts as a "black box" that may communicate with the WS end, the context properties or both. When it comes to specific implementations, each role may be assigned to separate nodes in the network, whereas it is also possible to assign the functionalities of more than one entities to the same node depending on the chosen architectural pattern. From the end-user point of view, the functionality of the privacy enforcer and the context awareness mechanism is seamlessly integrated with the WS provision leaving the user unaware of the details of the procedures that take place prior to the service delivery.

14.6 Combination with Context Adaptation Schemes

Taking into account the nature of Web Services, the presented configuration scheme can be combined with various context adaptation techniques

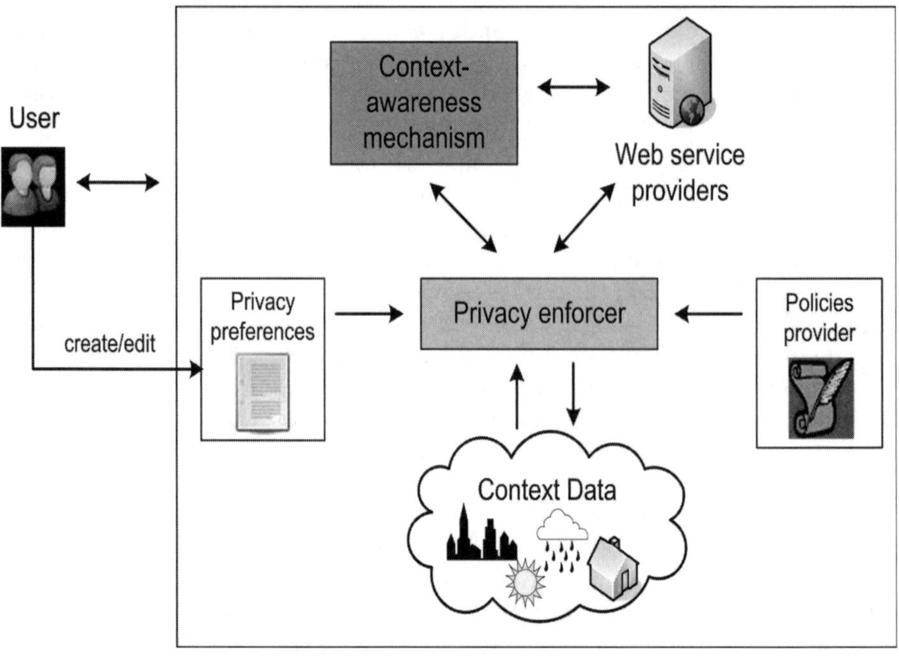

Figure 14.3: Main entities of the privacy enforcement framework.

(e.g., based on message interception or dedicated adaptation frameworks), regardless of whether the adaptation takes place on the server side, the client side or on a mediator entity. In general, the main decision that is taken in context-aware services is related with whether the procedure resulting in the service adaptation to context information takes place natively in the service main functionality or is linked to the service through external mechanisms that do not affect the main service logic.

In the former case, the context data acquisition is performed in the main service code through the WS itself or through external services that communicate directly with the initial service. In other words, the WS communicates with context sources and includes the appropriate mechanisms related with the privacy enforcer resulting to a common service provider/privacy enforcer entity. This case is feasible assuming that the service requester trusts the specific service provider and allows the service end to decide internally on the appropriate use of context information exploiting the functionalities offered by the user preferences and the policies provider.

On the other hand, the latter case leaves the main service logic intact and applies a number of external mechanisms that adapt the WS to the contextual situation. In this case, the privacy enforcer acts externally and is not incorporated in the service. In order to perform this kind of context injection to the service functionality in the field of WS message interception techniques are usually exploited. These techniques include the interception and the appropriate modification of the exchanged request and response messages, which usually take the form of SOAP envelopes.

As mentioned above, both of these cases are combined with the privacy enforcement principles at different levels in an attempt to guarantee that context awareness respects the user preferences and the privacy policies. In order to illustrate how this combination can be achieved some more specific directives are described next. These directives take the form of a framework for context-aware WS, which uses message interception. The framework architecture serves the provision of context-aware WS in a privacy preservation fashion, where access to context resources and the provision of the service follow the formally defined rules and preferences.

The communication among the main elements participating in the architecture is illustrated in Figure 14.4. The service request and response messages travel back and forth between the end-user and the service provider encapsulated in SOAP messages that include various information on the user necessary for the service adaptation to context. Through the introduction of appropriate message handler(s) these messages can be altered in order to reflect the contextual situation. However, this modification can be performed in a privacy-aware fashion by incorporating also the functionality of the privacy enforcer in the handler(s). The privacy enforcer portion of the handler(s) based on the information retrieved from the user preferences and the respective policies provider performs the tasks of context data acquisition through the supplementary context sources, and message adaptation based on the contextual space. After having modified the request message, it provides the service with the appropriate input fulfilling any constraints imposed by privacy-aware context sources. The same procedure can be followed for any modification that may be performed on the response message provided by the WS provider after the service execution.

The advantage of mechanisms that rely on message interception is that they remain totally independent of the main service logic. This way the WS can function without interruption with or without the existence of mechanisms for context-awareness and privacy guarantees. An example of research work on message interception and context awareness in WS can be found in (32).

14.7 Conclusions

Recently, the sphere of the electronic services' provision has been expanded. Among the underlying prominent technologies, the advent of context awareness creates impressive perspectives of rich, highly personalized and coherent

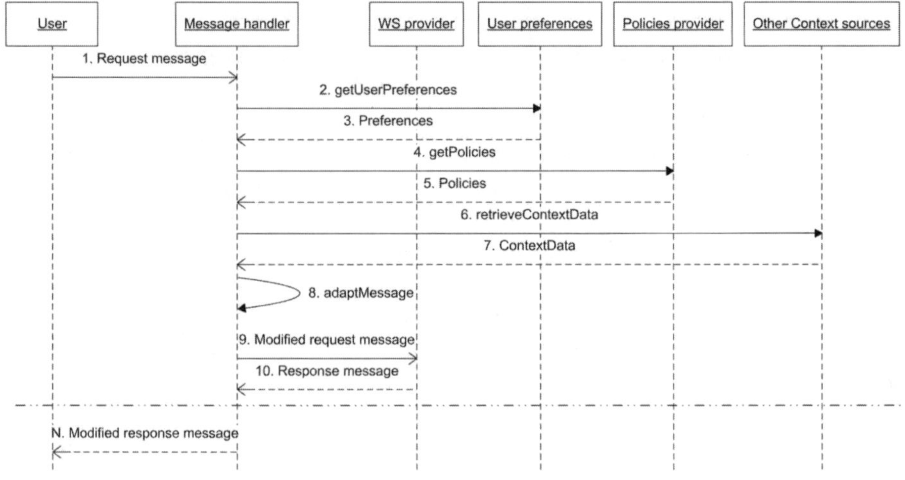

Figure 14.4: Sequence diagram for context adaptation and privacy enforcement through message interception.

services, information and computation ubiquity and, thus, spurs an information revolution that has the potential to bring significant improvements to the quality of life. Nevertheless, the risks for invasion of privacy constitute the flip side of context-aware services provision benefits. Ron Rivest's *reversal of defaults* seems to have started being realized: "what was once private is now public; what once was hard to copy is now trivial to duplicate; what was once easily forgotten is now stored forever." In that respect, the architectures fostering context awareness should be rethought so as to incorporate privacy preservation in their design and make the legal and regulatory provisions an integral part of services.

This chapter has delved into the need for privacy awareness in the provision of context-aware Web Services. It has presented a review of the legislation principles and the technical requirements that emerge and has presented some solution ideas that can be applied in different aspects of the provision cycle. From an architectural point of view, the main entity participating in this procedure is given in the form of a privacy enforcer responsible for the integration of privacy principles in the service function. Two privacy-related context sources are exploited: on the one hand, the user privacy preferences, specified for different context data and service properties, are becoming part of the user's contextual profile; on the other hand, a privacy policies provider offers new ground in terms of compliance with the legislation. The mechanisms described in this chapter can be implemented in different ways in specific context-aware architectures for providing privacy-aware functionality.

The issue of privacy protection is vital in context-aware services and the current analysis can serve as a useful reference for the potential developer and architect in the field. Nevertheless, there are several remaining challenges that have not been addressed by this chapter and constitute interesting research topics in the area. Among them, a prominent position is held by the issue of adapting the provision of a service to data that are incomplete. Indeed, as a result of privacy preferences enforcement, the available data can be reduced to an extent that makes them not sufficient and adequate for the execution of the underlying service's logic. That is, a dichotomy between privacy and service usability is created or, in other words, there is a trade-off between the enforcement of fair data practices and the provision of advanced context-aware WS. Therefore, an important aspect that should be considered concerns the adaptation of a service to operate with different amounts of information and information that is characterized by different levels of detail. For instance, a service can be adjusted to be functional under varying resolutions of location data or, with respect to accounting, to be able to exploit a number of alternative anonymous accounting mechanisms. These issues of data incompleteness and data-dependent service adaptation constitute our current research directions.

References

[1] 3GPP. 2007. Technical Specification 3rd Generation Partnership Project. Technical Specification Group Services and System Aspects. 3GPP Generic User Profile – Architecture, Stage 2, Sophia-Antipolis.

[2] Agrawal, R., Kiernan, J., Srikant, R. and Xu, Y. 2002. Hippocratic databases. *Proceedings of the 28th International Conference on Very Large Data Bases (VLDB'02).*

[3] Agrawal, R., Kiernan, J., Srikant, R. and Xu, Y. 2005. XPref: A preference language for P3P. *Computer Networks* 48(5):809-827.

[4] Ahmed, M., Anjomshoaa, A. and Tjoa, A M. 2008. Context-based privacy management of personal information using semantic desktop: SemanticLIFE case study. *Proceedings of the 10th International Conference on Information Integration and Web-based Applications and Services (iiWAS'08)*, 214-221.

[5] Ardagna, C. A., Cremonini, M., Capitani di Vimercati and S. De, Samarati, P. 2007. Privacy-enhanced location-based access control. In *Handbook of Database Security: Applications and Trends*, ed. M. Gertz, and S. Jajodia, Springer-Verlag.

[6] Ardagna, C. A., Cremonini, M., Damiani, E., Capitani di Vimercati and S. de, Samarati, P. 2007. Privacy-enhanced location services information. In *Digital Privacy: Theory, Technologies and Practices*, ed. Acquisti, De Capitani di Vimercati, Gritzalis, and Lambrinoudakis, Auerbach Publications (Taylor and Francis Group).

[7] Article 29 Data Protection Working Party. 2005. Opinion 5/2005 on the use of location data with a view to providing value-added services.

[8] Article 29 Data Protection Working Party. 2005. Working document on data protection issues related to RFID technology.

[9] Ashley, P., Hada, S., Karjoth, G., Powers, C. and Schunter, M. 2003. Enterprise Privacy Authorization Language (EPAL 1.2), `http://www.zurich.ibm.com/security/enterprise-privacy/epal/Specification` (accessed April 1, 2009).

[10] Backes, M., Pfitzmann, B. and Schunter, M. 2003. A Toolkit for Managing Enterprise Privacy Policies. *Proceedings of the 8th European Symposium on Research in Computer Security*, 162-180, Springer-Verlag.

[11] Bertino, E., Byun, J. and Li, N. 2005. Privacy-preserving database systems. In *Foundations of Security Analysis and Design III*, ed. A. Aldini, R. Gorrieri, and F. Martinelli, 178-206, Springer-Verlag.

[12] Birnhack, M. D. 2008. The EU data protection directive: An engine of a global regime. *Computer Law and Security Report* 24(6):508-520.

[13] Byun, J.-W., Bertino, E. and Li, N. 2005. Purpose based access control of complex data for privacy protection. *Proceedings of the 10th ACM Symposium on Access Control Models and Technologies*, 102-110. ACM Press.

[14] Casassa Mont and M. 2004. Dealing with privacy obligations: Important aspects and technical approaches. In *Trust and Privacy in Digital Business*, ed. S. Fischer-Huebner, S. Furnell and C. Lambrinoudakis, 120-131, Springer-Verlag.

[15] Casassa Mont, M. and Thyne, R. 2006. A systemic approach to automate privacy policy enforcement in enterprises. *Proceedings of the 6th Workshop on Privacy Enhancing Technologies*, 118-134.

[16] Cranor, L. F. 2004. I bidn't buy it for myself. In *Designing Personalized User Experiences in E-Commerce*, ed. C.-M. Karat, J.O. Blom, and J. Karat, 57-73, Norwell, MA, USA: Kluwer Academic Publishers.

[17] Damiani, E., Anisetti, M. and Bellandi. V. 2005. Toward exploiting location-based and video information in negotiated access control policies. *Proceedings of the 1st International Conference on Information Systems Security (ICISS'05)*, 21-35, Springer-Verlag.

[18] Eldin, A. A. and Wagenaar, R. 2006. A privacy preferences architecture for context aware applications. *Proceedings of the IEEE International Conference on Computer Systems and Applications*, 1110-1113, IEEE Computer Society Press.

[19] European Commission 2000. Commission Decision of 26 July 2000 pursuant to Directive 95/46/EC of the European Parliament and of the Council on the adequacy of the protection provided by the safe harbour privacy principles and related frequently asked questions issued by the U.S. Department of Commerce, *Official Journal of the European Communities* L 215:7-47.

[20] European Omnibus Survey "EOS Gallup Europe." 2003. "European Union companies' views about privacy." Flash Eurobarometer EB 147, `http://ec.europa.eu/public_opinion/flash/fl147_data_protect.pdf` (accessed June 6, 2009).

[21] European Parliament and Council. 1995. Directive 95/46/EC of the European Parliament and of the Council on the protection of individuals with regard to the processing of personal data and on the free movement of such data, *Official Journal of the European Communities* L 281:31-50.

[22] European Parliament and Council. 2002. Directive 2002/58/EC of the European Parliament and of the Council concerning the processing of personal data and the protection of privacy in the electronic communications sector (Directive on privacy and electronic communications). *Official Journal of the European Communities* L 201:37-47.

[23] European Parliament, Council, and Commission. 2000. Charter of fundamental rights of the European Union. *Official Journal of the European Communities* C 364:1-22.

[24] Ferraiolo, D.F., Sandhu, R., Gavrila, S., Kuhn, R.D. and Chandramouli, R. 2001. Proposed NIST Standard for role-based access control. *ACM Transactions on Information and System Security* 4(3):224-274.

[25] Ferrini, R., Bertino, E. 2009. Supporting RBAC with XACML+OWL. *Proceedings of the 14th ACM Symposium on Access Control Models and Technologies (SACMAT'09)*, 145-154, ACM Press.

[26] Finin, T., Joshi, A., Kagal, L., Niu, J., Sandhu, R., Winsborough, W. and Thuraisingham, B. 2008. ROWLBAC: Representing role based access control in OWL. *Proceedings of the 13th ACM Symposium on Access Control Models and Technologies*, 73-82, ACM Press.

[27] Gallup Organization. 2008. Data protection in the European Union: Citizens' perceptions – Analytical Report. Flash Eurobarometer 225, `http://ec.europa.eu/public_opinion/flash/fl_225_en.pdf` (accessed June 6, 2009).

[28] Gandon, F. L. and Sadeh, N. M. 2004. Context-awareness, privacy and mobile access: A web semantic and multiagent approach. *Proceedings of the 1st French-speaking conference on Mobility and ubiquity computing (UbiMob'04)*, 123-130, ACM Press.

[29] Gehring, S. and Gisch, M. 2008. The privacy badge revisited – Enhancement of a privacy-awareness user interface for small devices. *Proceedings of the Workshop on Security and Privacy Issues in Mobile Phone Use 2008 (SPMU'08)*.

[30] IST Simplicity project homepage, `http://www.ist-simplicity.org/`.

[31] Jorns, O. Jung, O. and Quirchmayr, G. 2008. A platform for the development of location-based mobile applications with privacy protection. *Proceedings of the 3rd International Conference on Communication Systems Software and Middleware and Workshops (COMSWARE'08)*, 120-127, IEEE Computer Society Press.

[32] Kapitsaki, G. M., Kateros, D. A. and Venieris, I. S. 2008. Architecture for provision of context-aware web applications based on web services. *Proceedings of the IEEE International Symposium on Personal, Indoor and Mobile Radio Communications (PIMRC'08)*, 1-5, IEEE Computer Society Press.

[33] Liberty Alliance Project. 2005. Liberty ID-SIS Personal Profile Service Specification, Version 1.1, http://www.projectliberty.org/liberty/content/download/1028/7146/file/liberty-idsis-pp-v1.1.pdf (accessed June 06, 2009).

[34] Lioudakis, G. V., Koutsoloukas, E. A., Dellas, N., Tselikas, N., Kapellaki, S., Prezerakos, G. N., Kaklamani, D. I. and Venieris, I. S. 2007. A middleware architecture for privacy protection. *Computer Networks* 51(16): 4679-4696.

[35] Lioudakis, G. V., Koutsoloukas, E. A., Dellas, N., Gaudino, F., Kaklamani, D. I. and Venieris, I. S. 2007. Technical enforcement of privacy legislation. *Proceedings of the IEEE International Symposium on Personal, Indoor and Mobile Radio Communications (PIMRC'07)*, 1-6, IEEE Computer Society Press.

[36] Lioudakis, G. V., Koutsoloukas, E. A., Dellas, N., Kapitsaki, G. M., Kaklamani, D. I. and Venieris, I. S. 2008. A Semantic Framework for Privacy-Aware Access Control. Proceedings of the *International Multiconference on Computer Science and Information Technology*, 813-820, IEEE Computer Society Press.

[37] McIlraith, S. A., Son, T. C. and Zeng, H. 2001. Semantic Web services. *IEEE Intelligent Systems* 16(2):46-53.

[38] Ni, Q., Trombetta, A., Bertino, E. and Lobo, J. 2007. Privacy aware role based access control. *Proceedings of the 12th ACM Symposium on Access Control Models and Technologies*, 41-50. ACM Press.

[39] Noorollahi Ravari, A., Amini, M. and Jalili, R. 2008. A Semantic Aware Access Control Model with Real Time Constraints on History of Accesses. Proceedings of the *International Multiconference on Computer Science and Information Technology*, 827-836, IEEE Computer Society Press.

[40] Organization for the Advancement of Structured Information Standards (OASIS). 2004. OASIS eXtensible Access Control Markup Language (XACML) TC. http://www.oasis-open.org/committees/xacml/ (accessed April 1, 2009).

[41] Organization for Economic Co-operation and Development. 1980. Guidelines on the Protection of Privacy and Transborder Flows of Personal Data, http://www.oecd.org/document/18/0,3343,en_2649_34255_1815186_1_1_1_1,00.html (accessed May 9, 2009).

[42] Paolucci, M. and Sycara, K. 2003. Autonomous semantic Web services. *IEEE Internet Computing* 7(5):34-41.

[43] Pletscher, T. 2005. Companies and the regulatory jungle. *Proceedings of the 27th International Conference of Data Protection and Privacy Commissioners.*

[44] Poullet, Y. 2006. The Directive 95/46/EC: Ten years after. *Computer Law and Security Report* 22(3):206-217.

[45] Rezgui, A., Ouzzani, M., Bouguettaya, A. and Medjahed, B. 2002. Preserving privacy in Web services. *Proceedings of the 4th International Workshop on Web Information and Data Management*, 56-62, ACM Press.

[46] Rukzio, E., Prezerakos, G. N., Cortese, G., Koutsoloukas, E. and Kapellaki, S. 2004. Context for simplicity: A basis for context-aware systems based on the 3GPP generic user profile. *Proceedings of the International Conference of Computational Intelligence (ICCI'04)*, 466-469.

[47] Rukzio, E., Schmidt, A. and Hussmann, H. 2004. Privacy-enhanced intelligent automatic form filling for context-aware services on mobile devices. *Proceedings of the Workshop Artificial Intelligence in Mobile Systems (AIMS'04).*

[48] Solove, D. J. 2006. A brief history of information privacy law. In *Proskauer on Privacy: A Guide to Privacy and Data Security Law in the Information Age*, ed. C. Wolf, 1-46, Practising Law Institute.

[49] Strimpakou, M. A., Roussaki, I. G., and Anagnostou, M. E. 2006. A context ontology for pervasive service provision. *Proceedings of the 20th international Conference on Advanced information Networking and Applications (AINA'06).*

[50] Tumer, A., Dogac, A., Toroslu and I. H. 2005. A semantic-based user privacy protection framework for Web services. *Intelligent Techniques for Web Personalization*, 289-305, Springer-Verlag.

[51] United Nations. 1948. Universal Declaration of Human Rights, `http://www.ohchr.org/EN/UDHR/Documents/60UDHR/bookleten.pdf` (accessed June 9, 2009).

[52] U.S. Public Law No. 93-579. Dec. 31, 1974. 5 U.S.C. 552a.

[53] W3C. 2002. Platform for Privacy Preferences 1.0 (P3P1.0) Specification. `http://www.w3.org/TR/2002/REC-P3P-20020416/` (accessed June 09, 2009).

[54] W3C. 2002. A P3P Preference Exchange Language 1.0 (APPEL1.0). `http://www.w3.org/TR/2002/WD-P3P-preferences-20020415` (accessed June 9, 2009).

[55] W3C. 2004. Web Ontology Language (OWL). `http://www.w3.org/2004/OWL/` (accessed June 09, 2009).

[56] Warren, S. D., Brandeis and L. D. 1890. The right to privacy. *Harvard Law Review* 4(5):193-220.

[57] Ying, X. and Fu-yuan, X. 2006. Research on context modeling based on ontology. *Proceedings of the International Conference on Computational Intelligence for Modelling, Control and Automation, and International Conference on Intelligent Agents, Web Technologies and Internet Commerce*, 188.

Chapter 15

A Knowledge-Based Framework for Web Service Adaptation to Context

Carlos Pedrinaci, Pierre Grenon, Stefania Galizia, Alessio Gugliotta, and John Domingue

Abstract. Web services provide a suitable level of abstraction and encapsulation for the development of applications out of reusable and distributed components. However, in order to increase the applicability of services as well as to fine tune their execution in particular situations, context-adaptive solutions are increasingly required. In this chapter we describe a generic knowledge-based framework for supporting the adaptation of Web services and Web service-based applications to context. The framework builds upon research in semantic Web services and Knowledge Engineering to support capturing contextual information in a way that is directly amenable to automated reasoning in order to support the recognition of relevant contextual information that is then used for selecting the most appropriate service or set of services to be executed.

15.1 Introduction

Since the appearance of Web services technologies and impelled by the global interest on Enterprise Application Integration (12) and Service-Oriented Computing (23), there has been much research and development devoted to better supporting the use of Web services as the core constituent of distributed applications. However, despite the appealing characteristics of service-orientation principles and technologies, their application remains mainly limited to large corporations and, effectively, we are still far from achieving a widespread application of service technologies over the Web. One aspect that is increasingly seen as a *condicio sine qua non* for achieving this is the capacity to dynamically adapt services based on contextual factors (17). These factors range from immediate concerns of location and language to legal issues and financial regulations. Swiftly accommodating to the context at hand will become increasingly important as the diversity of services expands with the global reach of the future Web of services.

Researchers focussing on Web services related technologies are beginning to consider aspects such as security, quality, trust and adaptability within the broader area of context-awareness (13, 16, 17). In fact given the broad meaning of context, characteristics such as low-level execution monitoring data or security and trust concerns are valuable and even necessary contextual information concerning services. Over time many different approaches to developing context-aware applications have been devised (4, 5, 14, 32). In most cases, applications were developed in an ad hoc manner and thus tailored to specific environments, domains, and purposes. As a consequence, from a general perspective the solutions proposed embed a certain set of trade-offs that prevent their systematic application across domains and environments. What can be distilled from this research is the complexity to design, develop and maintain context aware applications.

These difficulties have motivated the definition of frameworks and architectures for developing and supporting context-aware applications (5, 14, 32). These frameworks, however, need to support adapting services to particular contexts in a systematic manner in order to be adequate to the reality of the Web were one is likely to encounter a virtually infinite number of situations that need to be accommodated. We need, therefore, i) appropriate means for modelling contextual information based on a set of desirable requirements relating to the applicability of an information model, the support for comparing data, and the ability for inferring new data; and ii) a fully-fledged general purpose framework designed to provide an efficient, configurable, robust but nevertheless simple solution to context adaptation.

In this chapter we describe our approach for supporting the adaptation of services to context. We present a framework that is totally based on semantic technologies in order to provide a domain-independent solution. Application developers can use the framework for constructing their context-aware solutions by simply providing a set of domain-specific models that capture the degrees of flexibility the application should have with respect to different contexts and how it should react to changes. In the remainder of the chapter we shall first present an overall view of the framework describing in detail its core components. Additionally, a simple example is introduced in order to better illustrate the notions on which the framework builds upon.

15.2 Web Services Adaptation to Context: Overall approach

The contextual service adaptation framework described herein aims to provide a generic platform for supporting the adaptation of services to diverse contexts according to a virtually infinite variety of dimensions. In fact, our understanding of context is very much in line with perhaps the most widely agreed definition which is presented in (4):

> "Context is any information that can be used to characterize the situation of an entity. An entity is a person, place or object that is considered relevant to the interaction between a user and an application, including the user and the application themselves."

We here take a slightly wider understanding of context that is not limited to interactions between users and machines but also between machines themselves. After all, one of the main advantages service-oriented technologies provide is the capacity to compose existing services in order to provide more advanced ones.

In the light of this broad understanding of context, our approach to context adaptation is based on a set of conceptualisations providing the means for modelling contextual information of any kind, and a general purpose machinery able to manage contextual information, use it for recognising concrete contexts within particular situations, and apply this contextual knowledge for adapting services. Therefore, central to our approach, much like the definition of context itself, is the genericity of our framework so that it can cover the wide range of situations one is likely to encounter in a Web of billions of services.

Adapting services to particular contexts is envisaged as a process like the one illustrated in Figure 15.1. In a nutshell, this process involves gathering and deriving contextual information, recognising relevant contextual information given a concrete situation, and eventually adapting the execution of the service(s) based on the contextual knowledge gained. Although this process is essentially sequential, while trying to recognise contexts it might be required to acquire more information either from raw data or by deriving additional data on demand, and to trace the information obtained in order to determine how trustworthy it is. We shall not deal with the latter in this chapter.

Our approach to supporting the adaptation of services to contexts is based on previous research in the area of semantic Web services, Knowledge Engineering and Artificial Intelligence for modeling contextual information and for capturing the expertise for supporting Context Recognition and Service Adaptation in a domain-independent way. The kinds of adaptation currently contemplated are i) the use of *late binding* techniques for selecting at runtime the most suitable service to be executed based on contextual information; and

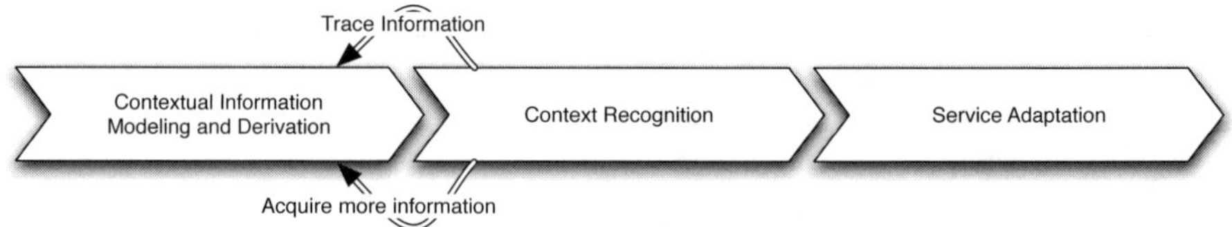

Figure 15.1: Web services adaptation process.

ii) the application of Parametric Design techniques to ensure that processes that use late binding for several activities are analysed such that services are locally contextualised and processes remain globally coherent and optimised. To this end, our framework is composed of three main generic services which support: i) Context Modeling and Derivation; ii) Context Recognition; and iii) Service Adaptation based on Context. In the remainder of this section we shall first describe IRS-III, the execution environment underlying our framework, and will then present the details of each of the services composing the adaptation framework.

15.2.1 IRS-III: a Semantic Execution Environment

Research in semantic Web services aims to increase the level of automation that can be achieved while discovering, selecting, composing, mediating, executing and monitoring Web services. Semantic Web services technologies make use of semantic annotations over existing services in order to support the application of automated reasoning to better support the aforementioned tasks. Several approaches have been proposed so far, the most prominent ones being WSMO (9), OWL-S (18) and WSDL-S (1) which was the basis for the only standard existing nowadays in the field, namely SAWSDL (7).

In our framework we build upon WSMO which provides ontological specifications for the core elements of semantic Web services. WSMO is based on four main building blocks, namely:

1. *Ontologies.* An ontology is an explicit formal shared conceptualization of a domain of discourse (11). Ontologies are used as data models throughout WSMO, so that all resource descriptions and the data exchanged during service execution are formally represented.

2. *Web Services.* Web services represent semantic models of traditional web services with formalised interfaces and capabilities.

3. *Goals.* Goals can be seen both as an abstraction over userÕs goals and as an abstraction over a set of Web Services. Web Services and Goals are the main abstractions relevant for the adaptation of services.

4. *Mediators.* Mediators support a proper integration –through data and process mediation– between any two WSMO elements while they ensure a strong decoupling between them.

Together with the conceptual work around WSMO, much effort has been devoted to implementing what we refer to as Semantic Execution Environments (22, 8). Semantic Execution Environments are systems that can process semantic Web services in order to support the (semi-)automated discovery, selection, composition, mediation, execution and monitoring of services. Work on Semantic Execution Environments is strongly based on WSMO as the conceptual model and is currently under standardisation within OASIS taking as starting point two reference implementations: WSMX (8) and IRS-III (6).

In our framework we build upon IRS-III, a framework for creating and executing semantic Web services, which takes a semantic broker-based approach to mediating between service requesters and service providers (6). IRS-III uses OCML for internal representation and incorporates at its core an OCML reasoner which supports both backward and forward-chaining reasoning (19). OCML is a frame-based language which includes support for defining classes, instances, relations, functions, procedures and rules. IRS-III includes its own implementation of WSMO and a set of core components for handling service discovery, invocation, choreography, orchestration and mediation. The reader is referred to (6) for more details.

A core design principle for IRS-III is to support capability-based invocation. A client sends a request which captures a desired outcome or goal and, using a set of semantic Web service descriptions, IRS-III will: a) discover potentially relevant Web services; b) select the set of Web services which best fit the incoming request; c) mediate any mismatches at the conceptual level; and d) invoke the selected Web services whilst adhering to any data, control flow and Web service invocation constraints. As the reader may realise, automating the transition from Goals to Web Services represents a relatively simple yet powerful technique for ensuring the most suitable solution is used taking contextual information into account. In the remainder of this chapter we shall describe the extensions that have been added to IRS-III in order to support the adaptation of services to contexts, covering context modeling and derivation, context recognition and finally Web services adaptation.

15.3 Context Modeling and Derivation

Context information is not any specific *kind* of information. Rather, for us, it is information that we assume is readily available or extractable from data and that is *contextually relevant for a task.* Thus the stake in context information gathering can be seen as holding onto efficient search space reduction.

This process has to be guided, however, and it is one role of ontologies to help structuring the information space and prepare for readily reduction. It is now widely recognised that context–aware computing can be supported in information systems using ontologies. (28, 2, 15). The role of ontologies in such systems however can be diverse.

We use ontologies for three conceptually separated purposes which are in truth related and complementary in their support to handling contextual information: i) expressing and structuring prospectively relevant information, ii) recollecting the selection criteria for a task, and iii) recollecting contextual information and generating contextual knowledge. These functionalities are supported by the application of problem solving methods in connection with ontological treatment of the relevant aspects feeding into these methods. The second point, in particular, ties into our use of heuristic classification (see section 15.4.1). Our framework handles these aspects modularly with i) ontologies having the general purpose of dealing with Web services (WSMO), temporal aspects and knowledge of that level, ii) dimensional ontologies dealing with quantities, units and their conversions, and other related aspects, and iii) a context modelling ontology dedicated to supporting ontological treatment of context and contextualistion. In the reminder of this section we describe the main ontological constructs needed to support our framework.

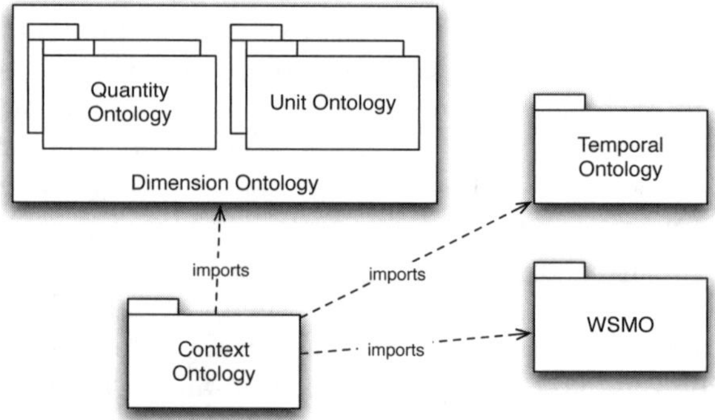

Figure 15.2: Modular ontologies each dedicated to different aspects and recollected within the context modelling ontology.

15.3.1 Background knowledge representation

Our system rests on knowledge–based representation of information about entities showing in the execution of web services. We thus apply generic ontological engineering techniques using OCML (19) as our representation language. As mentioned earlier, OCML is a frame–based language and it is used, typically, to define collective structures, *classes*, bearing slots (attributes) allowing to characterise *instances* of classes by associating values to slots. Thus in laying out the information about entities of interest (users, services and so on) we rely primarily on the slots & attribute-values mechanism. Against this background we focus on knowledge that can be collected by associating instances (for example, a particular user) with certain values (for example, an age on a given scale). Values in this sense are, of course, instances of certain classes and this forms the basis of our notion of dimension.

Our dimension ontology, shown in Figure 15.3, has at its core the class *dimension*. Figure 15.2 sketches the ontology in abstraction of its modularisation. There are different kinds of dimensions and in our dedicated ontology the class *dimension* has a number of specialisations. We can describe the typology of dimensions according to two main aspects: i) structural and ii) logical features. Structural aspects have to do with characteristics of the values that are members of dimensions. The most significant structural kinds for our purpose are, firstly, those of dimensions whose member values can be ordered and, secondly, those of dimensions whose member values are given in units. The definition of an ordered dimension is completed by the association of the dimension with an order relation that holds between its members. Because of this, a set of values can be common to two dimensions while the dimensions will remain distinct if the values are differently ordered. The definition of a dimension with values being given in units is less direct. Because dimensions are defined, partially, by a set of values, we define dimensional values *in abstracto* and independently of dimensions. The only thing we give to every dimensional value is a magnitude. We define a specialisation of *dimension value* that is instantiated by values whose magnitude is only meaningful in a unit. From such assemblies of values we can then define dimensions with units.

Thus, an instance of *dimension* is defined by: i) a number of (dimensional) values and ii) a slot that attaches these values to an entity[1]. The values in the dimension can be declared to belong in the dimension explicitly or they can be inferred when a test exists. Figure 15.4 shows an OCML listing defining the class *dimension* and examples of defined dimension for levels of trust. We can define two dimensions *LMH* and *MH* whose values are values of level of trust (for the sake of simplicity we ignore the associated slot). *LMH* and *MH* differ in that the former encompasses all levels of trusts that have been defined (hence we use as a membership test the unary-relation that holds of every levels of trust) and the latter is a more selective one for which the values are declared to include medium and high levels of trust only.

[1]The definition of a dimension 'mentions' the slot but it is a slot that characterises a class to which belong entities to be prospectively contextualised.

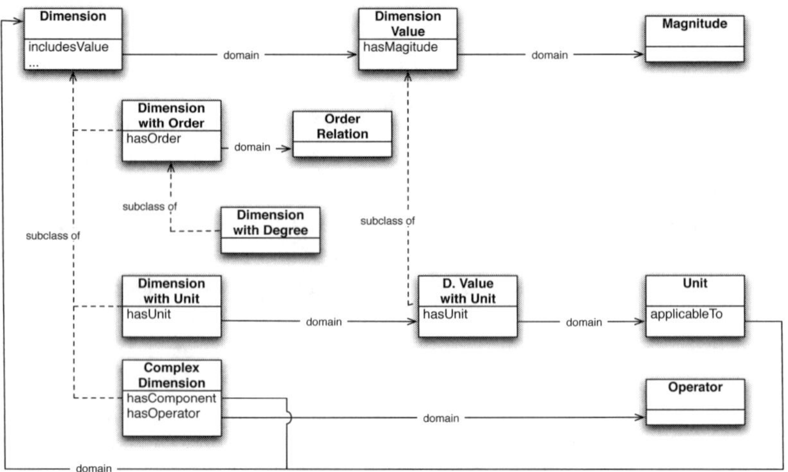

Figure 15.3: Main classes and slots with their domains in our dimension ontology—modularisation is not represented.

```
(def-class dimension ()
  ((dimension-includes-value :type dimension-value)
  (associated-slot :type slot :max-cardinality 1)
  (value-membership-test unary-relation)))

(def-class trust-dimension (dimension))

(def-class trust-value (dimension-value) ?x
  ((value-in-dimension-type :value trust-dimension))
  :iff-def (value-in-dimension-type ?x trust-dimension))

(def-instance high-trust trust-value
  ((has-magnitude high-degree)))

(def-instance medium-trust trust-value
  ((has-magnitude medium-degree)))

(def-instance low-trust trust-value
  ((has-magnitude low-degree)))

(def-instance LMH-trust-dimension trust-dimension
  ((value-membership-test trust-value)))

(def-instance MH-trust-dimension trust-dimension
  ((dimension-includes-value medium-trust high-trust)))
```

Figure 15.4: Examples of OCML definitions of dimensions.

The second aspect allowing to distinguish among dimensions is the logical one. The logical features involved here are those that have to do with the relative complexity and the relative convertibility of dimensions. By relative complexity of dimensions we refer to cases in which a dimension's values are obtained as a result of applying an *operator* to two or more *dimensions*. The operands are values in dimensions that are simple in relation to the dimension to which belong the result of the applied operator. A simple example of such operation is mere conjunction of two dimensions in which the values in the complex dimension are defined as conjunctions of values in two simpler dimension. Thus, for example, *security* could be seen as a complex dimension if a value in that dimension corresponded to a value in an *encryption level* dimension joint to a value in a *certification reputation* dimension. Convertibility is a somewhat similar affair but is between two dimensions with values given each in particular units such that there exist a conversion operation between these units (e.g., meters and inches). Figure 15.5 gives a simple illustration of a complex dimension of hit rates based on the combination of numbers of hits and time periods.

```
(def-class complex-dimension (dimension) ?x
  ((has-component-dimension :type dimension)
   (has-dimension-operator :type dimension-operator))
  :iff-def (exists ?y (has-component-dimension ?x ?y)))

(def-class dimension-operator (function) ?x)

(def-instance hit-number-dim-1 dimension
  ((dimension-includes-value 10-hit 100-hit 1000-hit)))

(def-instance time-span-dim-1 dimension-with-unit
  ((dimension-includes-value 1-day 1-week 1-month)))

(def-instance hit-rate-dim-1
  dimension-with-unit
  ((has-component-dimension hit-number-dim-1
                            time-span-dim-1)
   (has-dimension-operator divide)
   (has-unit hit-day)))

(def-instance 10-hit-week dimension-value
  ((has-magnitude 10/7)
   (has-unit hit-day)
   (value-in-dimension hit-rate-dim-1)))
```

Figure 15.5: The hit rate dimension is defined as a complex dimension combining two simpler ones.

So far, elements of our framework is reminiscent of the approach taken in (29) which also ties objects of interests to certain attributes using a notion of *aspect* to which resemble our notion of *dimension*. Our approach is somewhat finer grain insofar as it allows to build dimensions from sets of values while CoOL with its ASC model starts from what could be called in our terms generic dimension types and is preoccupied with convertibility between dimensions with units. This should be taken with a grain of salt and as an helper to intuition as we have not tried to establish the precise correspondence as we think it is not a direct one. Another difference with CoOL which is more significant is that while it is designed as a language for representing contextual information, it leaves open ways in which contextual knowledge is to be selected. We now provide the elements of a needed mechanism which is tied intimately with our treatment of dimensions.

15.3.2 Context reduction

Our context gathering mechanism relies on an ontological treatment of context information reduction to context knowledge – any contextual information that is relevant to the task at hand. We use a dedicated ontology for specifying the required elements that extends all that we need for background knowledge representation (Figure 15.2), namely the dimension ontologies where reside the elements just described but also WSMO which we use to model Web services.

We can contextualise any entity of interest in a systematic and uniform way. To do this, for any entity to be contextualised, we define a *context reduction* associated with this entity which, possibly after a number of processing steps, is able to recollect all the contextual knowledge relevant to the task at hand. While a context reduction is associated with only one entity, a contextualised entity may have different context reductions with which it is associated. Thus, we can also allow for navigating seamlessly between contexts.

The main element in our context ontology is the class of *context-reduction* of which an instance is an artifact allowing to recollect contextual information for an entity to be contextualised and to tie this information to the background activities triggering the delineation of the context in question (tasks and putative goals, in particular). Here we represent this tie as part of the definition of an instance of context reduction but for the sake of generality we could have used a relation rather than a slot.

```
(def-class context-reduction
  (time:TimeSpanningEntity)
  ((context-reduction-of :type ocml-thing)
   (context-reduction-in-dimension :type dimension)
   (has-context-value :type dimension-value)          ; individual values
   (has-context-value-set :type set)                  ; collecting set
   (has-associated-goal :type goal)                   ;  wsmo goal
```

Figure 15.6: OCML listing of the definition of the class *context-reduction.*

The slot context-reduction-of associates a context-reduction to an entity to be contextualsied by it. The slot context-reduction-in-dimension associates a *context-reduction* to *dimensions* in which values representing contextual information will be selected to generate contextual knowledge. The slot has context-value and its -set variant are collection slots for gathering contextual knowledge in the form of relevant values. As can be seen from the first line, context–reduction is a specialisation of a class in the temporal ontology (25) that allows to associate context-reduction with a time–span during which the reduction is considered valid. In our implementation we use a ternary relation in order to relate a contextualised entity (e.g., a user, a Goal, or a Web Service) to values of interest in a given context. This relation is populated through a call to a LISP function taking a context–reduction as an argument and computing the required values of the other arguments of the relation. It is this collected information, which provides the ground for applying context–parameterised heuristic classification.

15.4 Context Recognition

We understand context recognition as a knowledge intensive task that, given a body of information capturing contextual information about services and users, and a particular purpose for which a context has to be identified, returns the concrete context recognised. Our epistemological basis for context information derivation and context recognition is based on Heuristic Classification, a Knowledge Level (21) method that "relates data to a pre-enumerated set of solutions by abstraction, heuristic association, and refinement" (3), see Figure 15.7. In the remainder of this section we shall first present Heuristic Classification and then describe how we apply this method for context recognition.

15.4.1 Heuristic Classification

Heuristic Classification is a particular technique for classifying that was identified, thoroughly analysed and extensively described by Clancey in his seminal paper (3), (see Figure 15.7). For instance diagnosing diseases has often been approached in Artificial Intelligence as a process whereby basic observations about patients like temperature (e.g., 39ij C) are abstracted (e.g., temperature is high), matched against prototypical diseases (e.g., the flu), and refined so that the concrete diagnosis is the one that best explains the symptoms observed. Indeed, the previous scenario has been kept simple for clarity purposes. Realistic scenarios include a wide range of observations (e.g., fever, vomiting), which may have causal relations (e.g., high temperature might cause a certain disorientation), and all need to be reconciled to obtain a correct classification (e.g., fever and vomiting because the patient has gastroenteritis).

In most of the situations, the solution features will not be directly provided by the data observed. Instead they need to be abstracted from the raw data that is available. There are three different kinds of abstraction:

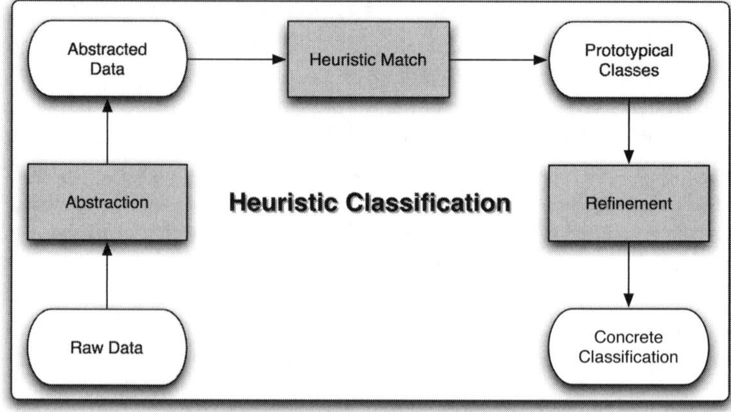

Figure 15.7: Heuristic Classification.

Definitional abstraction is based on the essential characteristics of the concept being analysed (e.g., a computer scientist knows about computers).

Qualitative abstraction is the derivation of some qualitative feature based on quantitative measures (e.g., if service A takes longer than 20 seconds it is performing slowly)

Generalization is based on the use of hierarchical information (e.g., a man is a mammal)

Refinement is a similar process but the direction of the inference is inverted. The goal is to, once a prototypical class has been identified, refine the solution trying to explain most of the observed facts. It is worth noting that refinement takes place within the target classification ontology trying to explain all the facts observed which are typically expressed in terms of a different data model.

The most distinctive feature of Heuristic Classification is the fact that a heuristic is used for achieving a non-hierarchical direct association between the abstracted data and the prototypical classes contemplated. The knowledge for relating problem situations to solutions is thus essentially experiential. A heuristic relation is uncertain, based on certain assumptions of commonality, and allows a reduction in search by skipping over intermediate relations. Heuristic Classification does not attempt to generate a solution that has not previously been identified and therefore, should the prototypical classes be too vague or the scope of the domain they capture too narrow, the solution obtained will also be affected.

15.4.2 Context Recognition as Heuristic Classification

In our framework we approach the task of determining a relevant context based on a task we want to achieve and a wide range of information concerning users and services, as a process involving the abstraction of information gathered, its heuristic classification with respect to general cases, and the refinement to the particular case at hand. In fact, it is this very process of heuristically classifying contextual information that produces contextual knowledge which can then support the appropriate adaptation of services. The reader should notice that we are here distinguishing between contextual information (i.e., all the contextual information gathered may it be useful in the situation at hand or not) and contextual knowledge which can support the system acting rationally (21).

This process is supported in our framework by a library of Heuristic Classification Problem-Solving Methods (PSM) developed by Motta et al (20). In a nutshell, PSMs are basically software components that encode sequences of inference steps for solving particular tasks in a domain-independent manner so that they can be applied for solving the same task within different domains. The library we use is based on the Task Method Domain Application (TMDA) framework (19). TMDA prescribes constructing Knowledge-Based Systems based on the definition of task ontologies that define classes of applications (e.g., diagnosis, classification), method ontologies that capture the knowledge requirements for specific methods (e.g., heuristic classification), domain ontologies that provide reusable task-independent models, and application ontologies for mapping domain models with domain-independent problem solvers.

The library defined in (20) includes a task ontology and a method ontology. Classification ontology characterises the task as the problem of finding the solution (a concept) which best explains a certain set of facts (*observables*) about some individual, according to some criterion. The concept *observables* refers to the known facts about the object (or event, or phenomenon) that we want to classify. Each *observable* is characterized as a pair of the form *(f, v)*, where f is a feature of the unknown object and v is its value. Feature is anything which can be used to characterize an object, such as a characteristic which can be directly observed, or derived by inference. As is common when characterizing classification problems–see, e.g., (31), we assume that each feature of an *observable* can only have one value.

The solution space specifies a set of predefined classes (*solutions*) under which an unknown object may fall. A *solution* itself can be described as a finite set of feature specifications, which is a pair of the form *(f, c)*, where f is a feature and c specifies a condition on the values that the feature can take. Thus, we can say that an *observable (f, v)* matches a feature specification *(f, c)* if v satisfies the condition c. The current implementation contemplates different solution criteria. For instance, we may accept any solution, which explains some data and is not inconsistent with any data. This criterion

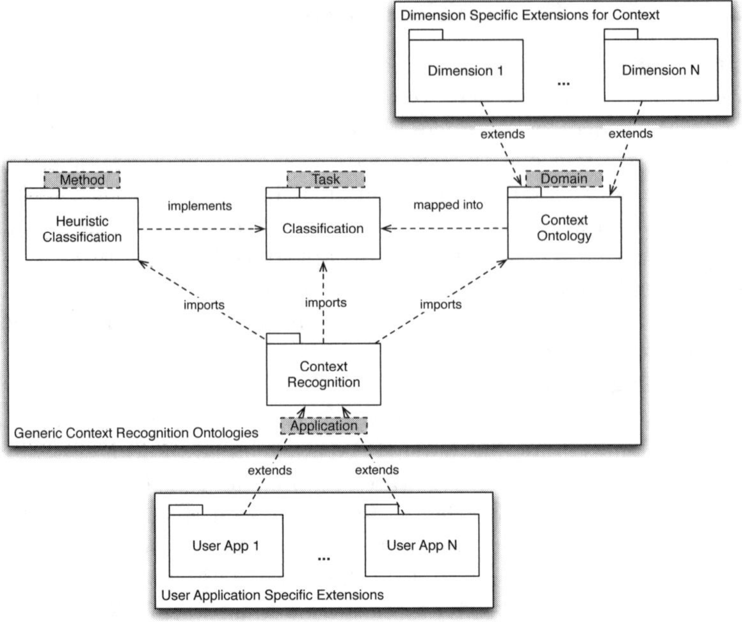

Figure 15.8: Context recognition ontology stack.

is called positive coverage (27). Alternatively, we may require a complete coverage - i.e., a solution is acceptable if and only if it explains all data and is not inconsistent with any data. Thus, the specification of a particular classification task to be performed needs to include a solution (admissibility) *criterion*. This in turn relies on a *match criterion*, i.e., a way of measuring the degree of matching between candidate *solutions* and a set of *observables*.

The library approaches classification as a data-driven search where *observables* try to be explained in accordance to some given criteria. Our classification library implements two different classification methods: *single-solution-classification*, and *optimal-classification*. The former implements a hill climbing algorithm with backtracking to find a suitable solution; the latter executes an exhaustive search for an optimal solution. Abstraction is supported by so-called *abstractors* which are basically functions that given a set of *observables* produce additional *observables*. Conversely, refinement is supported by so-called *refiners* which can be seen as the inverse of *abstractors*. *Refiners* themselves are also defined as functions that given the solution space refine it towards the solution. Coming back to our previous discussion concerning context derivation, it is quite clear that the context derivation techniques (metrics computation, derivation of values for composite dimensions, etc) are *abstractors* within our framework.

In order to use it for our particular purposes, i.e., recognising contexts, it is necessary to provide a concrete domain ontology, and an application ontology that performs the appropriate mappings between task concepts and domain concepts, and provides heuristic methods where appropriate, see Figure 15.8. Context Ontology (see Section 15.3) is provided as a general purpose domain ontology which users can extend in order to include particular dimensions of interest. Additionally, we provide as part of the framework a general purpose application ontology, namely Context Recognition ontology, which provides mappings between Context ontology and Classification ontology. In particular, the ontology maps the notion of *dimension-value* to the notion of *observable*, and the notion of *operator* to that of *abstractor*, see Section 15.3. In this way applying the framework for concrete applications only requires the provisioning of user-defined dimensions (or typical ones), the set of prototypical classes we want to match our context to, and domain-specific heuristics.

15.5 Web Service Adaptation

Once a relevant context or sets of contexts have been identified, services need to be adapted accordingly. The last infrastructural service provided by the framework is precisely in charge of this. Our framework provides the capability to adapt to particular contexts through two main complementary means. On the one hand we make use of the notion of *late binding* as supported by Semantic Execution Environments like IRS-III or WSMX (22). On the other hand, we enrich this feature with the capability to configure processes taking into account the possible adaptations of all the internal activities in order to maintain their compatibility and optimize the resulting compositions according to different criteria.

15.5.1 Goal-based Adaptation of Services

Like in real life, Goals can most often be achieved in several ways. The "best" one and even those that are suitable depend on a variety of aspects, many of which are merely contextual. Imagine for instance that we want to pay a certain object in an Internet shop. This could be achieved using different kinds of credit cards or even systems like Paypal. The suitability will depend on the kind of cards we have available, on whether we have a Paypal account, on the systems that we trust, etc. The "best" solution for us will depend on our preferences in terms of cost, security, simplicity, etc. These factors are all contextual since the essence of the service, that is to pay a certain amount of money, remains unchanged.

Our framework supports what we refer to as Goal-based adaptation of services by leveraging the capability-based invocation feature provided by IRS-III (see Section 15.2). This capacity for selecting the service to be executed at runtime is what is often referred to as *late binding.* Late binding support brings manageability to processes by specifying sets of suitable services by means of Goals (24). Additionally, this feature supports deferring the selection of the service to be utilised until runtime where concrete contextual information regarding the services (e.g., their latest performance, availability, price, etc) can be taken into account.

So far the selection of suitable services in IRS-III was uniquely driven by the functionality provided. As a consequence services providing compatible functionalities as specified in their capabilities were all considered suitable in no particular order (see (6, 9) for more details). There are however many non-functional properties (e.g., price, quality of service, trust) which should also be taken into account for fine-tuning the selection of services at runtime. IRS-III has been extended towards supporting context-adaptive capability-based invocation of services by taking into account contextual knowledge in order to restrict and rank the set of suitable services, and eventually choose the most appropriate one.

Somewhat implicit in the definition of context we introduced at the beginning of the chapter, is the fact that the relevance of contextual information is dependent on the actual application or service being provided. For instance, the encryption system used may well be irrelevant for a weather forecast service, whereas it is very important for payment services for example. The underlying essence is that what can be considered as relevant contextual information depends on the task being performed. In a scenario where a large number of services are provided and consumed by billions of users over the Web, the diversity of the tasks that will be supported will presumably be outstanding and there needs to be appropriate means for supporting adapting services to specific contexts in a generic yet scalable manner.

In order to cater for a context-adaptive execution of services we include in Goals a non-functional property (*contextually-sensitive-to*) specific to our framework. This non-functional property identifies context *dimensions* for which the Goal presents a certain *sensitivity* or, if you like, *flexibility.* In essence it identifies a set of context *dimensions* that can affect the late binding process. At runtime the framework can identify these degrees of flexibility, and use them for guiding a context reduction process whereby context knowledge is constructed out of the existing body of contextual information. Once this contextual knowledge is obtained, a Heuristic Classification problem-specification is generated and solved using the mechanisms previously introduced. During this process, an additional classification of services according to the contextual dimensions and heuristic methods are also necessary. The solution to this problem, which as we saw earlier can be restricted to the one that better fits the situation at hand, identifies the kinds of services that should be used for invocation, therefore bringing context-aware adaptability to the invocation of services.

15.5.2 Adaptation of Processes

Goals provide us with the required level of abstraction able to introduce contextual aspects within the execution of services. In simple cases, like the simple one outlined above, adapting services to a certain context is a matter of introducing within the execution infrastructure the capability to recognise a relevant context and use this information to restrict the suitable services and select the best option. In more complex cases like processes, which we define as orchestrations of Goals (6), contextual factors of one Goal may have implications over others. Imagine for instance that we want to contract an Internet connection and a VoIP solution with a certain quality of service (QoS). The kind of connection chosen might not support the QoS desired, yet an Internet connection must be in place before contracting the VoIP solution. Supporting this simple process in a completely automated way reveals to be more complex since context changes provoked by the first Goal (i.e., get an Internet connection) might prevent us from achieving the second Goal (i.e., get a VoIP with a certain QoS). Indeed, it is quite easy to envisage situations where the interdependencies are such that the complexity of the problem cannot be neglected.

In order to deal with this kind of situations and still maintain a level of flexibility allowing us to adapt the execution of processes and services to contexts, we envisage approaching this problem as a Parametric Design problem. This approach is similar to that of (30) although we aim to provide a general purpose infrastructural solution that will allow Semantic Execution Environments like the IRS-III or WSMX (8, 22) to execute contextualised semantic Web services in a generic way.

Parametric Design is a simplification of Configuration Design (19, 26), whereby the objects to be configured are assumed to have the same overall structure as preconfigured templates. For example, when configuring a computer, experts know that there will be a processor, a motherboard, a hard drive, etc. In these cases, design problem-solving consists of assigning values to the parameters defined in the preconfigured template taking into account a set of needs, constraints, and desires. This kind of problems, greatly decrease the complexity of the problem-solving task with respect to other types of designing and there exist tractable implementations (19).

One of the most comprehensive toolsets for solving Parametric Design problems was formalised and developed on top of the TMDA framework previously introduced (19). Therein Parametric Design is defined as a task that given a set of *parameters*, a set of *constraints*, a set of *requirements*, a *cost function*, a *cost algebra*, and a set of *preferences*, obtains a *design model. Design models*, that is the solutions to a Parametric Design problem, are defined as a set of parameters assignments of the form *(parameter, value)*. *Constraints* specify conditions that need to be satisfied by a design, and therefore restrict the solution space. *Requirements* on the other hand describe desired properties solutions should have. Requirements are therefore the positive counterpart of *constraints*. *Preferences* are basically the means for ranking the different solutions obtained. They are therefore different from *requirements* and *constraints* since they do not restrict the space of possible solutions but rather establish an order between the solutions obtained. Finally, the *cost function* and the *cost algebra* simply support the introduction of a global preference system for ordering solutions based on the combination of all the preferences provided.

In addition to the generic task definition, (19) includes a set of PSMs for parametric design. The library includes a general purpose search-based PSM based on the notions of *design state* and *design operators*. In this case Parametric Design is approached as the exploration of the possible design states which can be reached by applying the existing *design operators*. More advanced problem-solving strategies such as Propose & Backtrack or Propose & Revise are also supported in order to achieve further performance by including additional knowledge for guiding the exploration of the solution space. The trade-off indeed lays on the fact that more advanced problem-solving techniques require additional knowledge.

The adaptation of processes through Parametric Design is achieved by considering the overall control structure of the Goals orchestration as the preconfigured template. The Goals composing the orchestration represent the *parameters* and each of them can take a set of Web Services as value. Additional *constraints* or *preferences* between the different *dimensions* to which each of Goals are sensitive to can be specified so as to restrict unsuitable services combinations or to favour certain solutions over others. Solving the Parametric Design problem is a matter of selecting services (that is assigning values to the *parameters*) such that the *constraints* and *preferences* are honoured. The result obtained is an orchestration of services which has been adapted to context based on a set of constraints and preferences.

15.6 Application

So far we have presented the framework from a Knowledge-Level perspective focussing on the kinds of knowledge that are relevant and how they can be utilised for adapting services to context. In this section we present a simple application that aims to illustrate the notions presented so far. In this example, however, we shall limit ourselves to one single adaptation mechanism, namely Goal-based adaptation, for simplicity and space reasons.

The application is centred around the notion of *trust* which is indeed a very important aspect regarding services and their execution. The main issue of representing trust in all its facets lies in its contextual nature. Different users do not necessarily agree on which services are trustworthy. For example, a user may trust a Web service with a highly rated security certificate whenever she has to provide her credit card details, but may prefer an accurate service when precision is more important than security. Conversely, the same user weights the opinions of past users about a specific Web service in other situations, i.e., the evaluation of the Web service reputation is a priority in the current trust understanding of the user. Finally, distinct users may privilege different trust parameters in the same context; in this case, their priorities may depend on their different personal preferences.

Therefore, a trust-based mechanism for selecting Web services represents a compelling application of our framework. Following the principles introduced in the previous sections, we introduce an abstract model for trust that enables interacting participants–both Web service users and providers–to represent and utilize their own trust understanding with a high level of flexibility, and thus take the possible multiple interacting contexts into account.

15.6.1 The Trust Framework

Adapting service selection based on how trustworthy services are is here understood as a two steps process. Firstly, all participants specify their own requirements and guarantees about a set of trust parameters. Then, at runtime, our objective is to identify the class of Web services that matches the trust statements of involved participants, according to an established classification criterion; i.e. achieving Web service adaptation on the basis of trust-related parameters, see Section 15.5. In order to apply our framework to support the selection of services based on trust–and thus verify its benefits–we have extended the context model with trust-specific dimensions based on our previous work on Web Services Trust-management Ontology (WSTO) (10).

User preferences are the main elements on which Web service selection depends. Essentially, the user can decide which parameters should be considered in order to determine which class of Web services are trusted, in a given context. Therefore, in WSTO, the concepts *user*, *Web Service* and *Goal* are central ones, whereby *user* denotes the service requester and *Web Service* the service provider. Following the basic WSMO notions, a *Goal* represents the service requester's desire or intention. The user usually expresses different trust requirements in achieving different goals. For example, she can be interested in data accuracy when retrieving timetable information, and security issues when disclosing her bank accounts. On the other hand, the service aims to provide a set of trustworthy statements, in order to reassure the requester as well as to appear as attractive as possible.

The participants (*Web service* and *user*) are associated with trust profiles, represented in WSTO by the class *trust-participant-profile*. A profile is composed of a set of trust requirements and guarantees. Both *trust-guarantees* and *trust-requirements* are values within some dimension (e.g., *encryption-mechanism-dimension*), however their treatment within the adaptation process is different. Trust-guarantees are observables, i.e., contextual information that is available in the system, whereas *trust-requirements* are instead candidate solutions (pairs of feature and condition *(f, c)*).

We distinguish three logical elements in trust requirements: (i) a set of candidate solutions for expressing conditions on guarantees promised by the relevant parties; (ii) a candidate solution for requesting their reliability; and, (iii) a candidate solution for requesting their reputation evaluation. In a participant profile, the three elements are optional; choice depends strictly on the participant preferences in matter of trust. Similarly, the participant trust guarantees have three components: (i) a set of observables for representing the promised trust guarantees; (ii) an observable corresponding to the evaluation of the participant reliability; and, (iii) an observable for representing the reputation level of the participant. Whereas the promised trust guarantees are a set of promised values stated by the participant–such as (*execution-time, 0.9*) and (*data-freshness, 0.8*)–reliability and reputation guarantees are computed automatically based on data captured by monitoring software.

WSTO is comprehensive of all approaches usually adopted for representing trust (10). It embeds a policy-based trust management since the interacting participants express their trust policies in their semantically described profiles. The Semantic Execution Environment—in our case IRS-III—will behave as a Trusted Third Party (TTP) by storing participant profiles and reasoning on them. Moreover, the reputation module enables a Web service selection also based on ontological statements about reputation.

15.6.2 Case Study: A Trusted Virtual Travel Agent

The current prototype considers participant observables and needs, but it does not include the reputation module and the historical monitoring. As an illustration we implemented a simple virtual travel agent service whose goal is to find the train timetable, at any date, between two European cities. Origin and destination cities have to belong to the same country (European countries involved in our prototype are: *Germany*, *Austria*, *France* and *England*).

The client using this application publishes her *trust-profile*, with trust requirements and/or trust guarantees. In our prototype, we provide three different user profiles and three different Web services able to satisfy the user's goal, following the lines of Section 15.3 for the representation of knowledge. User profiles are expressed through trust requirements, without trust guarantees. All user requirements are performed in terms of security parameters: *encryption-algorithm*, *certification-authority* and *certification-authority-country*. These security parameters are three examples of possible *dimensions* for the trust domain. The listing of Figure 15.9 illustrates how users express a qualitative level of preference for each parameter.

For instance, *user4* would like to interact with a Web service that provides a high security level in terms of encryption algorithm, but she accepts medium value for certification authority and certification authority country. User requirements are represented in a qualitative way which is a more con-

```
USER4
(def-class trust-profile-USER4
  (trust-profile)
  ((has-trust-guarantee :value guarantee-USER4)
   (has-trust-requirement :value requirement-USER4)))

(def-instance requirement-USER4
  contextualised-trust-requirement
  ((has-encryption-algorithm high-trust-req)
   (has-certification-authority medium-trust-req)
   (has-certification-authority-country medium-trust-req)))

(def-instance user4-requirements-reduction
  context-reduction
  ((context-reduction-of requirement-USER4)
   (context-reduction-in-dimension
     required-datafreshness
     required-certification-authority
     required-certification-authority-country)
   (has-associate-goal get-train-timetable)))

USER5
...
(def-instance requirement-USER5
  contextualised-trust-requirement
  ((has-encryption-algorithm medium-trust-req)
   (has-certification-authority low-trust-req)
   (has-certification-authority-country low-trust-req)))

USER6
...
(def-instance requirement-USER6
  contextualised-trust-requirement
  ((has-encryption-algorithm low-trust-req)
   (has-certification-authority high-trust-req)
   (has-certification-authority-country high-trust-req)))
```

Figure 15.9: OCML listing of examples of user profiles and preferences.

venient way for users. As explained earlier, Heuristic Classification relies on an abstraction step that takes raw data and transforms it into higher-level information. The listing in Figure 15.10 shows an example of a particular abstractor that can take low-level information about the encryption algorithm used and transform it into a qualitative representation of its security[2].

```
(def-instance encryption-algorithm-abstractor
  abstractor
  ((has-body '(lambda (?obs)
   (in-environment ((?v . (observables-feature-value ?obs 'has-encryption-algorithm)))
   (cond
     ((== ?v DES) (list-of 'has-encryption-algorithm 'high-trust-req (list-of (list-of 'has-encryption-algorithm ?v))))
     ((== ?v AES) (list-of 'has-encryption-algorithm 'medium-trust-req (list-of (list-of 'has-encryption-algorithm ?v))))
     ((== ?v RSA) (list-of 'has-encryption-algorithm 'low-trust-req (list-of (list-of 'has-encryption-algorithm ?v))))))))
```

Figure 15.10: Encryption-Algorithm Heuristic OCML listing.

The heuristic *encryption-algorithm-abstractor* establishes that whenever the encryption algorithm adopted by a Web service provider is like *DES*, then its security level is considered high-trust-req. Whenever both User and Web service describe their profiles, they implicitly agree with the qualitative evaluation expressed in the heuristic. In turn, whenever the Web service provider makes use of an algorithm like *AES*, according to the heuristic in Figure 15.10, its encryption ability is deemed medium-trust-req, otherwise, if the adopted algorithm is like *RSA*, the security level is low-trust-req.

Other heuristics provide qualitative evaluations of *Certification Authorities (CA)*, and *CA countries*. For instance, security level of *globalsign-austria* is

[2]The categorisation we use is arbitrary and for the sake of illustration

Figure 15.11: Web application.

retained high, conversely German CAs are considered medium-secure[3]. The user can apply these heuristics, or define her own, sharing her expertise and knowledge with other users. Alternatively, the user can even express her requirements in a precise/quantitative way, by specifying the exact values expected from Web service guarantees, for example, the certification authority issuing security token has to be *VeriSign*.

We developed a user-friendly Web application to test our implementation, which is available at http://lhdl.open.ac.uk:8080/trusted-travel/trusted-query. The snapshot in Figure 15.11 shows the Web application interface. The user who would like to know the timetable for trains between two European cities enters the desired city names and date. The user owns a trust profile associated to her name: e.g. *dinar* is for instance associated to *user4*, *vanessa* to *user5* and *stefania* to *user6*.

Whenever the application starts, the system identifies the trust-profile based on the user name. Given several Web services semantically described in terms of WSMO, all with the same capability, but different trust profiles, the application is able to select the most trust-worthy service. Indeed, given the *trust-guarantees*, i.e., the *observables*, and the *trust-requirements*, i.e., the *candidate solutions*, selecting services based on trust is simply a matter of running the adaptation mechanism provided by the framework. For this application we applied the complete coverage criterion, i.e., every user requirement is explained (matches with a Web service trust guarantee) and none is inconsistent, and we used the optimal classification method which ensures the Web service selected is the most trustworthy one.

[3]Again, the reader should note that this simply arbitrary decisions

Trusted Travel

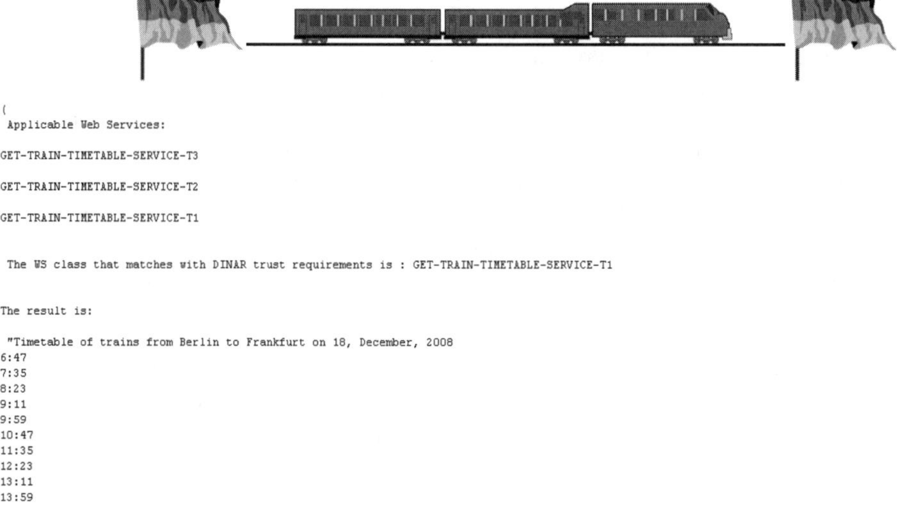

(
Applicable Web Services:
GET-TRAIN-TIMETABLE-SERVICE-T3
GET-TRAIN-TIMETABLE-SERVICE-T2
GET-TRAIN-TIMETABLE-SERVICE-T1
The WS class that matches with DINAR trust requirements is : GET-TRAIN-TIMETABLE-SERVICE-T1
The result is:
"Timetable of trains from Berlin to Frankfurt on 18, December, 2008
6:47
7:35
8:23
9:11
9:59
10:47
11:35
12:23
13:11
13:59

Figure 15.12: Web Application Output for the user "dinar".

Figure 15.12 is a snapshot of the resulted trusted VTA booking. The application returns the list of Web services able to satisfy the user goal, and the one that is invoked, which is the Web service that matches dinar trust requirements. It follows the Web service output, the requested timetable. Easily, the application can be tested with the other user instances implemented, *vanessa* and *stefania*. It can be noticed that *vanessa* trust profile matches with Web service class *get-train-timetable-service-T3*, while *stefania* with *get-train-timetable-service-T2*. A ""non-trusted" based version of the application is available at http://lhdl.open.ac.uk:8080/trusted-travel/untrusted-query for comparison purposes. This application implements a virtual travel agent based on the standard IRS-III goal invocation method. The output returns only the train timetable requested, without any trust based selection.

15.7 Conclusions

Service-Oriented Architecture is commonly lauded as a silver bullet for Enterprise Application Integration, inter-organizational business processes implementation, and even as a general solution for the development of all complex applications. However, despite the appealing characteristics of service-orientation principles and technologies, we are still far from achieving a widespread adoption of service technologies over the Web as initially predicted.

One aspect that is increasingly regarded as one of the main limitations of service technologies is the fact that current technologies do not support adapting services and processes behaviour to context. Gradually researchers focussing on Web services related technologies are beginning to consider aspects such as security, quality, trust and adaptability within the broader area of context-awareness. Even though appealing results have been achieved so far, most of the solutions created are rather ad hoc and can hardly cope with the road range of heterogeneous cases one could find should service technologies be widely adopted in the Web.

In this chapter we have presented a knowledge-based approach to adapting services in a generic automated and effective manner. The framework described is based on research on semantic Web services and Knowledge Engineering and provides a number of domain-independent services that can take contextual information and use it for supporting the adaptation of services and processes. The framework uses ontologies as its core backbone for context modeling and reasoning. Additionally it relies on its capacity for recognising relevant contexts by using Heuristic Classification and for adapting processes by configuring orchestrations of Goals using a Parametric Design engine.

The main advantage of the framework is that application developers can directly use it for constructing their context-aware solutions by simply providing a set of domain-specific models that capture the degrees of flexibility the application should have with respect to different contexts and how it should react to changes. Finally, as an illustration we have presented an application that uses the framework for supporting a trust-based selection of services dynamically.

Acknowledgements This work has been partially supported by the EU co-funded IST project SOA4All (FP7-215219). We would also like to thank Reto Krummenacher for his insightful comments.

References

[1] Rama Akkiraju, Joel Farrell, John Miller, Meenakshi Nagarajan, Marc-Thomas Schmidt, Amit Sheth, and Kunal Verma. Web service semantics - wsdl-s. http://www.w3.org/Submission/WSDL-S/, November 2005. W3C Member Submission.

[2] Matthias Baldauf, Schahram Dustdar, and Florian Rosenberg. A survey on context-aware systems. *International Journal of Ad Hoc and Ubiquitous Computing*, pages 263–277, June 2007.

[3] William J. Clancey. Heuristic classification. *Artificial Intelligence*, 27(3):289–350, 1985.

[4] Anind K. Dey. Understanding and using context. *Personal Ubiquitous Computing*, 5(1):4–7, 2001.

[5] Anind K. Dey, Gregory D. Abowd, and Daniel Salber. A Conceptual Framework and a Toolkit for Supporting the Rapid Prototyping of Context-Aware Applications. *Human-Computer Interaction*, 16(2):97 – 166, 2001.

[6] John Domingue, Liliana Cabral, Stefania Galizia, Vlad Tanasescu, Alessio Gugliotta, Barry Norton, and Carlos Pedrinaci. IRS-III: A broker-based approach to semantic Web services. *Web Semantics: Science, Services and Agents on the World Wide Web*, 6(2):109–132, 2008.

[7] Joel Farrell and Holger Lausen. Semantic Annotations for WSDL and XML Schema. http://www.w3.org/TR/sawsdl/, January 2007. W3C Candidate Recommendation 26 January 2007.

[8] Dieter Fensel, Mick Kerrigan, and Michal Zaremba, editors. *Implementing Semantic Web Services: The SESA Framework*. Springer, 2008.

[9] Dieter Fensel, Holger Lausen, Axel Polleres, Jos de Bruijn, Michael Stollberg, Dumitru Roman, and John Domingue. *Enabling Semantic Web Services: The Web Service Modeling Ontology*. Springer, 2007.

[10] S. Galizia. WSTO: A Classification-Based Ontology for Managing Trust in Semantic Web Services. In *3th International Semantic Web Conference (ESWC 2006)*, Budva, Montenegro, June 2006.

[11] T. R. Gruber. A translation approach to portable ontology specifications. *Knowledge Acquisition*, 5(2):199–220, 1993.

[12] Gregor Hohpe and Bobby Woolf. *Enterprise Integration Patterns: Designing, Building, and Deploying Messaging Solutions.* Addison-Wesley Longman Publishing Co., Inc., Boston, MA, USA, 2003.

[13] Markus Keidl and Alfons Kemper. Towards context-aware adaptable web services. In *WWW '04: Proceedings of the 13th international World Wide Web conference on Alternate track papers & posters*, pages 55–65, New York, NY, USA, 2004. ACM.

[14] Panu Korpipaa, Jani Mantyjarvi, Juha Kela, Heikki Keranen, and Esko-Juhani Malm. Managing Context Information in Mobile Devices. *IEEE Pervasive Computing*, 02(3):42–51, 2003.

[15] Reto Krummenacher, Holger Lausen, and Thomas Strang. Analyzing the Modeling of Context with Ontologies. In *International Workshop on Context-Awareness for Self-Managing Systems*, May 2007.

[16] Zakaria Maamar, Ghazi AlKhatib, Soraya Kouadri Mostefaoui, Mohammed B. Lahkim, and Wathiq Mansoor. Context-based Personalization of Web Services Composition and Provisioning. In *30th EUROMICRO Conference (EUROMICRO'04)*, pages 396–403, Los Alamitos, CA, USA, 2004. IEEE Computer Society.

[17] Zakaria Maamar, Djamal Benslimane, and Nanjangud C. Narendra. What can context do for web services? *Communications of the ACM*, 49(12):98–103, 2006.

[18] D. Martin, M. Burstein, Hobbs J., O. Lassila, D. McDermott, S. McIlraith, M. Paolucci, B. Parsia, T. Payne, E. Sirin, N. Srinivasan, and K. Sycara. OWL-S: Semantic Markup for Web Services. http://www.daml.org/services/owl-s/1.0/owl-s.pdf, 2004.

[19] Enrico Motta. *Reusable Components for Knowledge Modelling. Case Studies in Parametric Design Problem Solving*, volume 53 of *Frontiers in Artificial Intelligence and Applications.* IOS Press, 1999.

[20] Enrico Motta and Wenjin Lu. A Library of Components for Classification Problem Solving. Ibrow (ist-1999-19005) deliverable, The Open University, 2001.

[21] A. Newell. The Knowledge Level. *Artificial Intelligence*, 18(1):87–127, 1982.

[22] Barry Norton, Carlos Pedrinaci, John Domingue, and Michal Zaremba. Semantic Execution Environments for Semantics-Enabled SOA. *it - Methods and Applications of Informatics and Information Technology*, Special Issue in Service-Oriented Architectures:118–121, 2008.

[23] Michael P. Papazoglou, Paolo Traverso, Schahram Dustdar, and Frank Leymann. Service-oriented computing: State of the art and research challenges. *Computer*, 40(11):38–45, 2007.

[24] Carlos Pedrinaci, Christian Brelage, Tammo van Lessen, John Domingue, Dimka Karastoyanova, and Frank Leymann. Semantic business process management: Scaling up the management of business processes. In *Proceedings of the 2nd IEEE International Conference on Semantic Computing (ICSC) 2008*, Santa Clara, CA, USA, August 2008. IEEE Computer Society.

[25] Carlos Pedrinaci, John Domingue, and Ana Karla Alves de Medeiros. A Core Ontology for Business Process Analysis. In *5th European Semantic Web Conference*, 2008.

[26] A.Th. Schreiber and B.J. Wielinga. Configuration design problem solving. *IEEE Expert*, 12(2):49–56, April 1997.

[27] Mark Stefik. *Introduction to Knowledge Systems.* Morgan Kaufmann, San Francisco, CA, USA, 1995.

[28] Thomas Strang and Claudia Linnhoff-Popien. A Context Modeling Survey. In *First International Workshop on Advanced Context Modelling*, Nottingham, UK, 09 2004.

[29] Thomas Strang, Claudia Linnhoff-Popien, and Korbinian Frank. CoOL: A Context Ontology Language to Enable Contextual Interoperability. *Distributed Applications and Interoperable Systems*, pages 236–247, 2003.

[30] Annette ten Teije, Frank van Harmelen, and Bob J. Wielinga. Configuration of Web Services as Parametric Design. In Enrico Motta, Nigel Shadbolt, Arthur Stutt, and Nicholas Gibbins, editors, *EKAW*, volume 3257 of *LNCS*, pages 321–336, Whittlebury Hall, UK, October 2004. Springer.

[31] B. J. Wielinga, J.K. Akkermans, and G. Schreiber. A competence theory approach to problem solving method construction. *International Journal of Human-Computer Studies*, 49:315–338, 1998.

[32] Terry Winograd. Architectures for context. *Human-Computer Interaction*, 16(2):401 – 419, 2001.

Chapter 16

Ubiquitous Mobile Awareness from Sensor Networks

Theo Kanter, Stefan Forsstrom, Victor Kardeby, Jamie Walters, Patrik Osterberg, and Stefan Pettersson

Abstract Users require applications and services to be available everywhere, enabling users to focus on what is important to them. Therefore, context information from users (e.g., spatial data, preferences, available connectivity, and devices, etc.) needs to be accessible to systems that deliver services via a heterogeneous infrastructure. We present a novel approach to support ubiquitous sensing and availability of context to services and applications. This approach offers a scalable, distributed storage of context derived from sensor networks wirelessly attached to mobile phones and other devices. The support handles frequent updates of sensor information and is interoperable with presence services in 3G mobile systems, thus enabling ubiquitous sensing applications. We demonstrate these concepts and the principle operation in a sample ubiquitous mobile awareness service. The importance of this contribution, in comparison to earlier work, lies in the availability of real-time ubiquitous sensing to both applications on the Internet as well as applications in mobile systems.

16.1 Introduction

This chapter describes an approach and support for the provisioning of ubiquitous awareness of context information originating from sensors and wireless sensor networks via mobile systems. The approach is object-centric in that objects (individuals, artifacts, etc.) share sensor information on the Internet

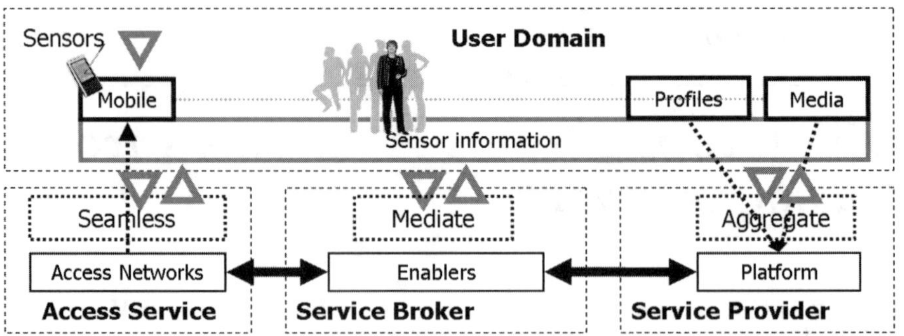

Figure 16.1: Overview of sensor information in the MediaSense open service framework.

both via Session Initiation Protocol for Instant Messaging and Presence Leveraging Extensions (SIMPLE) (15) and a novel Distributed Context eXchange Protocol (DCXP). In our approach, end-devices share sensor information both directly and via support in mobile systems, thus complementing each other. The research is mandated by the fact that access to context information both in 3G mobile systems and on the Internet provides ubiquitous real-time access to shared sensor information from distributed objects. This we can use to build distributed services that provide a better user experience, a better overall performance, and with entirely new behaviors. In Section 16.6 we provide such an example presenting a Ubiquitous Mobile Awareness service.

16.1.1 Motivations

Context-aware applications and services utilize information from users and sensors in order to present the users with application behavior that better matches their needs. As the proliferation of Web services increase across the Internet, there is a growing need to be able to provide more customized services based on users' context information. These will range from simple scenarios such as delivering location based advertisements to more complex tasks including digital communities, social interactions, and access to content and media in an intelligent manner. Any such delivery of Web services would require scalable systems capable of delivering vastly dynamic context data with minimal overhead and delay.

To this end, we propose an open services architecture as shown in Figure 16.1. This proposal originates from MediaSense, a large research project at Mid Sweden University (11). The aim of the project is to move users and their experiences to the center of multimedia services that may be delivered via various access networks. The open service architecture includes an open framework for the development and delivery of multimedia services utilizing context awareness created from a distributed sensing system. The services can be delivered via heterogeneous infrastructures. This motivates our search for a dissemination infrastructure that supports ubiquitous real-time access to shared sensor information from distributed objects, in which we preserve the association between the location of context sources and the end-points for service control.

DCXP is self-contained with respect to naming and address resolution of end-points, whereas Web services rely on DNS. Moreover, DCXP enforces a correspondence between the location and names of end-points, which is not the case with Web services, due to it reliance on DNS. Furthermore, DCXP constitutes a scalable system capable of delivering vastly dynamic context data with minimal overhead and delay. Thus, the novel DCXP creates a platform required for the real-time exchange of context information. By implementing a P2P structure along with a light-weight context-oriented information exchange format, we solve the problems of existing approaches including scalability, reliability, and availability. DCXP provides for the provisioning of context information in real time between mobile applications that run on devices attached to both mobile systems as well as the Internet including Web services.

16.1.2 Overview

The remainder of the chapter is organized as follows: Section 16.2 discusses related work. Section 16.3 presents the architecture and components for disseminating context in 3G mobile systems and the Internet. Section 16.4 covers DCXP and the underlying P2P infrastructure. Section 16.5 discusses a Bluetooth bridge to Wireless Sensor Networks for mobile devices. Section 16.6 presents a Ubiquitous Mobile Awareness application prototype. Section 16.7 summarizes the achievements along with a discussion of future work, followed by a concluding section with references.

16.2 Related Work

This section describes earlier results and approaches from related work. The concept of exchanging information from a sensor via a mobile phone is not new. Application clients running on mobile phones can access inbuilt GPS sensors and cameras or external sensor units, such as heart-rate monitoring via near field communication, etc. These clients can then make this information available to application servers on the Internet (17).

Earlier work has been focused on the brokering of sensor information by means of Web services via a 3G mobile system (12); as a network service (14); via Session Initiation Protocol (SIP) servers (2); with Web services on the Internet (8) (1) (6) and gateways to mobile systems for aggregating information from wireless sensor networks and making this available in services (13) (7).

The IST-EU-FP6 project, MobiLife, realized the provisioning and brokering of sensor information via mobile systems and services (12). This system is based on Web services that break the correspondence between the location of end-devices and the reachability of sensor information. In addition, exchanging application level information between clients and centralized servers indirectly imposes an upper limit on the freshness of information.

The IST-EU-FP6 project Ambient Networks created the ContextWare framework for the dynamic joining and leaving of context information networks. The framework provides Internet services via heterogeneous network infrastructures, supporting network composition, mobility, multiradio management etc. (14).

The Mobiscope approach federates distributed mobile sensors into a sensing system that achieves high-density sampling coverage over a wide area through mobility (1), through the use of a centralized repository.

CarTel is one example of providing and brokering sensor information via clients in end-devices that are opportunistically connected to an application server on the Internet (8). The approach is device centric with, as its objective, the survivability of the user application. End-devices negotiate for connectivity and compensate for random disruptions with delay-tolerant networking and information caching. As a consequence of this strategy, the system can provide no guarantees at all regarding the provisioning of sensor information between end-devices in real time.

Similar systems (18) and (3), present approaches for undertaking this task, however they utilize technologies that could impact negatively on scalability, reliability, and availability.

In the SenseWeb approach, applications can initiate and access sensor data streams from shared sensors across the entire Internet. The SenseWeb infrastructure provides support for an optimal sensor selection for each application and the efficient sharing of sensor streams among multiple applications based on Web services (6).

The solution presented in (13) uses gateways to mobile systems for collecting data from sensor networks (7). This eliminates the correspondence between an object and the sensor values that represent aspects of this object. Should other parties outside of this system need to interpret its sensor data in terms of objects, the system has to offer a correspondence function.

16.3 Enabling Ubiquitous Mobile Awareness

Ubiquitous real-time access to distributed context information from distributed objects, via end-devices attached to 3G mobile systems or on the Internet, is a challenge. Mobile services that utilize context information require parameters from individual sensors that are associated with the location of the end-device. This additional requirement is essential since it enables us to reason about object-oriented views of an Internet of Things where both the context and the sensor information are aspects of an object.

None of the approaches mentioned in Section 16.2 meets the requirements in a satisfactory way. This section examines what we consider to be the main requirements for an approach to enabling ubiquitous mobile awareness. It also examines our approach in relation to these requirements in making context information from sensor networks available in ubiquitous systems as well as the general operation of our solution. The section ends with a discussion about the advantages and disadvantages of our choices in comparison to previous results.

16.3.1 Associating Sensor Information with Objects

The research is mandated by the fact that services on the Internet benefit from having access to more information about the user's situation and intentions in order to support the user in achieving tasks. We also made the observation that through the proliferation of mobile systems (GSM, EDGE, UMTS, HSPA, etc.) there exists a growing, installed base system. This installed base makes the mobile phone (or other mobile devices attached to mobile systems) a viable option as a mediator of sensor information.

Earlier work for making information from wireless sensor networks available in mobile systems employed gateways or Web services, which have a disadvantage of breaking the implicit or explicit association between the sensor information and its owner or origin (13) (8) (12) (1) (6) (7). In choosing our approach we view the association between sensor information and its owner or origin as vitally important from the perspective that:

1. An association facilitates the management of privacy and trust when groups or communities share and exchange sensor information, avoiding the necessary policing in a nontransparent gateway.

2. An association is strongly related to concepts of identity and naming, which are the basis for building individual context profiles.

Furthermore, the services that we envision should take into account context about the situation of the user along with objects in the user's environment.

Therefore, decisions that coordinate multiple services, different service content, or the method of delivering the service, should be made in end-devices close to the user. As a result of this, we seek methods for the aggregation of sensor information from sensor networks and the provisioning of Context Information to end points such as mobile phones. The obvious reason for this requirement is that end-points correspond to a device with which the user is registered. User agents can use context information to invoke sessions for multiple services.

We believe that while the proposed solution does not explicitly address the matters of trust and privacy; the presented approach provides for an underlying architecture that will readily enable such an enhancement. This is achievable since the approach enforces the relationship between users and sensor information. This association guarantees sensor information sinks and dependent services of the source and by extension the validity of the sensor information.

16.3.2 Accommodating Sessions with Mobile Phones

In order for mobile devices to send information to the presence support system using the General Packet Radio Service (GPRS), mobile devices must first register with the home subscriber server of the mobile operator. These mobile devices can then request a radio link and obtain Packet Data Protocol (PDP) context from the Radio Network Controller for sending data via the radio link. This PDP context is valid as long as the mobile device is in an active state, i.e., when it actually transmits (sensor) data. Obtaining a PDP-context consumes device resources and incurs considerable latencies, which mandates the use of a Session Border Controller/Gateway for maintaining TCP sessions with the mobile device and avoid the loss of the PDP context.

16.3.3 Exchanging Sensor Information between Objects

We require real-time response times in the provisioning of Context Information, which means that we need to consider alternative methods for the exchanging of Context Information. This mandates the DCXP and an underlying P2P Infrastructure, which coexists and cooperates with IMS. The design of the DCXP protocol extends SIP naming and has a similar approach in that DCXP end-points signify context sources and sinks, which, when combined, run as a Context User Agent (CUA) in an end-device. This permit us to co-locate end-points for a context information network with the end-points for a service infrastructure where service decisions can be made.

We attach sensors or sensor networks (either wired, wirelessly, or hardware integrated) to mobile phones that are registered with a 3G mobile operator. We also consider other mobile or fixed end-devices, such as laptops or set-top boxes. Sensor agents running in these end-devices communicate with sensors or sensor networks that are attached to these devices. Sensor agents forward

Figure 16.2: Overview of provisioning of sensor information to the Internet and IMS.

sensor information both to a P2P infrastructure on the Internet for storage and provisioning of sensor information, as well as presence support that is available in IMS (see Figure 16.2).

16.3.4 Sharing Context with 3GPP IMS and DCXP

A seamless integration of both DCXP and IMS within our solution provides a novel approach to leverage context information residing in both domains. Legacy IMS applications would, with some modification be able to digest sensor information being propagated on DCXP. The converse would also be true, whereby applications existing within DCXP to exchange sensor information with applications running within IMS. Such inherent support within the proposed architecture would provide for homogeneity of sensor information while both systems are co-existing and enable a smoother, more gradual shift towards a distributed approach. The DCXP naming convention permits the interworking and co-location of DCXP and SIMPLE. A user on the P2P network can register with IMS using provided credentials and retrieve context information from presence and location services.

16.4 Distributed Context eXchange Protocol

The Distributed Context eXchange Protocol (DCXP) is an XML-based application level P2P protocol, which offers reliable communication among nodes that have joined the P2P network. Any end-device on the Internet that is DCXP capable may register with the P2P network and share context information. DCXP imposes a naming scheme similar in format to Uniform Resource Identifiers used in SIP (4). The DCXP naming scheme uses Universal Context Identifiers (UCIs) to refer to Context Information such as sensors that are stored in the DCXP network. The protocol currently employs XML as the preferred message format, mainly to benefit from the standardised properties of XML. While the use of XML could present some drawbacks to data parsing on resource constraint devices, the light structure of DCXP compensates for any parsing overheads incurred. Test results so far on mobile devices yielded parsing times of around 6 ms, which is acceptable for real-time oriented implementations.

16.4.1 Context Storage

A network that uses DCXP forms a Context Storage that utilizes a Distributed Hash Table (DHT) to map between UCIs and source addresses. The current DHT design choice is Chord, presented in (16), and it forms a logical ring structure of all the nodes in the network. An advantage of using a DHT is that entries can be found in log(N) time. In addition, the Context Storage also acts as a context exchange mechanism: clients query the Context Storage for a UCI to learn where the Context Information is located and the Context Storage returns the address of the desired Context Information. The Context Storage is able to resolve locations because it handles the resource registrations coming from the context sources. Thus, the Context Storage maintains a repository of UCI/source-address pairs and provides a resolution service to the clients via DCXP. With the exception of storing Context Information, the operation of the Context Storage is similar to the way that Dynamic Domain Name Server stores a mapping between a domain name and an Internet Protocol (IP) address.

16.4.2 Context User Agent

As the combined end-point for the origination and consumption of context information, Context User Agents (CUAs) are allowed to join a context network by registering with a Context Storage. A CUA corresponds to a node in the DHT ring that makes up the Context Storage. A CUA exposes functionality to enable applications and services to either resolve a UCI, get a UCI or register a UCI in the Context Storage.

16.4.3 DCXP Messages

DCXP is a SIMPLE-inspired protocol with five primitives outlined in Table 16.4.2. For more information on SIMPLE see (15). Figure 16.3 provides an example of the signalling involved in fetching a context value in the DCXP.

Message	Description
REGISTER_UCI	A CUA uses REGISTER to register the UCI of a CI with the Context Storage.
RESOLVE_UCI	In order to find where a CI is located, a CUA must send a RESOLVE to the Context Storage.
GET	Once the CUA receives the resolved location from the Context Storage it GETs the CI from the resolved location.
SUBSCRIBE	SUBSCRIBE enables the CUA to start a subscription to a specified CI, only receiving new information when the CI is updated.
NOTIFY	The source CUA provides notification about the latest information to subscribing CUAs every time an update occurs or if asked for an immediate update with GET.

Table 16.1: The primitive messages of DCXP.

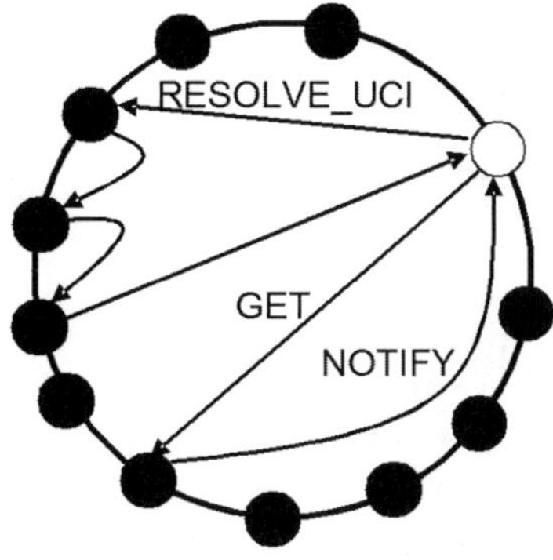

Figure 16.3: DCXP signaling.

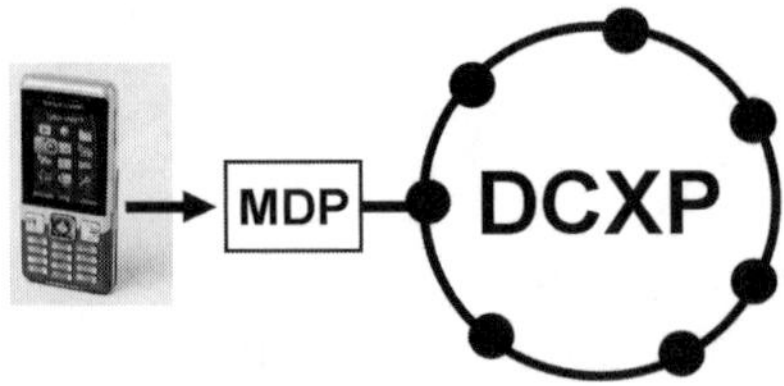

Figure 16.4: Mobile phone connected to DCXP via a MDP.

Each circle on the ring in the figure illustrates one CUA on the Context Storage. The white node initiates the retrieval process by sending a RESOLVE_UCI. This message is sent through the Context Storage until the node responsible for the UCI/source-address pair replies with the location of the context information. If the white node is only interested in retrieving the value once, it sends a GET message to the located node. If the white node is interested in a continuous subscription it sends a SUBSCRIBE message. The located node then responds with a NOTIFY message once for a GET or whenever the Context Information changes for a SUBSCRIBE.

16.4.4 Mobile DCXP Proxy

Proxy sensor agents running on mobile phones connect to and register with a Mobile DCXP Proxy (MDP) in order to communicate with the P2P network (see Figure 16.4). The reasons for this are twofold. Firstly, since mobile phones have limited resources it is desirable to limit its processing requirement through the use of the MDP. Secondly, mobile phones are attached to a 3G mobile system via a radio access network managed by a Radio NC. This inherently exposes the communication link to some of the unpredictable behaviors of radio communication such as packet loss, coverage, reliability, and interference. The MDP shields from this behavior and exposes a reliable connection to the P2P backbone. MDPs are part of a P2P network for disseminating context information.

16.4.5 DCXP Topology

The ring in Figure 16.5 symbolizes the DCXP ring. Each box on the ring symbolizes a node in the DCXP network and each node has a CUA service. There can be any number of nodes in a single ring, even a single node forming a ring with itself. The DB is the database, which serves both as the persistent storage and as the storage location for some meta information about the network, such as the users currently online. The bootstrap node is the first node to be started on the DCXP network. It is responsible for initializing and sustaining the network, and each DCXP ring requires one bootstrap node. Each node has some services running: a Database Agent (A), a Database

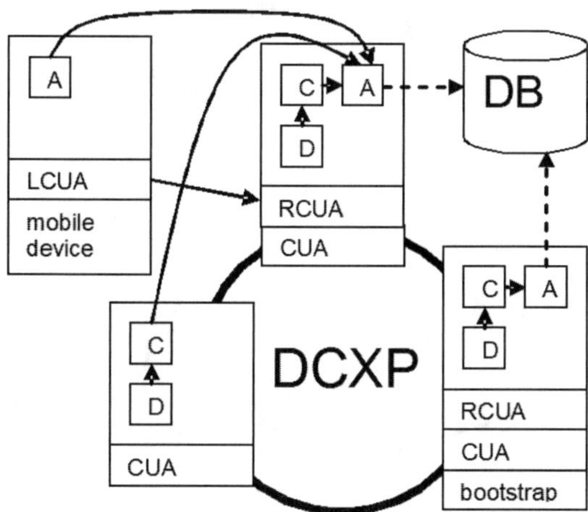

Figure 16.5: DCXP architecture.

Client (C) and a Dataminer (D). Database Agent is a service that listens to requests in the form of messages from Database Client and forwards them to the database and relays the response back to Database Client. Not all nodes in the DCXP ring need to have an A service. Dataminer gathers all information and sends it to the database for persistent storage. Dataminers are distributed across all the nodes in the DCXP network and this consequently distributes the load of subscribing to everything. The distribution is simple; since the DHT has already distributed the load of the UCIs, the Dataminer will assume the same distribution. It will subscribe to all UCIs that it's local CUA handles for the Context Storage.

The MDP is also shown in Figure 16.5. Its functionality is realized by two components, the Remote CUA (RCUA), which fulfills the server functionality of the MDP, and the Limited CUA (LCUA), which fulfills the client-side functionality. The RCUA is a service that runs on a node in the DCXP ring and it handles all the DCXP functionality for a mobile device. The LCUA offers all the functionality of DCXP to the mobile phones without them being actually incorporated into the DCXP ring, thus reducing the load on the mobile devices.

16.4.6 Super Nodes

A Super Node in our architecture is a CUA that assumes extra responsibilities. To be able to run DCXP as a Super Node, the computer or device needs ample resources, such as memory, storage, processing capacity, power, etc., to be able to handle the extra workload. The responsibilities and which nodes that are chosen to be a Super Node, differ depending on the network

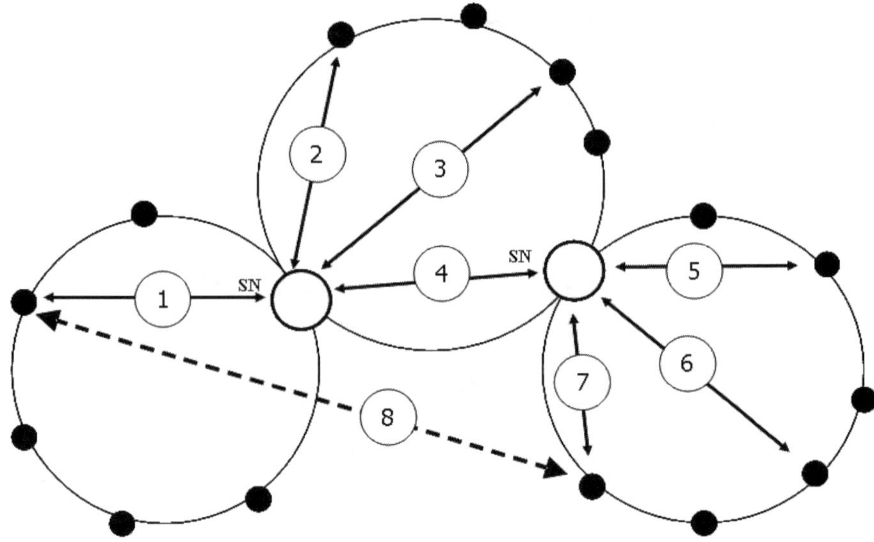

Figure 16.6: DCXP subnets connected through super nodes.

layout.

A Super Node can act as a first point of contact for other CUAs. This can be done by registering a DNS name to the IP of the Super Node. If Super Node acts as a static location for initial contact, we call this the Bootstrap Node of that specific DCXP network. The Bootstrap Node needs to contain all that is necessary for the network to start and when the Bootstrap Node is alone, it has a DCXP network that contains only itself.

If a node is a MDP, it is also considered a Super Node. This is because it has the responsibility for the mobile devices that it serves. In the same way that a SN can serve many mobile devices it can act as a mediator between many DCXP networks. The Super Node is then responsible for routing the DCXP messages between the subnets. By using a SN we provide for a way to partition and join DCXP networks. Figure 16.6 shows three joined DCXP networks, it also shows a numbered sequence, how the messaging will go when finding and transferring context information from a CUA in one network via two Super Nodes, to a CUA in the third network. The dashed line (8) denotes the resultant data exchange.

16.4.7 Universal Context Identifiers

DCXP identifies each Context Information by a Universal Context Identifier akin to a Uniform Resource Identifier, as described in (4). UCIs have the following syntax and interpretation:

```
dcxp://user[:password]@domain[/path[?options]]
```

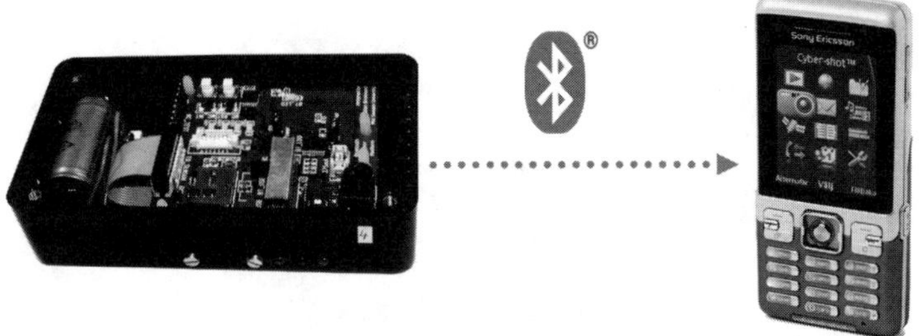

Figure 16.7: Bluetooth bridge to wireless sensor networks.

where *dcxp* is the protocol and *domain* is a Fully Qualified Domain Name relating to where the Context Information is located. The fields *user* and *password* provide optional means for authorization. The *path* field adheres to the context information namespace hierarchy, thus allowing for the organization and sorting of the items in a logical sense. The *options* facilitate further modifiers in the form of parameter=value. An example of a fully qualified UCI would be:

```
dcxp://alice@miun.se/weather/temperature?unit =celsius
```

16.5 Bluetooth Bridge to Wireless Sensor Networks

Figure 16.7 shows the prototype of the platform for associating Wireless Sensor Networks (WSNs) with a mobile phone or any other mobile device equipped with a Bluetooth interface. The box contains a class 1 Bluetooth radio, a sensor network controller, integrated humidity and temperature sensors as well as optional memory card storage for the storage of history information. This design enables further integration of additional sensors such as accelerometers, CO_2 or radiation, the design also has deep sleep capabilities that can lower the power consumption down to 40μA. The box is powered by a rechargeable lithium battery and the chosen form factor also allows for the inclusion of a higher capacity battery for a longer field deployment.

The current prototype employs the Bluetooth serial port profile to create a virtual serial port to communicate with other Bluetooth capable devices. The prototype publishes new sensor values every two seconds and communicates using a text-based protocol over the virtual serial port. Under this normal operation the power consumption is measured to 20mA. The current prototype employs a Bluetooth master that runs on the mobile phone and reads values from the sensors, forwarding these to both a presence server in IMS as well

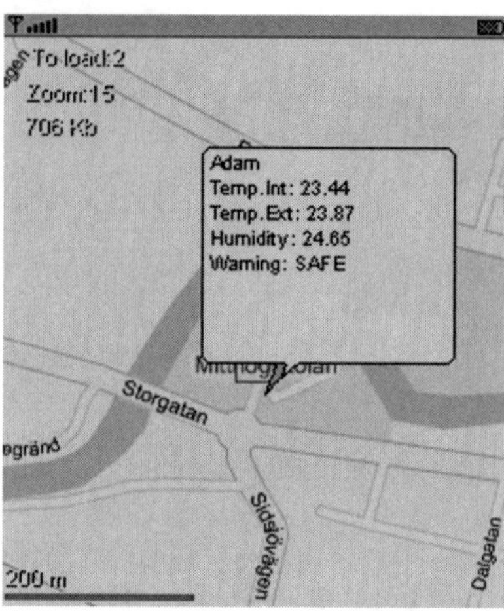

Figure 16.8: Ubiquitous mobile awareness application.

as a MDP, see Figure 16.2.

Bluetooth communication, in comparison to computing tasks and the sensing of humidity and temperature, consumes a considerable amount of power. However, subsequent generations of the prototype will use Bluetooth Low Energy Technology (9) that consumes much less power than the current version.

16.6 Ubiquitous Mobile Awareness Service

We built a prototype of our system and developed an Ubiquitous Mobile Awareness Service application on top of it, using the components and framework that we described earlier (10). The purpose of this application as a proof-of-concept is to:

1. Verify the principles of the intended operation of the framework along with that of its constituent components.
2. Demonstrate the feasibility of the approach and verify the principles of the intended operation of the framework and its constituent components.
3. Provide a testbed for relevant quantitative and qualitative measures of overall performance as well as performance of the constituent components.

We implemented the client application for the mobile device as a Java ME based MIDlet that uses the MDP to communicate with the P2P network and the Ericsson Mobile Java Communication Framework (5) for accessing IMS. The application, as demonstrated in Figure 16.8, is capable of receiving and manipulating real time sensor information about other users within the DCXP network. The sensor information obtained may be used, as was demonstrated, to display real-time information on a map on the mobile phone. It is capable of tracking and displaying the other user's location and additional context information while the user is actively mobile. The application derives this information from GPS sensors in the phones and other attached sensors via the Wireless Sensor Network Gateways.

The prototype realizes the framework's ability to utilize multiple real-time sensor information in order to provide a real-time view of the user's context in relation to other interesting users or objects. The demo application is capable of identifying a user's friends who are within close range and flag an alert displaying the friends' context information on the map. The application makes use of the MDP to offload computationally expensive tasks to a more powerful DCXP node. This is a seamless process within the application, with the user being unable to distinguish the fact that the underlying network connection is being proxied within the distributed architecture. The resulting application is compatible with any mobile phone with the Java ME environment and 3G Internet connectivity. For our measurements we used a Sony Ericsson C702 mobile phone.

16.7 Conclusions and Future Work

We have presented a solution for the scalable provisioning of Context Information derived from sensor networks wirelessly attached to mobile phones and other devices. The solution handles frequent updates of sensor information and is interoperable with presence services in 3G mobile systems, thus enabling ubiquitous sensing applications.

The importance of the Ubiquitous Mobile Awareness Service application lies in the fact that it demonstrates that simultaneous provisioning of Context Information to mobile systems and P2P infrastructure enables us to compensate for the limited capabilities of centralized presence systems. Other similar presence systems on the Internet are limited in the real-time, scalable provisioning of multidimensional data (2), it simply has never been a design objective. Yet, the provisioning of context via IMS Presence is still important in that it provides access to context information in mobile applications to a vast and growing population of mobile phones, which can be used for enabling ubiquitous coverage and sensing.

The reasons mandated the creation of a scalable P2P context network that is capable of client-side and server-side interfacing with IMS. In doing so we found that we could preserve the association between physical end-points and the logical end-point for sensing. This is not the case in the previous approaches that were based on Web services (8) (12) (1) (6). This also mandated the creation of the Distributed Context eXchange Protocol (DCXP); a simple protocol that adheres to the principles of SIP for locating end-points for session initiation. Another possible choice would have been the selection of P2P-SIP. We believe that our selection of DCXP is justified from the standpoint that SIP, while being standardized has evolved into a fairly complex architecture. This would have resulted in the use of a standard that satisfies several other requirements unrelated to context dissemination.

Scalable context networks that can interface with IMS raise questions about the need for synchronization as well as who might be in control. Here we believe that the answer lies in maintaining the initiative and control in end-points. If IMS orchestrates actions that fetch context values, then it must be on the initiative of a user or object.

Measurements of latencies in local area networks with DCXP showed latencies to be on par with TCP traffic. Therefore we may assume real-time capabilities for mobile phones. Initial testing with mobile phones also supports this statement.

Of equal importance, is the fact that the data models of DCXP networks allow us to visualize multidimensional view of Context Information, which through service layer agreements could include context records, mobile positioning, or other data that exists in mobile systems. This would enable reasoning and machine learning in order to provide applications with capabilities for more refined service decisions. This is addressed in the current work in which we exploit such knowledge to drive IPTV services or publication of personal networked media, in which the user is part of the control loop through the inclusion of mobile phones with sensors as described in this publication.

Through the use of P2P networks and persistent storage, we provide for resilient storage of Context Information. The choice of P2P infrastructure is essential since P2P networks such as Chord are vulnerable to attacks (16). We are thus investigating alternative topologies.

Other approaches have more specifically embraced gaining seamless connectivity via heterogeneous access networks as an issue. Although we did not prioritize this issue and specifically focused on complementing the capabilities of IMS, we claim that the principles of our approach still apply even when adding other access networks besides GSM, EDGE, W-CDMA, HSPA, etc. Furthermore, we based our work on IMS presence, which is based on IETF SIMPLE (15). We could move this functionality to the Internet, but what would have to remain in a 3G mobile operator's network is a Session Boarder Gateway in order to assist mobile devices in maintaining active sessions, minimizing power consumption, and avoiding latencies for acquiring a new PDP Context after a short period of inactivity.

Finally, we view transparency and reachability as key factors for future

Ubiquitous Mobile Awareness applications. Instead of gateways as aggregators and mediators such as in homes, we foresee the rethinking of mobile end-devices towards an Internet of Things as integrated functionalities of 3G radios, sensor communication, and local computation.

Acknowledgments

The research is partially supported by the regional EU target 2 funds, regional public sector, and industry such as Ericsson Research and Telia. In particular we acknowledge the efforts of Roger Norling for the implementation of the DCXP protocol and MDP; and finally Sebastian Brinkmann and Prof. Bengt Oelmann for design of the Bluetooth WSN platform.

References

[1] T Abdelzaher, Y. Anokwa, P. Boda, J. Burke, D. Estrin, L. Guibas, A. Kansal, S. Madden, and J. Reich. Mobiscopes for human spaces. *Pervasive Computing*, vol. 6:pp. 20–29, 2007.

[2] Carlos Angeles. Distribution of context information using the session initiation protocol (SIP). Master thesis, Royal Institute of Technology, Stockholm, Sweden, June 2008.

[3] Martin Bauer, Christian Becker, Jörg Hähner, and Gregor Schiele. Contextcube - providing context information ubiquitously. In *ICDCS Workshops*, pages 308–313. IEEE Computer Society, 2003.

[4] Tim Berners-Lee, Roy T. Fielding, and Larry Masinter. Uniform resource identifier (URI): Generic syntax. RFC 3986, Network Working Group, January 2005.

[5] Ericsson Labs Developer. Mobile java communication framework, November 2008.

[6] W.I. Grosky, A. Kansal, S. Nath, Jie Liu, and Feng Zhao. Senseweb: An infrastructure for shared sensing. *Multimedia*, vol. 14:pp. 8–13, 2007.

[7] J. Hjelm, T. Oda, A. Fasbender, S. Murakami, and A. Damola. Bringing IMS services to the DLNA connected home. In *Proceedings of the 6th International Conference on Pervasive Computing*, 2008.

[8] B. Hull, V. Bychkovsky, Y. Zhang, K. Chen, M. Goraczko, A. Miu, E. Shih, H. Balakrishnan, and S. Madden. Cartel: A distributed mobile sensor computing system. In *Proceedings of the 4th International Conference on Embedded Networked Sensor Systems*, pages 125–138, Boulder, Colorado, USA, 2006. ACM Press.

[9] N. Hunn. An introduction to wibree. White paper, EZURiO Ltd, 2006. http://www.tdksys.com/files/00616.pdf last visited 2009-08-14.

[10] Theo Kanter, Stefan Pettersson, Stefan Forsström, Victor Kardeby, and Patrik Österberg. Ubiquitous mobile awareness from sensor networks. In *Proceedings of the 2nd International ICST Conference on MOBILe Wireless MiddleWARE, Operating Systems, and Applications (MOBILWARE)*, Berlin, Germany, April 2009.

[11] Theo Kanter, Patrik Österberg, Jamie Walters, Victor Kardeby, Stefan Forsström, and Stefan Pettersson. The mediasense framework. In *Proceedings of th IARIA International Conference on Digital Telecommunications (ICDT)*, Colmar, France, July 2009.

[12] M. Klemettinen. *Enabling Technologies for Mobile Services: The MobiLife Book.* John Wiley and Sons Ltd, 2007.

[13] S. Krco, D. Cleary, and D. Parker. P2P mobile sensor networks. In *Proceedings of the 38th Annual Hawaii International Conference on System Sciences (HICSS)*, page 324c, 2005.

[14] Norbert Niebert, Andreas Schieder, Jens Zander, and Robert Hancock. *Ambient Networks: Co-operative Mobile Networking for the Wireless World.* Wiley Publishing, 2007.

[15] J. Rosenberg. SIMPLE made simple: An overview of the IETF specifications for instant messaging and presence using the session initiation protocol (SIP). Internet-draft, IETF, 2008.

[16] I. Stoica, R. Morris, D. Karger, F.M. Kaashoek, and H. Balakrishnan. Chord: A scalable peer-to-peer lookup service for internet applications. In *in Proceedings of the Conference on Applications, Technologies, Architectures, and Protocols for Computer Communications*, volume 31, pages 149–160, New Yourk, NY, USA, 2001. ACM Press.

[17] Esko Strömmer, Jouni Kaartinen, Juha Pärkkä, Arto Ylisaukko-oja, and Ilkka Korhonen. Application of near field communication for health monitoring in daily life. In *Proceedings of th IEEE International Conference Engineering in Medicine and Biology Society (EMBS)*, pages 3246–3249, New York, NY, USA, August 2006.

[18] Q. Wei, K. Farkas, P. Mendes, C. Prehofer, B. Plattner, and N. Nafisi. Context-aware handover based on active network technology. In *the International Working Conference on Active Networks.* Springer Verlag, 2003.

Chapter 17

Modeling and Storage of Context Data for Service Adaptation

Yazid Benazzouz, Philippe Beaune, Fano Ramaparany, and Olivier Boissier

Abstract Services need to be adapted to users and auto-configured for optimizing their functioning. Adaptation requires services to be aware of their contexts. However, making context-aware services costs many efforts from designers and a delicate setting. We propose to automate this process by identifying relevant contexts for services and then adapting these services accordingly. To accomplish our purpose, we explore a context data-driven approach. This approach refers to the process of collecting and storing data from a wide range of sources. Data are gathered in conformity with a context data model, stored in databases and then context recognition techniques are applied to identify relevant contexts. Our proposal differs from the existing approaches for context-aware applications, where context models and applications are closely related and ignore how context is derived from sources and interpreted. Our approach is supported by an overall architecture for service adaptation and intends to automatically adapt services without any prior knowledge about context models.

17.1 Introduction

The growing development of communication systems, the continuous miniaturization of electronic functions and the decrease of their cost, have favored the creation of diverse sensors. These sensors are becoming more autonomous and have high calculation capacity allowing to provide various environmental data. These data are used by applications or services, to be more efficient when situations change. These include:

- Adaptation to highly dynamic computing environment and unpredictable situation,
- Minimization of user's attention when executing tasks (e.g., attention that the system's administrator can devote to each configuration such as changing the status of a service),
- Anticipation of human activities and more generally exploiting information that should occur in the future (e.g., switching ON the heating system one hour before the person come back home, using stored information in the person agenda).

To date, Web services perform inefficient exchange of information and require human assistance. Web services need to be autonomous and perform tasks intelligently. For example, in a distributed peer to peer Web services system, it would be interesting for a peer to be aware of another peer. Therefore, the peer can adapt its communication capacities, or send its information according to the context of the other peers. In the same way, a reservation Web service might be aware about user preferences and his Internet consultations. Such adaptive services are called context-aware Web services for the reason that are capable of being configured or personalized according to their context.

C. Stefano et al. (8) present how context is used to reach adaptation in the Web. They cite four categories of adaptation:

1. Adaptability of contents and services delivered by accessed pages,
2. Adaptability of navigation through automatic navigation toward other pages,
3. Extendability of hypertext with context information for multi-channel access via Web and mobile applications,
4. Adaptability of presentation properties like application appearance.

In the same paper, the authors put forward a data-centric approach to achieve adaptation in the Web. They emphasize the need to identify and represent context data. Their modeling approach of context assumes that data supplying the context model are normalized with respect to some filters over raw data. However, the context model remains dependent on the application domain and on the adaptive goals. This approach neither allows automatic identification of relevant contexts for services nor takes benefit from machine learning techniques that can be applied to data.

In this chapter, we explore a data-driven approach for automatic service adaptation. Data gathered from different sources are stored in databases, then analysis algorithms are applied to recognize interesting patterns for services called *relevant contexts*. Relevant contexts are used by a variety of services to automatically adapt their published functions, or to reconfigure their parameters.

In several existing context-aware applications, context models and applications are closely related. Context models are predefined according to the goals behind the development of these applications. Such development of systems and context models does not promote reusability. Moreover, the development of adaptive services still based on predefined adaptation models and known contexts. In our approach, service adaptation aims to provide the service in a manner that matches the context being recognized. Contexts are automatically identified from environmental data and services can share their relevant contexts for the benefit of other services or applications.

The structure of this chapter is organized as follows: first, we explore the definition of context, then the role of context in service adaptation. We also discuss three basic components involved in service adaptation namely context data modeling, context storage and context recognition. We assist each component with technical choices. Finally, we present the overall architecture for service adaptation.

17.2 Context Definition

The term *context* appeared first in social science and psychology, then in linguistic science and artificial intelligence. Human sciences start from the standpoint that any human action lays within a context. In linguistics, context is a set of texts surrounding an element of the language that determines its meaning and value. In artificial intelligence, a first approach (28) (30) (29) considers context as an object used to resolve the problem of generality through reasoning and inference. A second approach (18) (19), defines context as a partial known set of facts.

To date, context has not reached a consensus on its definition in computer science (26). The commonly accepted definition of context is that of Dey (15) which defines the context as "any information that can be used to characterize the situation of an entity; an entity is a person, place, or object that is considered relevant to the interaction between a user and an application, including the user and applications themselves". In this definition, the interaction occurs between user and application, but interaction can be extended to applications and systems (e.g., in the World Wide Web, many services need other services to achieve their tasks).

Several existing approaches for the development of context-aware applications do not distinguish between context management and applications. One reason is the strong dependence between the context models and applications. In our approach, we disassociate context management from applications, and we use the following definitions for constructing service adaptation:

- **Context sources**: are computational or material entities able to perceive, acquire and provide data.
- **Context modeling**: is the process of producing a formal or semi-formal description of a set of data gathered from sensors and users present in a context-aware system (23). The context model can be specified by the context-aware application developer.
- **Context data modeling**: is the process of representing the context at its data level. Data include several types such as physical data (e.g., temperature, pressure, etc.), system data and network (e.g., devices and machines used, their state, quality of service, etc.), execution data (e.g., the current session, the date, the time, etc.), the user data (e.g., preferences, competencies, emotions, profile, etc.).
- **Context**: is a pattern that can be discovered from context data using data mining algorithms.
- **Adaptive service**: is a specific context-aware application that uses context data and identifies relevant patterns to adapt and reconfigure automatically. Adaptation can be achieved either according to the user or other applications needs.

The following section highlights the role of context in service adaptation.

17.3 Role of Context in Service Adaptation

Web services provide functionalities in the Web in the form of remotely executable services. Context-awareness can play an important role to improve the quality of these services, and make them able to operate efficiently in dynamic environments. Based on the context data modeling mentioned in section 17.2, we can present the use of context for service adaptation as follows: physical contexts are used to improve the performance of services; systems contexts increase the availability of services in the network; execution contexts allow services to construct users profiles and finally, user contexts enable services to provide contents according to user situation and needs.

Besides, we can illustrate how context can play a role in adapting services in several ways. For example:

- **Context-aware services for users**: The goal of such services is to satisfy users' requirements in accordance with their situations. Furthermore, these services can be personalized automatically by taking into account their historical uses.

- **Context-aware services for applications**: Services are usually adapted to environment in which users evolve. Context-awareness can be extended to services and applications themselves. For example, in the case of dynamic service composition, services should be aware of the availability and processing time of other services in the networks. As a result, a service selects more appropriate services for its composition.

- **Exchange of already learned contexts**: The exchange of already learned contexts eliminates the need for services to relearn context already known by other services. This exchange accelerates service adaptation. For example, if a user watches action movies every Saturday night, the video distribution service on the Internet can share this learned context with an advertising service. Therefore, the advertising service can propose to user movies that he might prefer.

- **Web services delivery in smart environments**: Web services have been often limited to the virtual Web environment, but they can also be extended to go beyond that and interact with the applications deployed in the smart environment. One way for smart environment applications and Web services to meet users requirements, is to share users context data. For example, a smart home application might need to know on which Web site a child is currently browsing. This information can be useful for parents to prevent their children against unsuitable content. In this way, the smart home application can decide to disconnect Internet or to inform parents. On the other hand, a Web site might need a description of the user location (e.g., is he surrounded by his children? is he in a meeting?). This information can be useful for an advertising service to adapt the publicity presented on the Web site.

17.4 Developing Context-Aware Services

Many efforts were focused on the developing of context-aware services for ubiquitous systems. H. Truong et al. (35) argue that it is not clear how context-awareness is useful and how to apply its techniques in the domain of Web services. They propose to go more in depth in the technical development of these systems. However, there is a need to describe the representation of context data, the storage of these data, and a need for techniques of adaptation for services.

In early context-aware services, M. Keidl et al. (24) present a context framework to facilitate the development and deployment of context-aware adaptable Web services. They defined the context as a data set about a service client and its environment, represented in XML format. Context information are added by the service client in the SOAP header message; therefore the Web service provider can interpret this data and make customized responses to the client. C. Yun et al. (11) propose a context-aware service oriented architecture for ubiquitous Web service discovery called CA-SOA in which both services requester and other services define their context in OWL-S. The service discovery mechanism is guided by context matching between both service requesters and other services contexts. This context matching strategy is used for finer granularity when discovering. T. Chaari et al. (10) propose a context-aware architecture providing a functional adaptation to context. The main idea is to separate application data and context data of a service. Adaptation is achieved by finding the best match between application data and context data at multiple level of services design (selection, compositions, etc.)

Activities on the development of context-aware services focus on implementation issues and do not take the problem of service adaptation in its globality. The proposed solutions do not predict situations or detect hidden contexts. In addition, used contexts are predefined and the addition, modification and deletion of contexts induce new implementation issues of context-aware services.

In the next section, we will detail three main parts that construct our approach namely context data model, storage solutions and context recognition techniques.

17.5 Context Data Modeling

Context data modeling is the process of representing context at its data level. The interest to context remains in the way to model and exploit data that contribute to the construction of context, and it is motivated by the following considerations:

- Ubiquitous applications are linked to various environmental factors. They are characterized by a huge quantity of data that need to be stored, managed and delivered. The environmental factors and data have to be coherently represented and interpreted, to enable ubiquitous applications to be adapted accordingly.

- Enabling easy ubiquitous applications development requires a context management system that represents data in order to be stored in an exploitable format.

- There is usually a significant gap between sensor output and the level of information that is useful to applications. This gap must be bridged by various kinds of processing (21). According to (16), there may be multiple layers that context data has to go though before reaching an application, due to the need for additional abstraction.

- Application-driven models fail to provide a context-independent data. The existing context representation models are application-dependent and do not support powerful interpretation and reasoning.

Several existing approaches of context-aware applications development have been interested in analyzing the representation of context at the application level. Applications are based on context models that are abstract representation of context conditions and situations that are relevant to applications (13). Context models are based on conceptual modeling theories such as UML (14) and Entity-Relation (ER) (38), and specified by the application developer. However, these approaches have some drawbacks when dealing with context modeling. This is due to the following points:

- The separation between application semantics and low level details of context acquisition from individual sensors leads to a loss of generality, making the sensors difficult to reuse in other applications and difficult to use simultaneously in multiple applications (16).

- Context model of a real application can vary depending on the application domain and the adaptively goals to be fulfilled. No universal context model exists that applies to all kinds of applications (8).

- Adaptive application makes use of a fixed context model and this limits the extendibility of the application process of new requirements of adaptation (36).

- The sensors might be physically distributed, and cannot be directly connected to a single machine (16).

We resume the above drawbacks in two main problems. A context model must be generic and flexible, and the necessity to separate context management from applications. Of course, these two problems are related and can be grouped into the independence between context management and context-aware applications. The following section discusses general requirements for context data modeling followed by a survey of the approaches that have been proposed for representing context data.

17.5.1 Requirements

According to the data-driven approach, the use of context for service adaptation requires primary that the context data model supports existing storage solutions and be usable by data analysis algorithms to identify relevant contexts. The context data model should meet the following requirements:

1. **Semantics**: Semantics is a potential requirement and refers to a common format for data drawn from diverse sources in order to be shared and reused across context-aware services. Therefore, context data model encompasses the data and its semantics.

2. **Interoperability**: The environment of services is characterized by its rapid change. Certain services might be deactivated or deleted and some others might appear in the environment. These new services should be able to use and exchange context data without new development efforts. Context model has the role to ensure the interoperability between context data and services.

3. **Support for time**: Time is an important characteristic of context because it denotes the date and time at which events occurred. Context model should be able to keep information about time to allow easy comparison of events and tracking of their progress over time.

4. **Support for context ontology**: Context data is represented according to the context data model using concepts that describe these data. Several relationships might exist among these context data (e.g., relationship between concepts, temporal relationships, etc.). These relationships can be used by data analysis algorithms to discover hidden patterns i.e., abstract descriptions of situations. Relationships between these concepts can be ensured by an ontology of context which cover service environments. Data analysis algorithms might need this ontology to calculate similarities between data concepts.

5. **Support for context data integration and fusion**: A key role for context data model is to permit context data merging based on conceptual, temporal or structural similarities.

6. **Support for flexibility**: Context model should adapt when changes occur in systems, architecture or applications.

7. **Support for quality**: context data model should take into account the uncertainty of context data. Context data is imperfect (2) (23) and can be incorrect if it fails to reflect the true state of the world or inconsistent if it is in contradiction with another context data. Certainty of context data can be described by functions which estimate the likelihood of confidence. For example, the quality of a context data about a user location can be characterized by its accuracy, measured by the standard error of the location system, and the freshness determined by the time when the event occurs.

8. **Support for queries**: Context data should support efficient queries due to the huge amount of context data stored in databases.

It remains some key points that are worth to mention. Firstly, the number of context data can become very heavy in space and computing time caused by data redundancy (e.g. a same data might appear several times at different instants). Finally, data collection poses security and ethical problems especially if these data are related to people lives.

17.5.2 Approaches for Context Data Modeling

T. Strang et al. (34) (9) presented an interesting study in which they detailed six approaches of context representation: Key-value modeling, Markup scheme modeling, Graphical modeling (UML, ER, etc), Object oriented modeling, Logic based modeling and Ontology based modeling approaches. The analysis of these models leads us to say that the ontologies based models are a promising approach for context modeling in the domain of ubiquitous computing.

However, this study does not distinguish between context modeling and context data modeling. Therefore, we focus our literature review on context data and their representation. This representation is described according to three types of data representation: numeric, symbolic and structural.

In its numerical form, context data is real value. These data do not have significance by themselves, but by the meaning we attribute to them. For example, the value (30) provided by a thermometer indicates a high temperature.

Context data in a symbolic form are concepts representing facts or signs. For example, the activity of a person (standing, sitting, walking, running, inside, outside etc.)(25) (27). This kind of data can not be issued from ordinary sensors, because they usually require low-level processing (digital learning algorithms or inference from rules) in order to give a signification to data.

In its structural form, context data is a composed data where its signification is carried by the whole structure. There are three techniques that popularly express structural context data: (1) attribute-value data, (2) object data and (3) semantic data.

1. **Attribute-value Data**: Context attribute-value data is organized in couple (name of the attribute/ value) in the form of rows or columns. The most well known data is that contained in the log files. Each column represents a physical implementation of relation between the name of the attribute and the data value and each row represents a physical implementation of relation between attributes themselves. Brown et al. (6) represented data structures in name/ value pairs, Chen and Kotz (20) present data structures as set of parameters/ attributes and as tags and corresponding fields in (5).

2. **Object Data**: Context object data is a structured representation mode for describing an element by its characteristics. For example the position provided by a camera tracker is an object composed of three values X, Y, and Z. Two reasons lead to the representation of context data as object data, first, to perform logical reasoning and inference and second, the manipulation of objects has become easier with the advent of the object oriented design supported by several techniques and programming languages.

3. **Semantic Data**: Context semantic data is the description of informational content composed of data and its signification, which means that the semantic of data is included in data itself. Several studies were interested in representing context using semantic models. Context data in (9) are represented by (subject, predicate, value, time, certainty). In (17), a context data is an RDF[1] assertion. In (8), context data is described in terms of properties and attributes expressed in Web modeling language. Other studies were interested in the Composite Capabilities / Preference Profile (CC/PP) model which is a W3C standard for expressing device capabilities and preference profile. What is common (1) to all semantic networks is a declarative graphic representation that can be used either to represent knowledge or to support automated systems for reasoning about knowledge.

Figure 17.1 summarizes the appropriateness of context data modeling representation based on three essential characteristics, the semantic, interoperability and flexibility.

We conclude that a semantic model for context data is more appropriate for data representation. Semantic models encompass all the above formats used for representing context data. They have very interesting facilities such as the capability to integrate the semantic of data, a rich language for expressing data and interoperability support.

[1] http://www.w3.org/RDF/

		Semantic	Data Exchange	Flexibility of the structure
Numerical context data		Not supported	Requires additional knowledge about the semantic and data format	Not flexible
Symbolic context data		Not supported	Requires additional knowledge about the semantic and data format	Not flexible
Structural context data	Attribute-value	Partially supported	The semantic is given by the attribute	Extension in tables
	Object	Partially supported	Requires additional knowledge about the access functions	Object Encapsulation
	semantic	Totally supported	Needs semantic parser	Description language

Figure 17.1: A comparative study between the three types of data representation

17.5.3 Our Context Data Nodel for Service Adaptation

In this section, we introduce our context data model. It is based on semantic Web technology precisely Resource Description Framework (RDF).

The RDF data model consists of the triple or statement (subject, predicate, object), where the subject represents the resource to be described, the predicate represents the property type applicable to this resource and the object represents the value of the property which can be another resource or a literal, such as string. RDF gives meaning to data through predicates. Also, RDF presents interesting features such as representing data and relations without ambiguity, resource manipulation and the possibility to represent them as graph.

In our approach, the context data model is a set of RDF triples. These triples concern data object, subject, source, quality and time. The object represents the entity concerned by the data, the subject represents the description of the situation as provided by the source. The equivalent graphical model for context data is presented in figure 17.2.

The source identifier allows us to assign a confidence index to the data, detect the malfunction of a device or its irrelevance. In case of malfunction, if the data provided by the infrared sensor does not contribute as much to the location of the person, this sensor can be removed or replaced by another device. In addition, temporality is introduced in the model to correlate events. Figure 17.2 shows the context data model, and figure 17.3 represents an example of a context data according to the context data model.

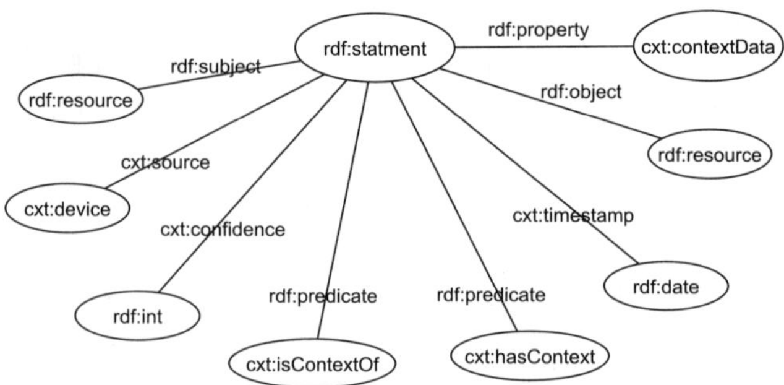

Figure 17.2: The context data model

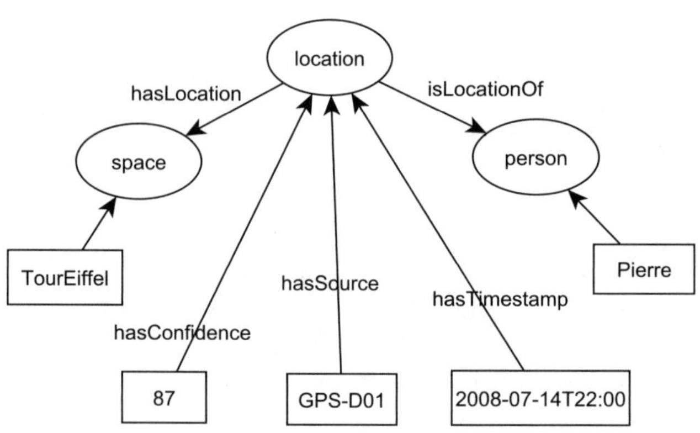

Figure 17.3: Example of a context data

17.5.4 Discussion

In literature, there are two descriptive languages that can be used to model context data. These languages are RDF and Topic Map[2]. Both are W3C standards. The basic Topic map data model consists of a set of topics that represent concepts, occurrences of a topic that represent pieces of information, and associations that are relations between topics. In comparison to RDF, topic map facilitates access to data, but it does not attribute meaning to data. Therefore, topic map is not appropriate for the description of context data in our approach.

17.6 Context Data Storage

A huge amount of data is generated by various sources disseminated in the running environment. To carry out efficient analysis (e.g., to identify groups or related clusters, find hidden patterns), it is necessary to maintain these data in databases. However, several solutions exist in the literature to maintain RDF data in databases. The following sections present respectively context data storage requirements, a comparison between fundamental database models used to store RDF data, and a discussion about our choice for context data storage.

17.6.1 Requirements

Our approach for services adaptation separates between context management and its usage by services. Context management is based on collection and storage of context data. The storage solution adopted in our approach should meet the following requirements:

- **Support for queries**: Services need to access to specific context data which are represented in our approach in RDF. The storage solution should be supported by an RDF query language. Graph database models can be a sound support for the design of an RDF model (32), particularly for RDF query languages. SQL requires fixed known column data types, and query languages based on SQL should overcome these difficulties.

- **Support for flexibility**: If any change occurs on our context data model, the storage solution should permit an easy access to the new model without affecting existing data. The query language should still be able to access new data.

[2]http://isotopicmaps.org/

Additionally, the storage solution depends on the access time and the amount of data. Nevertheless, an experimental evaluation is required.

17.6.2 Existing Approaches for Data Storage

A commonly accepted solution for mapping an RDF data model into a relational model is known as triple store. In a triple store, the statement table is composed of subject, predicate and objects columns. Additional tables may be needed to store long literals and URIs as the case of Jena (37).
Three storage types of RDF data were evaluated in (4) namely memory, persistent and native storage. According to this study, the native RDF storage systems sesame-native [3], Kowari [4] and Yars [5] provide solid performance on large data applications, but the DB based systems (sesame-db and Jena-db) are better for supporting inference. In object-oriented models, RDF resources are objects, RDF properties represent attributes, and RDF statements express the attribute values of objects. In a pure OODB approach, there is no need to program any mapping between the data stored in the database and the application data (7). According to the evaluation in (7), no OODBMS supports all the concrete specifications defined in the ODMG 3.0 standard, although ObjectStore has the biggest market while Objectivity system provides the most powerful functionalities.

The graph database model as defined by Renzo (33) "is a model in which the data structures for the schema and/or instances are modeled as a directed, possibly labeled, graph, or generalizations of the graph data structure". Since the RDF graph model is a directed graph, it is natural that it can be stored in a graph database. Renzo (33), compared multiple graph database models. We can deduce from this study that GDM (22) databases are more suitable for RDF storage because they have interesting characteristics for RDF storage such as labeled nodes and edges, schema support, derivation and inheritance and the ability to represent complex objects.

As a conclusion, the object oriented database models are strongly related to object oriented applications, thus OODBMs dispose little interoperability with other types of applications, also it is difficult to apply some interesting methods to these models e.g. learning algorithms, clustering, pattern recognition, etc. On the other hand, graph database models provide a sound support for the design of an RDF model and all graph properties but, they are relatively new and are not fully developed yet. In this study, we use the graph database models as promising solution for context data storage in our approach. Figure 17.4 summarizes the essential features related to database models.

[3]http://www.openrdf.org/
[4]http://www.kowari.org/
[5]http://sw.deri.org/2004/06/yars/

Figure 17.4: Comparison of database models

	Structure	Storage	Query language
Relational	- Tables. - Structure known in advance. - Fixed schema. - Integration of different schemas is neither easy nor automatizable.	- Best result with native storage of RDF.	- Only RQL supports RDF graph properties. - There is no object identification, but values. - Query languages cannot explore the relationship among data, paths neighborhoods, patterns
Object	- Object - All data conform to a predefined schema	- Pure OODB is close to the application. - It's possible to extend relational database model to OODB. - ObjectStore has the biggest market while Objectivity provides more powerful functionalities.	- OQL is the object query language. - Interaction via methods.
Graph	- Graph schema	- GDM proposes a transformation of RDF to bipartite graphs or hyper-graphs.	- G+, GraphLog support all graph properties.

17.6.3 Discussion on the Storage Strategy

The remainder of Section 17.6 provides a survey and analysis of a variety of context data storage techniques. We have mentioned that in our approach, graph databases are more appropriate for context data storage. There exist many topological forms that can be used for context data storage as graph. None of experimental analysis is done to determine the most appropriate topology. Here, we discuss two selected propositions that will be used in the future evaluation of our approach.

Storage of Context Data as Multiple Graphs

Here, the storage unit is a context data represented in an RDF model described previously. Each unit is identified by a unique identifier in the database. The advantage of this storage is to manipulate graphs as data. Hence, it is easy to retrieve, compare, merge and manipulate graphs as independent data. Nevertheless, this approach has two drawbacks, managing the list of identifiers and the large storage capacity due to data redundancy.

Storage of context data as single graph

A single graph of context data is a complex network. There are two topologies for such a graph, network of RDF nodes and network of RDF graphs.

An approach based on a network of RDF nodes

In this graph, each node is an RDF node and each arc is an RDF arc. In such a structure, context data sharing the same resources are linked together. Figure 17.5 shows one possible representation of information related to a single resource.

Figure 17.5: Network of RDF Nodes

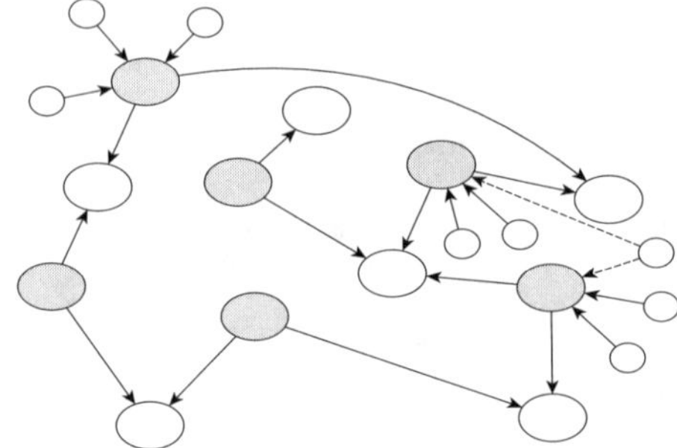

According to the context data model defined in the section 17.5.3, each gray ellipse is a statement. Big white ellipses represent subjects and objects. Small ones are confidence, source and timestamp nodes. Multiple statements can have the same subject and/or object which are defined by continued lines in the figure. The confidence, source, timestamp can be resources of several statements, as showed by discontinued lines in the figure 17.5.

An Approach Based on A Network of RDF Graphs

In this graph, each node is an RDF graph (context data). In such a structure, context data sharing the same meaning are interlinked. The meaning is given by the hascontext predicate in the RDF graph model. Figure 17.6 shows one possible representation of information related to a single context.

In figure 17.6, each node represents a context data model, and each edge in the graph represents the existence of a shared resource (object, subject, timestamp, confidence, and source). Dark gray circles represent the same subjects or objects in each linked nodes, while white gray circles describe the same timestamp, confidence and source in each linked nodes.

17.7 Context Recognition

The main purpose for using context is to construct an interpretation of what is happening in the environment and if possible to understand it in an automatic way.

Figure 17.6: Network of RDF Graphs

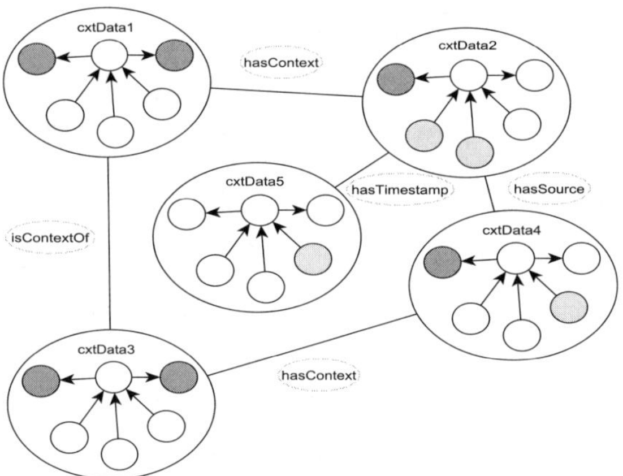

J. Mäntyjärvi et al (27) define context recognition as the process of extracting, merging and converting data from context sources to a representation to be utilized in the application. In our approach, we distinguish two types of context, low context and high context. Low context refers to a situation characterized by many connections between data, and a short duration. High context refers to a situation characterized by close connections between data, and a long period of time.

Context recognition when applied on RDF data has the advantage to take into account structural and relational information of data. This is not the case for numerical context data. Since our context data are represented as RDF graph, data mining techniques that can be applied to these data are called graph mining techniques (12). Graph mining techniques have been applied to video databases on the Internet, to mutagenesis databases (chemical data) and the Web. In our approach, graph mining techniques and especially the graph clustering, will help us to identify relevant contexts for services. In our case, the clustering consists of segmenting the data set into homogeneous subgroups called clusters (contexts). A cluster is called relevant context if the constructed group of data is related to a particular situation interesting for a service. For example, John likes watching classic movie on the Internet once a week is an interesting context for the video distribution service.

17.8 Software Infrastructure for Service Adaptation

In this section, we present our architecture for service adaptation based on context data-driven approach. Our architecture differs from the existing architectures discussed in (3). These architectures addressed more the problem of context data gathering and provisioning, but they do not tackle the recognition of contexts from these data and they do not demonstrate how context-aware applications can be adapted on the base of learning techniques on data. Our architecture encompasses the management and usage of context.

17.8.1 Amigo Context Management Service (CMS)

In our work, we used the CMS (31) for collecting and delivering context data. The CMS acquires data coming from various sources. The most salient features of the CMS include *(i)* its compliance to the Web service architecture both for interfacing to context consuming applications and integrating its sub components and *(ii)* the modeling of context data using a high level language with much expressiveness.

17.8.2 Our Architecture for Service Adaptation

Our system is implemented on top of the Amigo context management service. This design makes it possible to collect and store context data in order to be easily processed by data mining algorithms. The following diagram 17.7 introduces how our approach components are mapped to the home environment. Context data are collected and stored using the CMS component. After that, the learning module proceed in two phases to accomplish services adaptation. In the first one "offline phase", contexts and activated services are recognized using clustering methods and instance ontology matching. In the second phase, the system tries to detect learned contexts and to launch the corresponding adaptation model. The adaptation model should be robust to changes in the use of services.

17.8.3 Use Cases

In this section, we present some use cases that can be handled using our approach, followed by a summary example that illustrates how our approach can be applied to a use case in the home environment.

- **Intelligent reminder**: Let's suppose a person is watching TV and the sound is loud, it is not useful to remind the person with a phone call or audible alarm, a text message may be more interesting. The reminder system can also choose to decrease the sound of TV when receiving

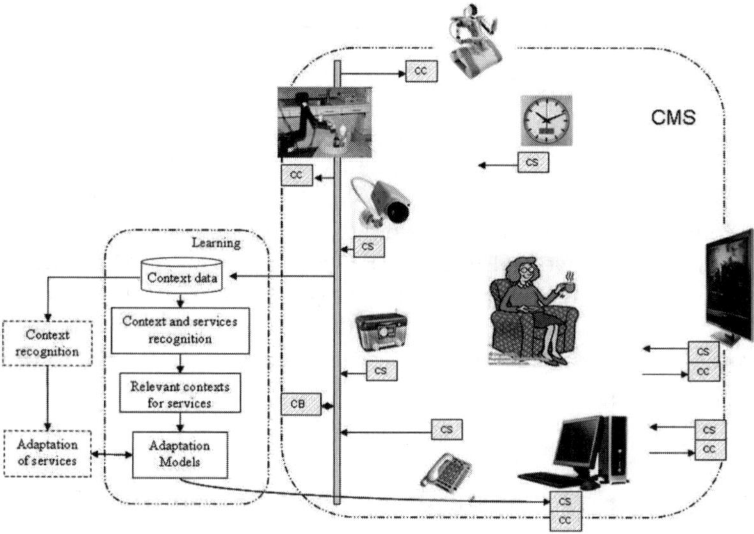

Figure 17.7: Our architecture mapped to home environment

an incoming call. Moreover, to communicate with other people, the reminder system can use different interaction devices and modalities (phone, mail, text, speech, sign, etc.). If the user is preparing something in the kitchen, and he can not hear an important incoming call because of the noise, then a message can be displayed to him in the kitchen communicator interface. The communicator could catch the user's voice response, encode it to text then send it by mail.

- **Service composition**: Presently, if a service requires other services to satisfy a request, and if these services are unavailable or temporarily absent, the composition fails. If services that can potentially be used by others were aware of the context of the services that might use them, they could adapt themselves accordingly. For example, a group of services required by a composite service in order to perform a task can decide to make themselves available to accommodate that service.

- **Search engine on the Web**: The user often makes requests on the Web using key words. There are many research efforts to improve search engines responses. By taking into account the user context such as: user location, language, previous search, search time, gender, etc. these services will provide appropriate answers to users' requests.

17.9 A Summary Example

We are interested in experimenting our approach in elderly people living domain. The improvement of life quality of elderly people has become a very active research field. Applications are increasingly numerous with very intense competition between professionals in health, telecom, device manufacturers, etc. We are involved in the MIDAS project [6]. The MIDAS proposal aims at demonstrating innovative and integrating semi-mature technologies. Techniques that will be developed intend to facilitate the deployment of usable products and services and to offer suitable solutions for the continuously growing community of elderly people.

In this section, we use the following scenario to illustrate our approach.

17.9.1 Scenario: Cultural and sport activities for elderly people

The existing solutions for cultural and sport activities reminder propose to elderly people some well known activities. The choice of an activity can be done according to the person situations. In our case, we will track some specific situations (the person hasn't had a social contact for 3 days, the person spends 5 hours in his bedroom between 8 AM and 4 PM), and we suggest some activities which can help him to improve his life quality, such as going to a meeting for elderly people 100 meters from his house.

Scenario: John is a senior who likes cooking and lives alone since 3 years ago. Recently, John has equipped his house with an intelligent reminder system. The reminder system detects that John spends 5 hours in his bedroom between 8 AM and 4 PM. The reminder system learns that John needs a social contact. It checks elderly people that have similar profiles to John. When done, the system proposes to John to contact Marry who lives in the same city, and likes cooking. John and Marry become friends, and prepare cooking together frequently.

[6]http://www.midas-project.com

Context data

Context parameter	Subject	Object	Predicate subject	Predicate object	Identifier subject	Identifier object	Confidence	Source	Timestamp
Location parameter 1	Room	Person	hasLocation	isLocationOf	bedRoom	John	90	infrared 1	20/10/2008: 09:30:00
Location parameter 2	Room	Person	hasLocation	isLocationOf	Corridor	John	85	infrared 2	20/10/2008: 14:35:00
Location parameter 3	Room	Person	hasLocation	isLocationOf	bedRoom	John	80	infrared 1	21/10/2008: 08:30:00
Location parameter 4	Room	Person	hasLocation	isLocationOf	Corridor	John	90	infrared 2	21/10/2008: 13:30:00
Location parameter 5	Room	Person	hasLocation	isLocationOf	bedRoom	John	90	infrared 1	22/10/2008: 10:30:00
Location parameter 6	Room	Person	hasLocation	isLocationOf	Corridor	John	80	infrared 2	22/10/2008: 15:30:00
Location parameter 7	Room	Person	hasLocation	isLocationOf	livingroom	John	95	infrared 3	23/10/2008: 09:00:00
TV parameter 1	Status	Device	hasStatus	isStatusOf	ON	TV	95	Electrical Sensor 1	23/10/2008: 09:00:00

Relevant contexts

Context	Description
Context 1	John spends 5 hours in the bedroom
Context 2	John is watching TV

Adaptation

Service	Adaptation
Reminder	Display Marry's profile on John's TV at 23/10/2008: 09:00:00

Figure 17.8: Running example on the scenario

Figure 17.8 shows an example of context data with respect to the above scenario. Each row in the table represents nodes and relations of a context data graph according to our model discussed in section 17.5.3. The reminder system recognizes two hidden context which are ("John spends 5 hours in the bedroom", "John is watching TV"). These contexts are respectively deduced from the location parameters (1,2),(3,4),(5,6), and (location parameter 7, TV parameter 1). When the system recognizes the context (1) , it searches a social contact for John, and after recognizing the context (2), it proposes the contact "Marry" then displays her profile on the TV.

17.10 Conclusion

In this chapter, we addressed the use of context-awareness for service adaptation. We proposed a data-driven approach which includes three main parts, context data modeling, storage and context recognition. We argued our technical choices through the presentation of requirements and discussion issues related to each part. We also emphasized the importance of context recognition for service adaptation. The overall architecture of our approach illustrates how the above parts can be integrated together to fulfill service adaptation. At the end of this chapter, we show by a summary example a description of relevant contexts that can be extracted from data and a description of a possible adaptation for service. We are still working on the implementation of our approach, and experimentations with multiple context data.

References

[1] H. P. Alesso and C. F. Smith. *Thinking on the Web: Berners-Lee, Gödel and Turing*. Wiley-Interscience, 2006.

[2] T. B. An, L. Young-Koo, and L. Sung-Young. Modeling and reasoning about uncertainty in context-aware systems. In *ICEBE 05 Proceedings of the IEEE International Conference on e-Business Engineering*, pages 102–109, Washington, DC, USA, 2005. IEEE Computer Society.

[3] M. Baldauf, S. Dustdar, and F. Rosenberg. A survey on context-aware systems. *International Journal of Ad Hoc Ubiquitous Computing*, 2(4):263–277, 2007.

[4] L. Baolin and B. Hu. An evaluation of rdf storage systems for large data applications. In *Proceedings of the International Conference on Semantics, Knowledge and Grid*, Los Alamitos, CA, USA, 2005.

[5] P. J. Brown. The stick-e document: a framework for creating context-aware applications. In *Proceedings of the International Conference on Electronic Publishing and Document Manipulation*, pages 259–272, 1996.

[6] P. J. Brown and G. J. F. Jones. Exploiting contextual change in context-aware retrieval. In *Proceedings of the 2002 ACM symposium on Applied computing*, pages 650–656, New York, NY, USA, 2002. ACM.

[7] L. Caixue. Contents object-oriented database systems: A survey. Technical report, Computer Science Department, University of California at Santa Cruz, 2003.

[8] S. Ceri, F. Daniel, M. Matera, and F. M. Facca. Model-driven development of context-aware web applications. *ACM Transaction on Internet Technology*, 7(1), 2007.

[9] T. Chaari, E. Dejene, F. Laforest, and V. M. Scuturici. Modeling and using context in adapting applications to pervasive environments. In *ICPS'06 : IEEE International Conference on Pervasive Services*, 2006.

[10] T. Chaari, F. Laforest, and A. Celantano. Design of context-aware applications based on web services. Technical report, RR-LIRIS-2004-033, France, 2004.

[11] I. Y. Chen, S. J. H. Yang, and J. Zhang. Ubiquitous provision of context aware web services. In *IEEE International Conference on Services Computing*, pages 60–68, 2006.

[12] D. J. Cook and L. B. Holder. *Mining Graph Data*. John Wiley & Sons, 2006.

[13] P. D. Costa. *Architectural Support for Context-Aware Applications: From Context Models to Services Platforms. CTIT Ph.D.-Thesis Series, No. 07-108, Telematica Instituut Fundamental Research Series, No. 021 (TI/FRS/021), Enschede, The Netherlands, 2007, ISBN 978-90-75176-45-2*. PhD thesis, 2007.

[14] E. Degene, S. Marian, and B. Lionel. Semantic approach to context management and reasoning in ubiquitous context-aware systems. In *2nd International Conference on Digital Information Management, ICDIM07*, pages 500–505, 2007.

[15] A. K. Dey. Understanding and using context. *Personal Ubiquitous Computing*, 5(1):4–7, 2001.

[16] A. K. Dey., G. D. Abowd., and D. Salber. A conceptual framework and a toolkit for supporting the rapid prototyping of context-aware applications. *Human-Computer Interaction*, 16(2):97–166, 2001.

[17] J. Euzenat, J. Pierson, and F. Ramparany. Dynamic context management for pervasive applications. *Knowledge Engineering Review*, 23(1):21–49, 2008.

[18] F. Giunchiglia. Contextual reasoning epistemologia. *Special Issue on I Linguaggi e le Macchine*, XVI:345–364, 1993.

[19] F. Giunchiglia and C. Ghidini. Local models semantics, or contextual reasoning = locality + compatibility. *Artificial Intelligence*, 127, 2001.

[20] C. Guanling and D. Kotz. Context aggregation and dissemination in ubiquitous computing systems. In *Proceedings of the Fourth IEEE Workshop on Mobile Computing Systems and Applications*, page 105, Washington, DC, USA, 2002. IEEE Computer Society.

[21] K. Henricksen, J. Indulska, and A. Rakotonirainy. Modeling context information in pervasive computing systems. In *Pervasive '02: Proceedings of the First International Conference on Pervasive Computing*, pages 167–180, London, UK, 2002. Springer-Verlag.

[22] J. Hidders. Typing graph-manipulation operations. In *Proceedings of the International Conference on Database Theory*, pages 391–406, 2003.

[23] J. Indulska and K. Henricksen. *The Engineering Handbook on Smart Technology for Aging, Disability and Independence*, chapter 31, pages 585–605. John Wiley and Sons, Inc., Publication, 2008.

[24] M. Keidl and A. Kemper. A framework for context-aware adaptable web services. In *Proceedings of the International Conference on Extending Database Technology*, pages 826–829, 2004.

[25] K. V. Laerhoven. Combining the selforganizing map and k-means clustering for online classification of sensor data. In *International Conference on Artificial Neural Networks*, pages 464–469, 2001.

[26] R. Macgregor and K. In-Young. Representing contextualized data using semantic web tools. In *Workshop "Practical and Scalable Semantic Systems" of 2nd International Semantic Web Conference*, 2003.

[27] J. Mantyjarvi, J. Himberg, P. Kangas, U. Tuomela, and P. Huuskonen. Sensor signal data set for exploring context recognition of mobile devices. In *Workshop "Benchmarks and a database for context recognition" of 2nd International Conference on Pervasive Computing*, 2004.

[28] J. McCarthy. Generality in artificial intelligence. *Communications of the ACM*, 30(12):1030–1035, 1987.

[29] J. McCarthy and S. Buvac. Formalizing context. In *Computing Natural Language, CSLI Lecture Notes*, pages 13–50, 1998.

[30] G. Ramanathan. *Contexts: a formalization and some applications.* PhD thesis, Stanford, CA, USA, 1992.

[31] F. Ramparany, R. Poortinga, M. Stikic, J. Schmalenstroer, and T. Prante. An open context information management infrastructure the ist-amigo project. In *Proceedings of the 3rd International Conference on Intelligent Environments*, pages 398–403, 2007.

[32] A. Renzo and G. Claudio. *The Semantic Web: Research and Applications*, chapter Querying RDF Data from a Graph Database Perspective, pages 346–360. Springer, 2005.

[33] A. Renzo and G. Claudio. Survey of graph database models. *ACM Computing Surveys*, 40(1):1–39, 2008.

[34] T. Strang and C. L. Popien. A context modeling survey. In *Workshop "Advanced Context Modelling, Reasoning and Management" of the Sixth International Conference on Ubiquitous Computing*, 2004.

[35] H.-L. Truong and S. Dustdar. A survey on context-aware web service systems. *International Journal of Web Information Systems*, 5(1):5–31, 2009.

[36] R. D. Virgilio and R. Torlone. Modeling heterogeneous context information in adaptive web based applications. In *ICWE '06: Proceedings of the 6th international conference on Web engineering*, pages 56–63, New York, NY, USA, 2006. ACM.

[37] K. Wilkinson, C. Sayers, H. A. Kuno, and D. Reynolds. Efficient rdf storage and retrieval in jena2. In *EXPLOITING HYPERLINKS 349*, pages 35–43.

[38] H. Wu, M. Siegel, and S. Ablay. Sensor fusion for context understanding. In *Proceedings of the 19th IEEE Instrumentation and Measurement Technology Conference*, pages 13–17, 2002.

Chapter 18

Research Challenges in Mobile Web Services

Chii Chang, Sea Ling, and Shonali Krishnaswamy

Abstract Mobile devices such as smart phones, personal digital assistants (PDA), or ultra-portable computers are used as the medium for accessing heterogeneous distributed services in smart environment or mobile Internet. As the Web becomes mobile, it raises various challenges that make it unique and distinct to the traditional Web service approaches. Moreover, since mobile devices evolved, hosting Web services in mobile devices has become a reality. It brings about many possibilities such as using a mobile device as a moveable context information collector. However, it also raises many challenges derived from various areas such as resource constraints, interoperability, lack of standards, availability of tools, and other management-related issues. This chapter describes these challenges and indicates possible research directions particularly the area of mobile peer-to-peer Web services and enabling context-aware mobility support for Web services.

18.1 Introduction

Since the emergence of ubiquitous computing, various contributions have been proposed to realize it. The ultimate goal is to provide users a convenient smart environment where information and facilities are available for easy access and interaction without needing the users to play the active role in discovering them. Such an environment can be realized today with distributed computing technologies where the environmental entities and services such as smart devices and facilities, advertisements, shopping, and tourists guide can be implemented as services and are available to be seamlessly accessed by users. However, to realize the human-computer interaction ubiquitously, an adaptive medium is required. Mobile devices such as smart phones, personal digital assistants (PDA), or ultra-portable computers are used as the medium since they are carried by users in their daily lives almost anytime, anywhere. These devices have provided various benefits such as fast booting time compared to traditional computers; available wireless network connections such as Bluetooth, WiFi in local wireless network, and General Packet Radio Service (GPRS) (40), 3G (42) in mobile Internet; embedded Radio-Frequency IDentification (RFID) reader enabling mobile devices to sense environmental information (21); camera for photograph taking, video recording, and moreover, for touching graffiti information (24). Mobile devices have gradually been used as means to access services. Furthermore, mobile devices can be designed as service providers in order to switch their role from passive to active to enable autonomous interaction with the environmental services. Examples include automatically providing information to the unsubscribed mobile users who arrive in the environment for the first time.

Services that have been designed specifically to serve mobile devices are known as mobile services. Specifically, mobile services have been defined as Web services that are accessed via mobile devices (3, 32, 48, 49, 52). We refer the term — mobile Web service (MWS) (13, 17, 20, 11, 16, 51) to represent such a service. A typical MWS should encompass the features of traditional Web services, that is, supporting service-oriented architecture, platform independence, loose coupling, ability to register, ability to be discovered, and so on.

In recent years, research aspects on MWS concern the limitation of hardware such as the screen size of the device, performance, memory, storage, battery life, and network issues such as connectivity, speed, costs, stability, and reliability in mobile Internet. These limitations and issues raised many challenges. Moreover, a primary objective of MWS is to provide an adaptive service to requesters based on their needs. A major aspect is on how the services can be provided adaptively based on factors such as user preference, device specification, network conditions, etc.

This chapter is organized as follows. In Section 2, we classify MWS infrastructures into two categories and summarize the features and limitations of each. Follow up with the foundation of MWS is described. In order to enhance the capability of MWS, context-aware MWS and context-aware dynamic composite MWS have been applied in a number of works and is introduced. We also describe existing technologies that enable MWS and a discussion on MWS management approaches. We identify research challenges in Section 3.

18.2 Enabling Mobile Web Services: State of the Art

18.2.1 Mobile Web Services: Two Models of Hosting

The term MWS has been described discordantly. On the one hand, MWS indicates a mobile application accessing a remotely hosted Web service deployed on a stable, resource-rich stationary Web server. On the other hand, MWS also implies a mobile device hosting/running a Web service that is accessible by mobile and nonmobile applications. We have reviewed a number of works in each model and discuss their features and limitations in the following sections.

Static-Hosted Mobile Web Service (SHMWS)

In (17, 46, 51, 53), mobile Web services have been described as the Web services that are designed particularly for mobile clients such as smart phones and PDAs. The service providers of this infrastructure are no different from the static Web service systems. We call these MWS as static-hosted MWS (SHMWS) in which mobile devices are the service consumers requesting services from a remote static host. These works focused mainly on how to deliver services to the mobile clients efficiently. We summarize the features and limitations of SHMWS infrastructure below:

Features:

- It is easier for the implementation and maintenance to be performed within a centralized service control infrastructure managed by static Web service systems.
- Due to the existence of relevant standards such as Universal Description Discovery and Integration (UDDI), it simplifies the service discovery process.
- Static mobile Web services provide better quality of service because stationary servers can have higher resource specification. The servers are capable of handling high amount of interaction between client and server. Existing security technologies can also be directly applied.
- Benefiting from storage capability, static mobile Web services are able to store clients' data on the server side machines.

Limitations:

- Since every interaction needs to go through the static server, the process will be slower and will cost more than that for a direct interaction between mobile peer to peer (in the case when such a request is made).

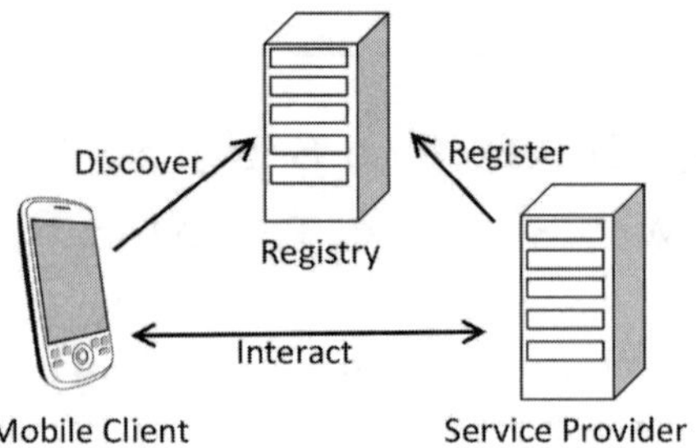

Figure 18.1: Static-Hosted Mobile Web Service (SHMWS).

- Pure peer-to-peer connectivity is no longer an option. This means that if the central service is down, the communication between mobile peers cannot be performed.
- Due to the Web service not being hosted by the mobile device, the device itself cannot be utilized as a mobile context information collector in a smart environment.

Mobile-Hosted Mobile Web Service (MHMWS)

Mobile-hosted mobile Web service also known as Nomadic Mobile Service (NMS) (36), has been defined as the service that is *"hosted on the mobile host such as a handheld device, mobile phone or any type of embedded device capable of connecting to the Internet using wireless network."* (35). In the infrastructures described by (13), (20), (34), and (44), MWS denotes the Web services that have been hosted on the mobile devices to provide Web services to other mobile or stationary clients. Hence, mobile devices can be designated as both service requesters and providers enabling the pure mobile peer-to-peer interoperability and Web service-based mobile ad hoc network ubiquitously. Essentially, hosting Web services on mobile devices is a highly challenging task, due to resource limitation such as physical hardware and network bandwidth. However, the capability of hardware (such as CPU, RAM, storage, etc.) has been improved in recent years and the network bandwidth can be provided at 2.5Gbps to this day.[1] It is anticipated that many previous challenges can then be solved in the near future. We summarize the features and limitations of MHMWS:

Features:

- In the central static service infrastructure, when two mobile users need to interact, they have to upload data via the central service controller

[1]NTT DoCoMo Achieves World's First 2.5Gbps Packet Transmission in 4G Field Experiment <http://www.nttdocomo.com/pr/2006/000722.html>

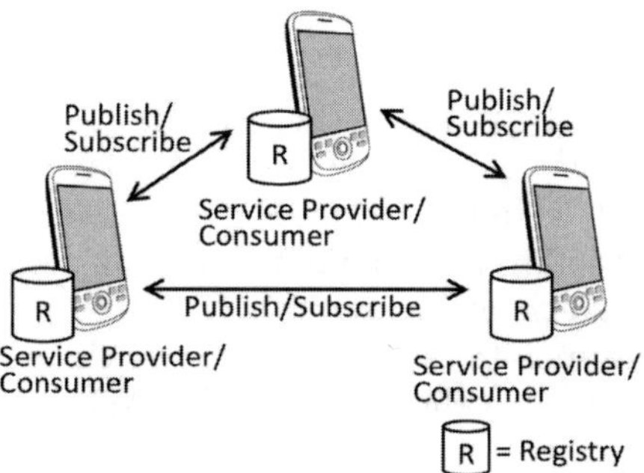

Figure 18.2: Mobile Hosted Mobile Web Service (MHMWS).

where the controller charges a service fee for each transaction. This means that both mobile users (mobile service provider and the consumer) have to pay the transaction fee. However, if the mobile service provider node also hosts the server on its mobile device, the service consumer can directly interact with the service without going through the central controller. Hence, the cost can be reduced.

- MHMWS allows mobile devices to establish more readily Web service-based adhoc network that realizes the wireless virtual community environment, due to its more direct peer-to-peer nature.

- Context retrieval can be more efficiently performed by a MHMWS. For example, a MHMWS node can be requested by a mobile node (which is located in another part of the world) to provide its current location, environment context or moreover, the user's activity and situation.

Limitations:

- Hosting Web services on mobile devices may be limited by the capacity of hardware resources that include CPU, RAM, storage, and battery life. Due to the limitation, it can only serve a restricted number of clients at a time.

- Quality of service may be an issue due to the various limitations on bandwidth, connectivity, and current hardware resources that are unable to handle a high amount of interaction asynchronously.

- It is a challenge to ensure the mobile service provider node stays connected during the service invocation period.

- Available standards and frameworks are limited. Compared to the traditional Web services, MHMWS lacks industrial standards and open-source frameworks.

Rationale

Both SHMWS and MHMWS have their strategies in different domains. We describe two scenarios for comparing these two MWS styles. For a scenario based on an m-Health system, the patient's mobile device is utilized as a context collector interacting with environmental services to send the collected context data to the end-point in hospital. If the system is built in SHMWS infrastructure, the end-point is unable to request the patient's mobile device for a particular data, due to the fact that the Web service is not hosted on the patient's mobile device. The only option is to rely on the patient's mobile device recurrently submitting all the data it has collected to the SHMWS, which is costly and inexpedient. On the other hand, such scenarios can be performed more efficiently in the MHMWS-based infrastructure.

The second scenario is based on a ubiquitous shopping mall where the stores provide SHMWSs for consumers to search information (items, locations, customer services, etc.) via their mobile devices. In this scenario, consumers are in active roles who have to search services manually. On the other hand, if MHMWSs are hosted on consumers' devices together with user preference profiles that can be retrieved through MHMWSs. The role of consumers can be switched to passive in which store services are able to promote services to consumers automatically based on user preference and location.

The above scenarios indicated a key strategy of MHMWS that is to enhance the autonomy and interoperability of ubiquitous systems. Although MHMWS can be hosted on mobile devices, the issues of resource constraint are major concerns. A MHMWS is usually of low throughput (compared to SHMWS) and high latency (due to high speed mobile Internet connections are expensive). The issue of battery life also restricts the availability of MHMWS. Not to mention, the lack of available MHMWS tools (to support Web service standards) enforces MHMWSs being implemented as highly coupled systems.

18.2.2 Mobile Web Service Foundation

Recently, various MWS approaches have been introduced in both SHMWS and MHMWS infrastructures. Numerous works applied middleware technologies toward resolving the challenges. They include:

- Enhancing the overall system performance in the resource constraint mobile device (2).
- Providing MWS discovery in the cellular network or mobile broadband (11).
- Improving Quality of Service (QoS) in the MWS interaction affected by resource limitation such as network latency and the throughput of MHMWS provider (43).

- Providing service mobility to or from mobile peers including MHMWS providers and clients whose availability and stability are influenced by dynamic factors such as available network connection and protocol in mobile user's current location (36).

Intuitively, resolving these fundamental challenges relates to the basic design principles of MWS infrastructure.

A traditional Web service infrastructure encompasses three basic elements: service provider, service registry, and service requester. Service providers are required to register themselves to the service registry in order to make themselves visible to service requesters who are capable of discovering service providers from the service registry. The registration process usually relies on submitting a standard-formatted service description that provides information of the service provider including basic information such as address and functions. In SHMWS domain, such a process can be directly inherited since service provider and registry are implemented in a stable and stationary node. However, in MHMWS domain, service providers are movable objects with different connectivities and protocols depending on their device specifications. A service provider can be interoperable this minute but disappear in the next minute due to numerous factors such as out-of-network range or it's a change in IP address when switching to another domain. Although enabling static IP address to mobile peers is possible if the mobile Internet service provider is willing to provide such a service, it does not address service discovery and stability issues. Hence, numerous researchers introduced open-source mobile peer-to-peer frameworks enabling dynamic service discovery for MHMWS. Dynamic service discovery mechanism supports runtime service registration based on advertisement or publish/subscribe scenario in which a mobile node (either a service provider or requester) is capable of discovering or be discovered when it enters the MWS-enabled network. Existing technologies such as JXTA2 (23), Knopflerfish OSGi (10), Jini (34) are capable of supporting MWS systems realizing such mechanism.

Indeed, the MWS foundation is not only about discovering and interacting services. To accomplish the best practice in MWS, researchers such as (4) and (43) have designed their MWS infrastructure based on Enterprise Service Bus (ESB) architecture (39). ESB is software infrastructure that enables a Service-Oriented Architecture (SOA) by providing the "Link" between service providers and requesters. A service is no longer specifically an XML Web service or a Java Enterprise Servlet or a SOAP-based Web service. A service in ESB is derived from any type of software component implemented in any programming language and on any platform. For example, in a ubiquitous computing environment, environmental facilities are utilizable via their corresponding components that were implemented in different forms. ESB provides a runtime environment enabling these components to be registered as manageable ESB service providers — Bus Service Providers (BSP), which are usually utilized as XML Web service, with a common service description format (usually in XML format such as WSDL) letting service requesters interact with them without having a single knowledge about the service provider. Within the ESB infrastructure, both service provider and requester are ESB's clients and are able to register themselves to ESB. A registered ESB requester — Bus Service Requester (BSR) does not need to be aware which BSP is capable of handling its request. The request only needs to be submitted to ESB with a clear description of what the goal is and perhaps with additional information describing its criteria, policy, and QoS requirements. ESB will handle the request by routing the request message to corresponding BSP who is capable of accomplishing the request.

18.2.3 Context-Aware MWS Approaches

Context awareness was introduced by (38). Context was described as the environmental phenomenon influencing the system process and enhancing the adaptive autonomous actions of the system. Later on, (9) defined context as: *"Context is any information that can be used to characterize the situation of an entity. An entity is a person, place, or object that is considered relevant to the interaction between a user and an application, including the user and applications themselves"*.

Context awareness computing has been an interested research field in the Web services area for many years. However, more factors need to be considered in the mobile environment. In general, context can be categorized into two types: static context and dynamic context (28). Static context is a user-customizable context that considers a user profile containing the user's personal information and preferences. Dynamic context can be considered as environmental factors in which the user is passive and is unable or has less in control. The system relies on external entities to be observed, sensed, and to derive meaningful information.

In SHMWS, a number of approaches (15, 22, 17, 46, 41) have applied different contexts as request parameters in order to provide feasible service to the service consumer. In (15), applications supported by the device and data transmission delay have been considered. Base on these contexts, the central controller service will redirect the request to the specific service provider in order to provide quality content to the request. In (22, 17), the primary objective of the service is to push service to subscribed clients autonomously based on their current context. (46) have mainly focused on QoS-Aware service provision in which the service broker has been placed between service provider and consumer. The service broker handles a client's request with QoS requirement described by WS-QoS and discovering the feasible service provider who matches the requirement. The broker directly interacts with the service provider and retrieves the result for the client. The approach proposed by (41) used a multi-agent technique to handle the client's context and acting as client's proxy that interact with Web service providers.

In MHMWS, (11) aim to support context-aware service direction/discovery in which the approach considers user-related context, an available resource, a location, and time context. This work involved a semantic model that structures the service direction recommendation based on both client and server-side context.

Different to traditional Web service environments where the service provider and clients are usually connecting to the stable LAN or WAN, mobile nodes in MWS environments are usually dynamically connected to the wireless network such as Bluetooth, WiFi, wireless broadband, or satellite Internet depending on the user's location and available network. In order to provide the best network connection for MWS nodes, context-aware vertical handover middleware (36) was proposed for supporting the prediction of the vertical handover. The middleware is capable of sensing the MWS provider node's movement, direction, and surrounding network. It enables the MHMWS provider node to measure the network availability and performing the autonomous decision-making based on the utility algorithm for the vertical handover. Moreover, the system is capable of ensuring that the MHMWS client-service interaction will not be interrupted due to network switching.

18.2.4 Mobile Web Service Composition

Service composition enables multiple services to be invoked as a single, customizable adaptive service, to best accomplish a client's request. It usually involves orchestration. Orchestration, using workflow-based approaches such as BPEL4WS (26), mainly describes the interoperation between services. It involves the following elements:

Goal achievement — Composite service is customizable. It provides adaptive service provisioning to the requester based on the goal of the request query (54). A goal denotes the objective of the request, which in the case of the SOA-based environment, is usually based on orchestration, involving multiple tasks performed by different service invocations.

Invocation accomplishment — Composite service is capable of analyzing the request query and performs semantic service discovery, executing tasks by invoking relevant services to achieve the goal. The invocation relies highly on the decision making of the composite service to select the best service to perform the tasks, which is based on the QoS, availability, reliability, stability, etc. (1).

Fault handling — Recently, approaches that involve service composition are usually dynamic in nature due to the potential failure of predefined service providers at runtime. Dynamic composition, which denotes defining the invoked partner components at runtime, enhances the capability of fault handling by the composite service (7). Failure can be replaced by substitution, which can be accomplished by numerous ap-

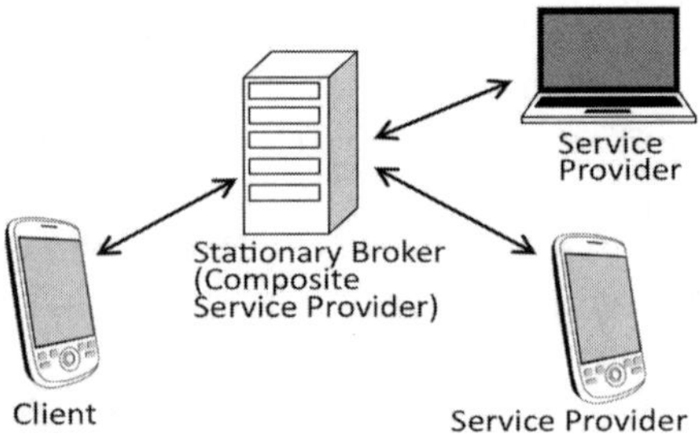

Figure 18.3: Infrastructure-based MWS composition.

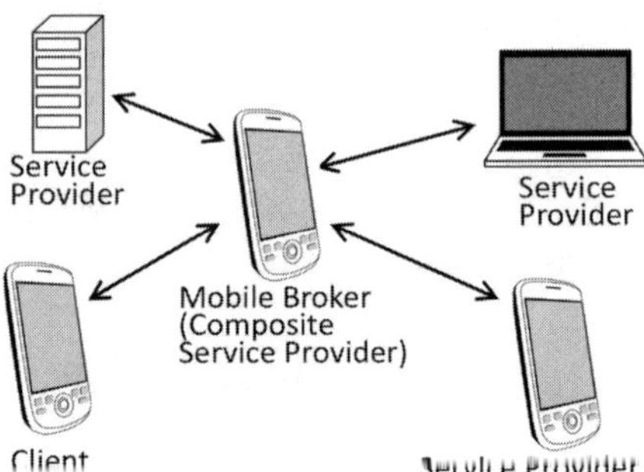

Figure 18.4: Infrastructure-less MWS composition.

proaches such as policy-based (12), context-aware rule-based (54), or semantic-based substitution such as ontology (18).

Service composition in MWS can be classified into two different approaches: infrastructure-based and infrastructure-less.

Infrastructure-based MWS composition has been applied in both SHMWS and MHMWS. It relies on an "always available" stable node (usually a stationary node) placed within a MWS network. Such a node is also known as a broker. Since the broker is located at the stationary node, it encompasses numerous benefits such as: availability of various industrial tools for development (e.g., ServiceMix[2], Open ESB[3]) and ease in maintenance and QoS control. Furthermore, the stationary broker can be administrated by I.T. professionals all the time. However, a single point of failure may cause the entire system to fail and consumers may not been keen to pay for the service (since the brokers are maintained by I.T. professions, it requires extra service fees to use the system).

[2]Apache ServiceMix. <http://servicemix.apache.org/home.html>

[3]Open ESB. <https://open-esb.dev.java.net/>

On the other hand, infrastructure-less service composition (5, 14) is mainly used in mobile ad hoc network environments in which mobile devices interact with other WS providers, as well as MWS provided by other mobile device users and WS provided by environmental mobile or static hosts such as sensor embedded environmental facilities that are intractable via WS interface. A set of tasks that involves a number of devices that are communicable via WS in a specific environment can be accomplished via the composite MHMWS from a requester that performs distance control. An infrastructure-less composite service can avoid the single point of failure and reduce the cost derived from the stationary node maintenance. However, numerous limitations such as the lack of available industrial tools support, QoS and security constraints due to mobile nodes being maintained by regular mobile users, resource constraints such as the battery life of mobile devices and wireless network speed, contribute to the challenge of infrastructure-less service composition.

Context-Aware Dynamic Service Composition for MWS

In order to provide customizable, adaptive service provisions based on contexts, a number of researchers have applied context-aware capabilities towards supporting dynamic service composition in a mobile service area. The fundamental service composition in a mobile service environment is based on two major principles:

- Dynamism: The environment is purely dynamic in which mobile peers are joining and leaving the network freely. Hence, *"compositions cannot be predefied, but must be created and recomposed at runtime."* (29).
- Decentralization: It not only involves the distribution of an entire system, it must also consider a central, trusted party for managing and governing the composition can be nonexistent (in an infrastructure-less environment).

In this section, we bring forward a number of dynamic service composition approaches that have been designed specifically for mobile computing and pervasive computing. These works have applied context-awareness towards enabling the adaptive service composition in numerous areas.

In the earlier contribution of the context-aware dynamic service composition in mobile computing, MobiPADS (6) emphasized a situation-aware adaptive composite service provision based on network and service conditions. The approach relies on the predefined rule and policy to react to a runtime event in order to provide stable service to its mobile clients. However, such an approach does not address many context factors that have been illustrated as major concerns in today's mobile and pervasive environments. Such context factors include a mobile user's current location and movement that have however been considered in Plumber Service (21). Plumber Service is an RFID-based context-aware composite service that utilized the RFID reader embedded in mobile devices to sense the user's current environment. Based on the user's location and movement, the composite service analyzes the situation and then dynamically performs recommendations to the mobile user.

Recently, workflow-based dynamic composition approaches have been proposed in numerous contributions. (8) have proposed the workflow-based approach based on context factors. The adaptive composition process is based on location, identity, time, activity, and connectivity. Each context type is retrieved by its corresponding component and to analyzed by the system in order to avoid service mismatching. The composition process is based on a structural mapping model that was defined in policy.

The other two workflow-based approaches are COCOA (27) and (50). These two works have numerous similarities including using OWL-S[4] for the semantic service matchmaking and both are SOA Web service-based systems. By using OWL-S, structural service mapping is built on an open-standard technology in which the scalability is enhanced (comparing to (8)). COCOA is a QoS-Aware system in which the quality related contexts such as Service availability, latency, cost have been considered in the interaction parameter. Distinctive to COCOA, an approach proposed by (50) is intended to support adaptive service selection for composition at runtime in which the service will be dynamically defined based on a user's current location, preference, and previous service interaction record (for the quality ranking)

The above approaches have a common limitation, *they do not support an infrastructure-less environment in which both service provider and consumer are moving objects.* Although (16) and (37) have addressed dynamic service composition in infrastructure-less environments, they focus mainly on the methodology of dynamic service composition in which the involved technology was not explicitly described. These two works have a common adaptive capability in which the system is capable of predicting the stability of the service provider. In an infrastructure-less environment, the mobility support of the service provider is extremely important. In (37), the prediction is performed at the time of service selection (design time) in order to choose a stable service for the client. Kalasapur's prediction is performed at runtime in which the potential failed operation (caused by the invocation failure) will be partially replaced by a dynamic defined operation (with a stable service provider defined at runtime).

The algorithms proposed by (16) have considered both the mobile and non-mobile service provision and adaptation. However, to accomplish the objective of dynamic service composition for MHMWS in an infrastructure-less environment, there are numerous context factors that were not covered by Kalasapur's work (16), such as QoS-related context (e.g., throughput, latency) and hardware-related context (e.g., battery life). The speed of message transmis-

[4]OWL-S: Semantic Markup for Web Services. <http://www.w3.org/Submission/OWL-S/>.

sion also highly influences the overall performance. Moreover, (16) did not provide a structural semantic service matchmaking model.

Besides the approach of service composition, realizing dynamic service composition in an infrastructure-less environment requires a feasible middleware technology. Most existing frameworks (6, 21, 8, 27, 50) were designed for infrastructure-based environments. Infrastructure-less approaches such as (16) and (37) did not propose a generic open-standard framework (such as COCOA) that is capable of leveraging heterogeneous resource.

18.2.5 Mobile Web Service Development and Management

In recent years, various solutions have been introduced enabling Web servers to be hosted on mobile phones, such as Nokia Mobile Web Server[5] that hosts Apache HTTP server in S60 system[6] enabling users provide Web sites on their S60 mobile phones to share contents for social purposes; i-Jetty[7] adapt Jetty Web server in Google Android enabling the phone providing Java Servlet based services; and the ServersMan[8] developed by FreeBit turning iPhone/i-Pod into a Web server with a similar functionality as the Nokia Mobile Web Server.

Although these tools only have simple capabilities (e.g., sharing contents via the Internet), they can accomplish some useful tasks. For example, in an M-health system, a patient's mobile device can host the MobileSite (e.g., a Web site deployed on a Nokia Mobile Web Server) with extra standard Web service client applications and an embedded context middleware. As a Web service client, it is capable of retrieving environmental context information (e.g., current weather and temperature) from environmental services and user-related context (e.g., the patient's movement) from the embedded context sensor (e.g., the patient's coordinate from GPS service). This data is stored on the mobile device and retrievable from the MobileSite on the device.

On the other hand, if a more complex process is required, MobileSites are not entirely satisfactory. The complex process requires the MWS provider to interact with various distributed resources to provide composite service to end-users. It is needed when the MWS provider is required to complete a sequence of activities that involve a number of operations derived from various resources. A MobileSite is unable to satisfy such a scenario because the MobileSite can only provide Web pages. In this case, SOA-based Web service is required to be hosted by the mobile device.

---

[5]Nokia Research Center — Mobile Web Server. <http://research.nokia.com/research/projects/mobile-web-server/>

[6]Symbian. <http://www.symbian.org/>

[7]I-Jetty: Web server for the android mobile platform. <http://code.google.com/p/i-jetty/>

[8]Turn your iPhone into a Web server. <http://blogs.zdnet.com/Apple/?p=2983>

Management of MWS aims to improve the overall system performance and Quality of Services (QoS). Existing Web service standards for management such as Web Services Distributed Management (WSDM), does not address the management issues for MWS (47). (47) defined a number of principles for MWS management. Some of the defined principles are generic to common context-aware Web service as well. We summarize them by separating the principles into two categories:

Generic Web service management principles:

- Enabling the publishing, monitoring, analyzing, and updating of context information (e.g., location) of a MWS and using context information for monitoring and controlling activities in the management of MWS and its compositions.
- Supporting reconfiguration based on Service Level Agreement (SLA), policy, constraints, and dynamic-changed context factors.
- Maintaining MWS status and providing such information to clients.
- Enabling catching and pre-fetching of the operation results for fault handling.
- Supporting application-level invocation retries to a disconnected MWS.
- Enabling failures to be replaced by replicas/substitutions.

Mobile Web service-related principles:

- Requiring minimal resource usage. Such resources include processing power, memory, electric power, and wireless network communication needs.
- Implementing mechanisms for supporting efficiency and reducing system overheads.

The two MWS-related principles are the most important concerns in a MHMWS domain. A number of approaches have been proposed towards supporting the efficiency and reducing overheads of MWS specifically. In the message exchange level, (31) have proposed a mechanism to optimize the SOAP message for reducing the time of message exchange. (2) have proposed an architecture in which the remote stationary server has been used as the broker between the MWS client and MHMWS provider in order to reduce the burden on the MHMWS provider. (36) have introduced context-aware vertical handover to support dynamic wireless network connection changes for optimizing the connectivity of the MHMWS provider.

18.3 Research Challenges

18.3.1 Challenges in Mobile Web Service Foundation

Enterprise Service Bus has been introduced for several years and various contributions and solutions have been proposed. However, not much work on ESB covers the MWS domain. Implementing ESB in the MWS-based environment can raise research challenges that have not yet been explored thoroughly. Approaches and solutions designed for traditional static Web service environments cannot be readily applied to the MWS environment. The major challenges are described in the following subsections.

Dynamic-configuration

Existing industrial ESB solutions support various interactions for components implemented on different platforms. However, these solutions do not consider mobile devices as service requesters or providers. ESB enables MWS providers support dynamic configuration at runtime by examining the status of MWS nodes that include network condition, device performance, specification, battery life, runtime environment, available applications, and so on. Moreover, the approach should also cater to the unstable connectivity that can potentially occur in MWS nodes. Additional mechanisms should be regarded to support mobile client reconnection in order to continue from the previous process that involves the interaction of service providers that were invoked by the ESB. Such issues can be resolved by using a mobile software agent technique, which has been proposed in the SHMWS domain (53). However, the highly coupled approach has limited interoperability due to the fact that common mobile devices and Web service providers do not have such capability initially. On the other hand, in the MHMWS domain, the invoked service provided by a mobile host may be potentially disconnected from the network forever. Such situation may require adaptive substitution of service to be performed.

Semantic-service-discovery

In a common MWS infrastructure where service providers are registered in the centralized service repository, MWS requesters discover services in the repository and interact with the service provider directly. However, in the ESB-based environment, the functionalities provided by the invoked service can be integrated with other MWS providers. Such a multilevel interoperation requires a semantic service discovery mechanism in which the service discovery mechanism is based on the task and related contexts such as the user, device, application, and network condition. It should also be considered together with other service-provider-related contexts such as the stability of the service provider and the involved third-party services, their security level, and the description of QoS characteristics.

18.3.2 Challenges in QoS-Aware Interaction

QoS-aware service involves the overall performance of runtime invocation. QoS commonly involves throughput and latency. Typical QoS approach in MWS domain such as the work proposed by (46) combines with message compression technology to reduce the size of SOAP messages and enhance the message exchange speed in the SHMWS environment, while (36) applied context-awareness to enhance the overall performance in the MHMWS domain. These works were applied to the client server-based architecture. Performance testing was not considered in the multilevel service interaction based architecture where the invocation may involve multiple third-party services that require extra mechanisms to ensure satisfactory QoS.

When the service provision involves invoking multiple external services, each service having its policy, agreement, security level and condition, it is essential that a management party is available within the network. In a traditional infrastructure-based Web service environment, a central trusted management party can be applied to control the overall QoS. However, in infrastructure-less MHMWS environment, it is a challenge to control QoS and security due to the nonavailability of the centralized service for managing the QoS-related information of MHMWS providers.

18.3.3 Challenges in Mobile Web Service Composition

Common aims in both infrastructure-based and infrastructure-less service composition include self-configuring, self-optimizing, self-healing, and self-adapting capabilities. In MWS, dynamic composition should consider contexts including: mobility-related contexts (mobile user's movement, location, coordinate, etc.), resource-related contexts (device performance, the battery life of device, network connectivity, network speed etc.), and QoS-related contexts that involves service level agreement and constraints, the latency of the MWS provider and requester, and the throughput of the MWS provider in its current condition to the MWS requester who is also influenced by its context.

Generally, as mentioned previously, most existing works in dynamic service composition were designed for an infrastructure-based environment. We believe, in the near future, the service composition for an infrastructure-less environment will be one of the major aspects in a mobile Web service provision.

One of the primary purposes to use a composite service is to provide a dynamic, autonomous service that is capable of reacting to events at runtime based on contexts. A composition process for MHMWS in an infrastructure-less environment can potentially fail because of the following events:

- The external service provider is unavailable at runtime during the composition process because of the sudden disconnection caused by a hardware or network issue.
- The throughput and latency of the external service provider is not satisfactory due to system overhead in which the operation is unable to be completed in time.
- The service consumer has switched its network protocol before the response is delivered due to the user moving to a different zone.

As mentioned in the previous section, most existing frameworks in context-aware service composition do not consider service providers as mobile entities. When the service provider becomes mobile, changes become dynamic and influence the composition process. Predefined operations can fail because of mobility-related changes occurring in either the service provider node or service consumer node. Hence, the composite NMWS needs to be able to re-configure at runtime to adapt to the new situations dynamically.

18.3.4 Challenges in Mobile Web Service Development and Management

Developing MWS is not a straightforward task. There is no complete industrial released framework available (30). Platform-dependent and language-specific approaches cannot fulfill the loosely coupled SOA in MWS. In traditional desktop-based distributed systems, developing a SOA can be done by utilizing exiting standard technologies, adapting a message-oriented Web service to enable a loosely coupled system that is capable of leveraging resources from different platforms. Some researchers (45, 10) proposed SOA-based MWS. More work needs to be done towards the MHMWS environment. For example, Java ME does not contain key libraries to support hosting Web services like Axis,[9] which cannot be directly ported in the Java ME environment. Java-based ESB framework — ServiceMix used in (43) is also not compatible to the Java ME environment. Works like (10) using Knopflerfish OSGi for context sharing in mobile peer-to-peer environment, have mentioned that the AXIS-OSGI package for Knopflerfish OSGi enables OSGi services to be published as a standard Web service. However, the AXIS-OSGI package requires some Java SE specific libraries that are missing in the Java ME environment. Although it is possible to import the missing libraries to enable Axis in the Java ME environment, it also increases the burden of a resource-constrained mobile device.

Improving the efficiency and reducing the burden of the MHMWS provider is one of the major research aspects in managing MWS. Existing approaches

[9]Web Services — Axis. <http://ws.apache.org/axis/>

such as (2) and (47) have used a stationary/nonmobile service broker for effectively handling complex processes. However, such a design may not be suitable for a purely mobile environment, such as a decentralized mobile ad hoc network because a stationary node is unavailable. Approaches in a mobile peer-to-peer domain such as (13) and (19) used the mobile-hosted MWS broker as a service directory in which the MWS broker manages the publish/-subscribe processes for mobile peers. However, QoS issues still need to be explored.

An efficient MWS interaction is influenced by MWS node configuration and runtime factors. The configuration of a MWS node should take into account:

- Hardware resource-related issues such as operation performance, the display capability, and available applications for presentation.

- Network-related issues such as the supported network connections (Bluetooth, Wi-Fi, 3G, etc.) of the device, the communicable protocol (SOAP, RPC[10], REST[11], etc.) of the device platform.

- QoS-related issues such as SLA, QoS requirements, communication criteria, and the user-preference profiles.

The runtime factors include the current location of the mobile user, available network connection, available services and any context that may influence the user's task, the operation of the task, and the decision making by the autonomous mechanisms. All these issues need to be considered in the MWS interaction. An MWS provider should be able to change its configuration and behavior to fulfill user's requests.

According to (33), the management of SOA involves self-healing and self-protecting management services. A self-healing mechanism is capable of observing the operation of each single interaction between the service provider and requester and performs the reactive failure recovering mechanism without disrupting runtime operations. A self-protecting mechanism is capable of observing the nodes of interaction and predicting events for immunity to threats, blocking unauthorized accesses, and immunity to malicious attacks. Existing SOA-based MWS management solutions, especially in a MHMWS infrastructure such as (47), have not specifically addressed these two areas.

Due to the page limits, we are unable to provide much detail about trust and security related topics in this chapter. However, a recommended literature survey proposed by (25) will fulfill such a need.

[10]XML-RPC. <http://www.xmlrpc.com/>

[11]RESTful Web Services. <http://java.sun.com/developer/technicalArticles/WebServices/rest

18.4 Summary

In this chapter, we have discussed a number of approaches to enable MWS and have described the challenges in this research area. We have declared and compared different types of MWS including static-hosted mobile Web service and mobile-hosted mobile Web service. In order to apply Service-Oriented Architecture to enhance the interoperability of MWS, Enterprise Service Bus based approaches have been described. To support the customized, adaptive service provision, the composite service-based approach is a stream. We have compared infrastructure-based and infrastructure-less approaches for enabling service composition in MWS. Furthermore, due to traditional Web service frameworks cannot be readily applied to MWS development, a number of feasible solutions have been described in this chapter to enable mobile Web service. Finally, to support the stability, connectivity, and to address resource constraint, MWS management principles and approaches have been discussed. Ultimately, we have raised a number of challenges as future research direction in the area of enhancing the efficiency of MWS using context awareness.

References

[1] P. Ahluwalia and U. Varshney. Composite quality of service and decision making perspectives in wireless networks. *Decis. Support Syst.*, 46(2):542–551, 2009.

[2] M. Asif, S. Majumdar, and R. Dragnea. Partitioning the ws execution environment for hosting mobile web services. In *Proceedings of the 2008 IEEE International Conference on Services Computing (SCC '08)*, pages 315–322, Washington, DC, USA, 2008. IEEE Computer Society.

[3] G. Broll, S. Siorpaes, E. Rukzio, M. Paolucci, J. Hamard, M. Wagner, and A. Schmidt. Supporting mobile service usage through physical mobile interaction. In *Proceedings of the 5th IEEE International Conference on Pervasive Computing and Communications (PERCOM '07)*, pages 262–271, Washington, DC, USA, 2007. IEEE Computer Society.

[4] H. Cao, R. Ma, and C. Zhu. Reasearch on architecture of battlefield mobiserv of cv and implementing method. In *Vehicle Power and Propulsion Conference, 2008. (VPPC '08). IEEE*, pages 1–4, Sep. 2008.

[5] D. Chakraborty, A. Joshi, T. Finin, and Y. Yesha. Service composition for mobile environments. *Mob. Netw. Appl.*, 10(4):435–451, 2005.

[6] A.T.S. Chan and S. Chuang. Mobipads: A reflective middleware for context-aware mobile computing. *IEEE Transactions on Software Engineering*, 29(12):1072–1085, 2003.

[7] W. Chen, Z. He, G. Ren, and W. Sun. Service recovery for composite service in manets. In *Proceedings of the 4th International Conference on Wireless Communications, Networking and Mobile Computing (WiCOM '08)*, pages 1–4, Oct. 2008.

[8] J. Cubo, C. Canal, and E. Pimentel. Towards a model-based approach for context-aware composition and adaptation: A case study using wf/.net. In *Proceedings of the 2008 5th International Workshop on Model-based Methodologies for Pervasive and Embedded Software (MOMPES '08)*, pages 3–13, Washington, DC, USA, 2008. IEEE Computer Society.

[9] A.K. Dey. Understanding and using context. *Personal Ubiquitous Computing*, 5(1):4–7, 2001.

[10] C. Dorn and S. Dustdar. Sharing hierarchical context for mobile web services. *Distrib. Parallel Databases*, 21(1):85–111, 2007.

[11] C. Doulkeridis and M. Vazirgiannis. Casd: Management of a context-aware service directory. *Pervasive Mob. Comput.*, 4(5):737–754, 2008.

[12] A. Erradi, P. Maheshwari, and V. Tosic. Recovery policies for enhancing web services reliability. In *Proceedings of the IEEE International Conference on Web Services (ICWS '06)*, pages 189–196, Washington, DC, USA, 2006. IEEE Computer Society.

[13] G. Gehlen and L. Pham. Mobile web services for peer-to-peer applications. In *Proceedings of the 2nd IEEE Consumer Communications and Networking Conference (CCNC '05)*, pages 427–433, Jan. 2005.

[14] G. De Giacomo, M. De Leoni, M. Mecella, and F. Patrizi. Automatic workflows composition of mobile services. In *Proceeding of the IEEE International Conference on Web Services (ICWS 2007)*, pages 823–830, Jul. 2007.

[15] B. Han, W. Jia, J. Shen, and M. Yuen. Context-awareness in mobile web services. In *Parallel and Distributed Processing and Applications*, pages 519–528. Springer Berlin, Heidelberg, 2005.

[16] S. Kalasapur, M. Kumar, and B.A. Shirazi. Dynamic service composition in pervasive computing. *IEEE Trans. Parallel Distrib. Syst.*, 18(7):907–918, 2007.

[17] E. Kang, D. Park, and Y. Lim. Content aware selecting method for reducing the response time of an adaptive mobile web service. In *Ubiquitous Intelligence and Computing*, pages 748–757. Springer Berlin, Heidelberg, 2007.

[18] Y. Kim, E. Kim, J. Kim, E. Song, and I. Ko. Ontology based software reconfiguration in a ubiquitous computing environment. In *Proceeding of the 6th IEEE International Conference on Computer and Information Technology (CIT '06)*, page 260, Sep. 2006.

[19] Y. Kim and K. Lee. A light-weight framework for hosting web services on mobile devices. In *Proceedings of the 5th European Conference on Web Services (ECOWS '07)*, pages 255–263, Washington, DC, USA, 2007. IEEE Computer Society.

[20] T. Koskela, N. Kostamo, O. Kassinen, J. Ohtonen, and M. Ylianttila. Towards context-aware mobile web 2.0 service architecture. In *Proceedings of the International Conference on Mobile Ubiquitous Computing, Systems, Services and Technologies (UBICOMM '07)*, pages 41–48, Washington, DC, USA, 2007. IEEE Computer Society.

[21] C. Lee, S. Ko, S. Lee, W. Lee, and S. Helal. Context-aware service composition for mobile network environments. In *Ubiquitous Intelligence and Computing*, pages 941–952. Springer Berlin, Heidelberg, 2007.

[22] H.J. Lee, J.Y. Choi, and S.J. Park. Context-aware recommendations on the mobile web. In *On the Move to Meaningful Internet Systems 2005: OTM Workshops*, pages 142–151. Springer Berlin, Heidelberg, 2005.

[23] S. Li. *JXTA 2: A high-performance, massively scalable P2P network*, Nov. 2003. Available at http://www.ibm.com/developerworks/java/library/j-jxta2/.

[24] D. LopezdeIpina, J.I. Vazquez, and J. Abaitua. A web 2.0 platform to enable context-aware mobile mash-ups. In *Ambient Intelligence*, pages 266–286. Springer Berlin, Heidelberg, 2007.

[25] J. Merwe, D. Dawoud, and S. McDonald. A survey on peer-to-peer key management for mobile ad hoc networks. *ACM Comput. Surv.*, 39(1):1, 2007.

[26] Microsoft, IBM, Siebel, BEA, and SAP. *Business Process Execution Language for Web Services Version 1.1*, May 2003. Available at http://www-106.ibm.com/developerworks/library/ws-bpel/.

[27] S. Ben Mokhtar, N. Georgantas, and Valérie Issarny. Cocoa: Conversation-based service composition in pervasive computing environments with qos support. *J. Syst. Softw.*, 80(12):1941–1955, 2007.

[28] P. Moore, B. Hu, and J. Wan. Smart-context: A context ontology for pervasive mobile computing. *The Computer Journal Advance Access*, pages 1–17, Mar. 2008.

[29] V. Myllärniemi, C. Prehofer, M. Raatikainen, J. Gurp, and T. Männistö. Approach for dynamically composing decentralised service architectures with cross-cutting constraints. In *Proceedings of the 2nd European Conference on Software Architecture (ECSA '08)*, pages 180–195, Berlin, Heidelberg, 2008. Springer-Verlag.

[30] T. Nguyen, I. Jorstad, and D. Thanh. Security and performance of mobile xml web services. In *Proceedings of the 4th International Conference on Networking and Services (ICNS '08)*, pages 261–265, Washington, DC, USA, 2008. IEEE Computer Society.

[31] S. Oh and G.C. Fox. Optimizing web service messaging performance in mobile computing. *Future Gener. Comput. Syst.*, 23(4):623–632, 2007.

[32] S. Panagiotakis and A. Alonistioti. Context-aware composition of mobile services. *IT Professional*, 8(4):38–43, 2006.

[33] M.P. Papazoglou, P. Traverso, S. Dustdar, and F. Leymann. Service-oriented computing: State of the art and research challenges. *Computer*, 40(11):38–45, Nov. 2007.

[34] P. Pawar, J. Subercaze, P. Maret, B.J.F. van Beijnum, and D. Konstantas. Towards business model and technical platform for the service oriented context-aware mobile virtual communities. In *Proceedings of IEEE Symposium on Computers and Communications, 2008. (ISCC 2008)*, pages 103–110, Jul. 2008.

[35] P. Pawar, B.-J. van Beijnum, M. van Sinderen, A. Aggarwal, P. Maret, and F. De Clercq. Performance evaluation of the context-aware han-

dover mechanism for the nomadic mobile services in remote patient monitoring. *Comput. Commun.*, 31(16):3831–3842, 2008.

[36] P. Pawar, K. Wac, B.-J. van Beijnum, P. Maret, A. van Halteren, and H. Hermens. Context-aware middleware architecture for vertical handover support to multi-homed nomadic mobile services. In *Proceedings of the 2008 ACM Symposium on Applied Computing (SAC '08)*, pages 481–488, New York, NY, USA, 2008. ACM.

[37] L.D. Prete and L. Capra. Reliable discovery and selection of composite services in mobile environments. In *Proceedings of the 2008 12th International IEEE Enterprise Distributed Object Computing Conference (EDOC '08)*, pages 171–180, Washington, DC, USA, 2008. IEEE Computer Society.

[38] B. Schilit, N. Adams, and R. Want. Context-aware computing applications. In *Proceedings of the 1994 First Workshop on Mobile Computing Systems and Applications (WMCSA '94)*, pages 85–90, Washington, DC, USA, 1994. IEEE Computer Society.

[39] H. Schmidt, A. Köhrer, and F.J. Hauck. Soapme: a lightweight java me web service container. In *Proceedings of the 3rd workshop on Middleware for service oriented computing (MW4SOC '08)*, pages 13–18, New York, NY, USA, 2008. ACM.

[40] E. Seurre, P. Savelli, and P. Pietri. *GPRS for Mobile Internet.* Artech House Publishers, 2003.

[41] Q.Z. Sheng, B. Benatallah, and Z. Maamar. User-centric services provisioning in wireless environments. *Commun. ACM*, 51(11):130–135, 2008.

[42] C. Smith and D. Collins. *3G Wireless Networks.* McGraw-Hill, Inc., 2001.

[43] S.N. Srirama. *Mobile Hosts in enterprise service integration (PhD-thesis).* RWTH Aachen University, 2008.

[44] S.N. Srirama, M. Jarke, and W. Prinz. Mobile web service provisioning. In *Proceedings of the Advanced International Conference on Telecommunications and Int'l Conference on Internet and Web Applications and Services (AICT-ICIW '06)*, page 120, Washington, DC, USA, 2006. IEEE Computer Society.

[45] S.N. Srirama, M. Jarke, and W. Prinz. Mwsmf: a mediation framework realizing scalable mobile web service provisioning. In *Proceedings of the 1st International Conference on Mobile Wireless MiddleWARE, Operating Systems, and Applications (MOBILWARE '08)*, pages 1–7, ICST, Brussels, Belgium, Belgium, 2007. ICST (Institute for Computer Sciences, Social-Informatics and Telecommunications Engineering).

[46] M. Tian, A. Gramm, H. Ritter, J.H. Schiller, and T. Voigt. Adaptive qos for mobile web services through cross-layer communication. *Computer*, 40(2):59–63, 2007.

[47] V. Tosic, H. Lutfiyya, and Y. Tang. A management infrastructure for mobile/embedded xml web services. In *Network Operations and Management Symposium, 2006. NOMS 2006. 10th IEEE/IFIP*, pages 1–4, Apr. 2006.

[48] A. van Halteren and P. Pawar. Mobile service platform: A middleware for nomadic mobile service provisioning. In *Proceedings of the IEEE International Conference on Wireless and Mobile Computing, Networking and Communications 2006 (WiMob '06)*, pages 292–299, Jun. 2006.

[49] N. WeiBenberg, R. Gartmann, and A. Voisard. An ontology-based approach to personalized situation-aware mobile service supply. *Geoinformatica*, 10(1):55–90, 2006.

[50] Y. Yamato and H. Sunaga. Context-aware service composition and component change-over using semantic web techniques. In *Proceedings of the IEEE International Conference on Web Services (ICWS 2007)*, pages 687–694, Jul. 2007.

[51] T. Yoshikawa, K. Ohta, T. Nakagawa, and S. Kurakake. Mobile web service platform for robust, responsive distributed application. In *Proceedings of the 14th International Workshop on Database and Expert Systems Applications (DEXA '03)*, page 144, Washington, DC, USA, 2003. IEEE Computer Society.

[52] M. Younas, K.-M. Chao, P. Wang, and C.-L. Huang. Qos-aware mobile service transactions in a wireless environment: Research articles. *Concurr. Comput.: Pract. Exper.*, 19(8):1219–1236, 2007.

[53] W. Zahreddine and Q.H. Mahmoud. An agent-based approach to composite mobile web services. In *Proceedings of the 19th International Conference on Advanced Information Networking and Applications (AINA '05)*, pages 189–192, Washington, DC, USA, 2005. IEEE Computer Society.

[54] L. Zeng, A.H. Ngu, B. Benatallah, R. Podorozhny, and H. Lei. Dynamic composition and optimization of web services. *Distrib. Parallel Databases*, 24(1-3):45–72, 2008.

Index

E